European Studies in Philosophy of Science

Volume 3

This new series results from the synergy of EPSA - European Philosophy of Science Association - and PSE - Philosophy of Science in a European Perspective: ESF Networking Programme (2008–2013). It continues the aims of the Springer series "The Philosophy of Science in a European Perspective" and is meant to give a new impetus to European research in the philosophy of science. The main purpose of the series is to provide a publication platform to young researchers working in Europe, who will thus be encouraged to publish in English and make their work internationally known and available. In addition, the series will host the EPSA conference proceedings, selected papers coming from workshops, edited volumes on specific issues in the philosophy of science, monographs and outstanding Ph.D. dissertations. There will be a special emphasis on philosophy of science originating from Europe. In all cases there will be a commitment to high standards of quality. The Editors will be assisted by an Editorial Board of renowned scholars, who will advise on the selection of manuscripts to be considered for publication.

More information about this series at http://www.springer.com/series/13909

Edward Slowik

The Deep Metaphysics of Space

An Alternative History and Ontology Beyond
Substantivalism and Relationism

 Springer

Edward Slowik
Department of Philosophy
Winona State University
Winona, MN, USA

ISSN 2365-4228   ISSN 2365-4236   (electronic)
European Studies in Philosophy of Science
ISBN 978-3-319-83153-4   ISBN 978-3-319-44868-8   (eBook)
DOI 10.1007/978-3-319-44868-8

Printed on acid-free paper

This Springer imprint is published by Springer Nature
The registered company is Springer International Publishing AG
The registered company address is: Gewerbestrasse 11, 6330 Cham, Switzerland

Preface

The origins of this project can be traced back to a discussion among Newton scholars at the 2006 Paris meeting of the International Society for the History of Philosophy of Science. The conversation focused on Newton's ontology of space and if there are any modern treatments that actually capture the details of Newton's views. After many years, this book is an attempt to meet that challenge, as well as for Leibniz' equally unconventional spatial ontology, all the while dovetailing the discussion with analogous issues and elements in the modern spacetime ontology debate. Long before I began researching Newton and Leibniz' views on space and motion in more detail, I was intrigued by the confusions and uncertainties involved in the attempts to extend the standard substantivalist versus relationist scheme to both general relativity, the theory that replaced Newtonian gravitation, and the new breed of quantum gravity hypotheses that espouse spacetime emergence (in order to link quantum mechanics and general relativity). I had planned to cover both issues, the historical (Newton, Leibniz, and others) and modern spatial ontologies in contemporary physical theory, in separate works, but the results of the historical investigation revealed an intriguing host of parallel concerns and concepts, in particular, the relationship between a grounding entity or theory (God and monads, in the seventeenth century, and usually quantum mechanics, in the modern setting) and the familiar macrolevel world of bodies and their interactions, such as gravity. After bracketing away the theological and immaterialist implications of the earlier theories (which play no role in this study other than as a necessary part of seventeenth-century spatial ontologies), it turns out that in both cases, the seventeenth-century and the modern, spatial ontologists advanced a similar array of hypotheses that strived to relate the spatial properties of the foundational entity/entities with the spatial properties at the emergent level of matter, thus suggesting an entirely new way of approaching the spatial ontology debate, both historically and for contemporary theories. Given that seventeenth-century natural philosophers often conceive space as akin to a property of God, although formulated in many different guises, and that emergent spacetime theories also exhibit many property-like features, I was eventually led to examine that long-neglected theme in the philosophy of space and time, the property theory of space, in conjunction with spatial emergence. A further aspect of these historical theories pertains to the Platonist versus nominalist divide as regards spatial geometry, an issue that fits naturally with a discussion of properties and which proved to be quite useful in untangling many of the seemingly conflicting claims made by seventeenth-century thinkers about space. As a result, contemporary alternatives to substantivalism and relationism that either tacitly or openly involve these additional elements, properties and nominalism, were ultimately incorporated into the investigation alongside the property theory, i.e., the definitional approach and spacetime structuralism. Needless to say, this investigation does not feign to procure the "one true view" on spatial ontology, assuming there is such a thing, but it is hoped that this (somewhat unorthodox) approach, which culminates in the new taxonomy developed in later chapters, will demonstrate its superiority over the standard dichotomy that is still in use, namely, substantivalism and relationism.

While the conclusions reached, and the ideas developed, are entirely my own (unless otherwise noted), there are numerous people that I would like to thank for helpful discussions and comments on earlier work and for their support over the years (and I apologize to the many people that I forgot to include). Among scholars in the history and philosophy of science and/or Early Modern philosophy: Peter Machamer, Dana Jalobeanu, Peter Anstey, Doina-Cristina Rusu, Doug Jesseph, Michael Futch, Eric Schliesser, Zvi Biener, Katherine Dunlop, Roger Ariew, Raffaella De Rosa, Maarten Van Dyck, Andrew Janiak, Joe Zepeda, Mihnea Dobre, Ori Belkind, Jasper Reid, Antonia LoLordo, Anja Jauernig, Tad Schmaltz, Ben Hill, Vincenzo De Risi, Paul Lodge, Margaret Atherton, Cees Leijenhorst, Christoph Lüthy, and David Miller. Among scholars in the philosophy of space and time and/or philosophy of physics: Rob Rynasiewicz, Mauro Dorato, Tim Maudlin, Graham Nerlich, John Earman, Jonathan Bain, Laura Ruetsche, Jeremy Butterfield, Carl Hoefer, Vincent Lam, Elaine Landry, John Norton, Chris Wüthrich, Michael Friedman, Howard Stein, Erik Curiel, and Mark Wilson. For their generous assistance regarding specific issues and/or philosophers, I would especially like to thank the following: Ric Arthur, Dan Garber, and Don Rutherford (Leibniz); Ted McGuire (Newton); Marius Stan (Huygens and Kant); Nick Huggett, Gordon Belot, Oliver Pooley, Rob DiSalle, Dean Rickles, and Dennis Dieks (substantivalism vs. relationism and quantum gravity); and Chris Pincock, Katherine Brading, and Michael Esfeld (philosophy of mathematics or structuralism in physics). Last, but certainly not least, I would like to thank Geoff Gorham for his help and encouragement over the many years of this project.

I would like to thank: the American Philosophical Society for a Sabbatical Fellowship Award for the academic year 2008–2009; the Center for Philosophy of Science, University of Pittsburgh, for a visiting fellowship during the 2007–2008 academic year; the Minnesota Center for Philosophy of Science, University of Minnesota, Twin Cities, for a visiting fellowship during the fall term, 2008; and Princeton University, for a visiting fellowship during the spring term, 2009.

I would also like the thank the editors of the following journals for allowing me to include various portions of previously published articles, although the material has been greatly revised and augmented: *Studies in History and Philosophy of Modern Physics, European Journal for Philosophy of Science, HOPOS: The Journal of the International Society for the History of Philosophy of Science, International Studies in the Philosophy of Science, Perspectives on Science, Intellectual History Review, Foundations of Science, Journal of Early Modern Studies, Journal for General Philosophy of Science, History of Philosophy Quarterly, Early Science and Medicine, Studies in History and Philosophy of Science, Philosophia Scientiae.*

Contents

Abbreviations and Other Conventions

The symbol "§" will refer to chapter and sections in this book: e.g., "§2.1.1" refers to chapter 2, section 1, sub-section 1. References to most texts will list author, year of publication, and page number. For more frequently cited primary sources, the following abbreviations will be used (listed in alphabetical order), along with the standard pagination convention for that author (unless otherwise noted):

A   Leibniz, G. W. (1923). *Sämtliche Schriften und Briefe*, ed. by Akademie der Wissenschaften der DDR. Darmstadt and Berlin: Akademie-Verlag. Referenced with series, volume, and page number.

AAA   More, H. (1997 [1655]). *An Antidote Against Atheism*, 2nd ed. Bristol: Thommes Press. Referenced with page number.

AG   Leibniz, G. W. (1989). *Leibniz: Philosophical Essays*, ed. and trans. by R. Ariew and D. Garber. Indianapolis: Hackett. Referenced with page number.

AT   Descartes, R. (1976). *Oeuvres de Descartes*, ed. by C. Adams and P. Tannery. Paris: J. Vrin. Referenced with volume, page number.

C   Clarke, S. (2000). *Leibniz and Clarke Correspondence*, ed. and trans. by R. Ariew. Indianapolis: Hackett. Referenced with letter number, and section.

CAP   Philoponus, J. (1992). *Commentaries on Aristotle's Physics*, trans. by J. O. Urmson, in *Corollaries on Place and Void*. Ithaca, New York: Cornell University Press. Referenced with the standard pagination from *Commentaria in Aristotelem Graeca*.

CPQ   Newton, I. (1983). *Certain Philosophical Questions: Newton's Trinity Notebook*, ed. and trans. by J. E. McGuire and M. Tamny. Cambridge: Cambridge University Press. Referenced with page number.

CPR   Kant, I. (1998). *Critique of Pure Reason*, trans and ed. by P. Guyer and A. Wood. Cambridge: Cambridge University Press. Referenced with edition, A or B, and line number.

CSM I   Descartes, R. (1985). *The Philosophical Writings of Descartes, Vol.1*, ed.
            and trans. by J. Cottingham, R. Stoothoff, D. Murdoch. Cambridge:
            Cambridge University Press. Referenced with page number.
CSM II   Descartes, R. (1984). *The Philosophical Writings of Descartes, Vol.2*, ed.
            and trans. by J. Cottingham, R. Stoothoff, D. Murdoch. Cambridge:
            Cambridge University Press. Referenced with page number.
CSMK   Descartes, R. (1991). *The Philosophical Writings of Descartes, Vol.3,
            The Correspondence*, eds. and trans. by J. Cottingham, et al. Cambridge:
            Cambridge University Press. Referenced with page number.
CWA   Aristotle. (1984). *The Complete Works of Aristotle*, ed. by J. Barnes.
            Princeton: Princeton University Press. Referenced with Bekker number.
DLM   Pardies, I. (1670). *A Discourse of Local Motion (Discours du mouvement
            local)*, trans. by A. M. London: Moses Pitt. Referenced with page
            number.
E   Locke, J. (1975 [1690]). *An Essay Concerning Human Understanding*,
            ed. by P. Nidditch. Oxford: Oxford University Press. Referenced with
            part, chapter, section.
EM   More, H. (1995 [1679]). *Henry More's Manual of Metaphysics: A
            Translation of the* Enchiridium Metaphysicum, Parts I and II, trans. by
            A. Jacob. Hildesheim: Olms. Referenced with page number.
EW   Hobbes, T. (1839). *The English Works of Thomas Hobbes of Malmesbury*,
            vol. 1, ed. by W. Molesworth. London: John Bohn. Referenced with page
            number.
G   Leibniz, G. W. (1965). *Die Philosophischen Schriften von Leibniz*, ed. by
            C. I. Gerhardt. C. I. Hildesheim: Olms. Referenced with volume, page
            number.
GM   Leibniz, G. W. (1962). *Leibnizens Mathematische Schriften*, ed. by C. I.
            Gerhardt. Hildesheim: Olms. Referenced with volume, page number.
H   Huygens, C. (1993). *Codex Huygens7a* in *Penetralia Motus: la fondazi-
            one relativistica della meccanica nell'opera di Chr. Huygens*, ed. by
            G. Mormino. Firenze: Nova Italia. Trans. by M. Stan. Referenced with
            page number of the translation.
IS   More, H. (1997 [1659]). *The Immortality of the Soul*. Bristol: Thommes
            Press. Referenced with page number.
L   Leibniz, G. W. (2000). *Leibniz and Clarke Correspondence*, ed. and
            trans. by R. Ariew. Indianapolis: Hackett. Referenced with letter number,
            and section.
LDB   Leibniz, G. W. (2007). *The Leibniz-Des Bosses Correspondence*, ed. and
            trans. by B. Look and D. Rutherford. New Haven: Yale University Press.
            Referenced with page number.
LoC   Leibniz, G. W. (2001). *The Labyrinth of the Continuum: Writings of 1672
            to 1686*, ed. and trans. by R. Arthur. New Haven: Yale University Press.
            Referenced with page number.

Lm   Leibniz, G. W. (1969). *Leibniz: Philosophical Letters and Papers*, 2nd ed., ed. and trans. by L. E. Loemker. Dordrecht: Kluwer. Referenced with page number.

LG   Barrow, I. (1976). *Lectiones Geometricae*, 4–15, trans. by M. Capek, in *The Concepts of Space and Time*, ed. by M. Capek. Dordrecht: Reidel, 203–208. Referenced with page number.

MF   Kant, I. (2004). *Metaphysical Foundations of Natural Science*, trans. and ed. by M. Friedman. Cambridge: Cambridge University Press. Referenced by volume, followed by a colon, and page number, from the standard Academy edition of Kant's works.

MP   Leibniz, G. W. (1995). *Leibniz: Philosophical Writings*, ed. and trans. by M. Morris and G. H. R. Parkinson. Rutland, Vermont: C. Tuttle. Referenced with page number.

MWB   Barrow, I. (1860). *The Mathematical Works of Isaac Barrow D. D.*, ed. by W. Whewell. Cambridge: Cambridge University Press. Referenced with page number.

N   Newton, I. (2004). *Philosophical Writings*, trans. and ed. by A. Janiak and C. Johnson. Cambridge: Cambridge University Press. Referenced with page number.

NE   Leibniz, G. W. (1996). *New Essay on Human Understanding*, ed. and trans. by P. Remnant and J. Bennett. Cambridge: Cambridge University Press. Referenced with book, chapter, section.

NS   Kant, I. (2012). *Natural Science*, ed. by E. Watkins, trans. by L. W. Beck, J. B. Edwards, O. Reinhardt, M. Schönfeld, and E. Watkins. Cambridge: Cambridge University Press. Referenced by volume, followed by a colon, and page number, from the standard Academy edition of Kant's works.

O   Oresme, N. (1968). *Le Livre Du Ciel et Du Monde*, ed. and trans. by A. D. Menut and A.J. Denomy. Madison: University of Wisconsin Press, 1968. Referenced with page number.

PFM   Kant, I. (1997). *Prolegomena to Any Future Metaphysics*, trans. and ed. by G. Hatfield. Cambridge: Cambridge University Press. Referenced by volume, followed by a colon, and page number, from the standard Academy edition of Kant's works.

Pr   Descartes, R. (1983). *Principles of Philosophy*, trans. by V. R. Miller and R. P. Miller. Dordrecht: Kluwer Academic Publishers. Referenced with part, section.

PS   Patrizi, F. (1943). "On Physical Space, Francesco Patrizi", ed. and trans. by B. Brickman, *Journal of the History of Ideas*, 4, 224–225. Referenced with page number.

RIV   Gassendi, P. (1976). "The Reality of Infinite Void", trans. by M. Capek and W. Emge, in *The Concepts of Space and Time*. Dordrecht: D. Reidel, 1976, 91–96. Referenced with page number.

SM   Mach, E. (1960, [1893]). *The Science of Mechanics*, 6th ed., tans. by T. McCormack. London: Open Court. Referenced with page number.

ST   Aquinas, T. (1964). *Summa Theologiae*, Blackfriars edition, 61 volumes. New York: McGraw-Hill. Referenced part number, question number, article number.

SWG   Gassendi, P. (1972). *The Selected Works of Pierre Gassendi*, ed. and trans. by C. G. Brush. New York: Johnson. Referenced with page number.

T   Hume, D. (2000, [1739]). *A Treatise of Human Nature,* ed. by D. F. Norton and M. J. Norton. Oxford: Oxford University Press. Referenced with book, part, section, and paragraph.

TeL   Newton, I. (1978). "Tempus et Locus", in "Newton on Place, Time, and God: An Unpublished Source", trans. and ed. by J. E. McGuire, *British Journal for the History of Science,* 11, 114–129. Referenced with page number.

TP   Kant, I. (1992). *Theoretical Philosophy: 1755–1770*, trans. and ed. by D. Walford and M. Meerbote. Cambridge: Cambridge University Press. Referenced by volume, followed by a colon, and page number, from the standard Academy edition of Kant's works.

UML   Barrow, I. (1734). *The Usefulness of Mathematical Learning Explained and Demonstrated*, trans. by J. Kirkby. London: Stephen Austin. Referenced with page number.

W   Aristotle (1996). *Physics*, trans. by R. Waterford. Oxford: Oxford University Press. Referenced with Bekker number.

WGB   Berkeley, G. (1843). *The Works of George Berkeley, D. D.*, ed. by G. Wright. London: Thomas Tegg. Referenced with volume, page number.

All italics are from the original texts unless otherwise noted.

Introduction

In contemporary scholarship devoted to the philosophy of space and time, two chief rivals dominate the field: substantivalism (or absolutism) and relationism. Providing a precise definition of these opposing viewpoints is quite difficult, but a fairly straightforward reading, albeit simplistic, is that substantivalists reckon space or spacetime to be an entity of some sort that can exist independently of material entities, whereas relationists contend that space or spacetime is a mere relation among material things and is thus neither an entity nor independent of matter.[1] There are numerous difficulties with this modern or "standard dichotomy," as we will often designate the debate between substantivalists and relationists, but our investigation will focus on two specific issues as a means of examining and developing alternative ontological conceptions of space that go beyond the limitations imposed by the standard dichotomy. First, while Newton and Leibniz (or occasionally Descartes) are often upheld as the progenitors of, respectively, substantivalism and relationism, their own work in the natural philosophy of space often contradicts the central tenets of the modern dichotomy. Part I of our investigation will largely be devoted to the historical examination of the spatial hypotheses of Newton and Leibniz in order to demonstrate this point, that is, to reveal the important discrepancies manifest between the modern dichotomy and the details of Newton and Leibniz' views on space. Second, while the modern substantivalist-relationist dichotomy is to some extent functional within the setting of Newtonian mechanics, it has proved extremely problematic when transferred to the setting of modern field theories, in particular, general relativity, but also as regards the more recent quantum gravity hypotheses. Part II and III of our investigation will highlight the attempts to apply the modern dichotomy to these contemporary physical theories, although additional aspects of the spatial ontologies of Newton and Leibniz, and other metaphysical themes, will also be developed and discussed. In place of the standard substantivalist and relationist ontologies, an alternative set of concepts and distinctions will be developed

---

[1] Throughout the remainder of our investigation, unless otherwise noted, the terms "space" and "spatial" will be used to designate both the standard three-dimensional space conception and the four-dimensional spacetime conception.

that (1) more accurately fits the details of Newton and Leibniz' distinct ontologies of space, (2) avoids the pitfalls that beset the modern dichotomy in the context of general relativity and quantum gravity, and (3) secures successful analogies between Newton, Leibniz, and other seventeenth-century natural philosophers with these contemporary (and often nonclassical) physical theories. Along the way, a number of alternative approaches and concepts will be explored that offer insights into the spatial ontologies of the seventeenth-century and contemporary physical theories, and thus the examination of these concepts can be deemed important subsidiary goals of the investigation as well: in particular, (4) the supervenience or emergence of space from a deeper ontology that is either nonspatial or has a drastically different geometric structure than classical physics, sometimes labeled the "emergent spacetime" hypothesis, (5) versions of platonism and nominalism as they pertain to spatial geometry, and (6) several alternative spatial ontologies—the structural realist, definitional approach, and property theories of space—as rivals to the standard dichotomy.

The motivating idea behind this project can be stated quite simply: the contemporary substantival-relational distinction, which claims to be based on Newton and Leibniz' spatial concepts, actually fails to capture the metaphysical elements that undergird these seventeenth-century spatial hypotheses, and this failure helps to explain both the deficiencies in their attempted analogies between seventeenth-century and contemporary spatial theories, but also, more generally, the inadequacy of substantivalist and relationist assessments of contemporary theories in physics, namely, general relativity and quantum gravity hypotheses. The aspect of seventeenth-century spatial hypotheses that the exponents of the modern dichotomy have neglected is, oddly enough, the deep metaphysical entity that underlies and actually brings about space for such natural philosophers as Newton and Leibniz, i.e., God, although Leibniz' monadological ontology is an additional factor of primary importance. By ignoring this central feature of seventeenth-century spatial ontologies—i.e., the deep metaphysics of space—the modern practitioners of the traditional dichotomy have unwittingly deprived themselves of a set of valuable conceptual resources that, if pursued, offer a potential way out of the limitations and uncertainties that currently plague their appraisals of spatial ontology. It is important to note, however, that our investigation of the role that God and Leibniz' monads play in seventeenth-century natural philosophy is based on purely metaphysical and scientific, and not theological, motivations. The intrusion of supernatural and theological dogma into the domain of modern science is, from our perspective, both scurrilous and pseudoscientific, and all such misguided attempts should be utterly repudiated. Rather, the goal of this investigation is to demonstrate that, in the seventeenth century, the complete ontological account of space includes a metaphysical entity or entities (God/monads) as a crucial component—and, if this component is accorded its proper status within the larger ontology, then the structure and conceptual elements of the overall theory exhibit a form that differs drastically from the appraisals furnished using the modern dichotomy. But it is not just historical accuracy that prompts our analysis: when this metaphysical grounding role is incorporated into a more general scientifically informed, and thus non-theological, account of spatial ontology, a host of conceptual alternatives are brought into play that have the potential to transform the current (quite limited and inadequate) state of onto-logical speculation based on the standard dichotomy into a more fruitful and effec-tive scheme that is not saddled with the dichotomy's limitations. In particular, the grounding relationship between a foundational level and derived level of ontology, a feature that typifies the vast majority of seventeenth-century spatial theories, is closely analogous to the grounding relationship postulated between the microlevel structures and processes and the resulting macrolevel phenomena in numerous cur-rent quantum gravity hypotheses. Given our more accurate evaluation of seventeenth-century spatial theories, consequently, entirely new avenues are opened for the evaluation of spatial ontologies, both in historical and metaphysical terms.

As noted above, Part I of our investigation will be concerned with the historical details of Newton and Leibniz' spatial ontologies, with special emphasis placed on how contemporary substantivalisms and relationisms often fail to accurately repre-sent their respective hypotheses. Prior to the examination of the seventeenth centu-ry's natural philosophy of space, however, Chap. 1 will first explore the contemporary interpretation of substantivalism (absolutism) and relationism, the various influ-ences that have shaped its current form, and how this dichotomy may have evolved from the late Scholastic period up to Newton and Leibniz' time. Even though the complexity of the historical record almost certainly militates against any exact, grand narrative of the evolution of the traditional dichotomy, Descartes' spatial the-ory will be offered as a prime historical example of an approach to space that lies much closer to its ancient and medieval predecessors than any contemporary assess-ment founded on the standard dichotomy. This chapter will also provide an update and reevaluation of several of the key issues raised, and arguments presented, in Slowik (2002). Turning to one of the central figures in our analysis, the anti-substantivalist reading of Newton's spatial hypotheses that has been pioneered by Stein, DiSalle, and several earlier philosophers will be given a thorough historical examination in Chap. 2. While some of the anti-substantivalist conclusions reached by Stein and DiSalle will be upheld, the more contentious epistemologically based interpretation that they seem to offer, a positivist-leaning approach that treats spatial hypotheses as definitional axioms for his physics, will be shown to be in conflict with the historical evidence. In Chap. 3, the complex maze that comprises the spa-tial hypotheses of our other key figure, Leibniz, will be navigated in order to dem-onstrate its numerous, and profound, divergences from relational orthodoxy. Many crucial components of Leibniz' ontology will only be taken up in the second half of our inquiry, but Chap. 3 will lay out the main case for an anti-relationist interpreta-tion as well as present many of the central components of his spatial ontology. Rounding out Part I, a detailed examination of the relationship between Leibniz' non-extended foundational entities, monads, and material bodies, as well as the many difficulties associated with Leibniz' account of motion, will be the topics of Chap. 4. Despite emphasizing monads and motion, rather than space, the hypothe-ses put forward in this chapter will compliment many of the conclusions that will be reached elsewhere in our investigation, especially as regards supervenience and emergence. In Chap. 4, furthermore, Leibniz' largely neglected notion of ubiety, which he describes as the way a being is located in (or related to) space, will figure prominently, for it will assume a central role in later chapters, especially in Part III. On the whole, the conclusions that will be drawn from our close examination of seventeenth-century theories of spatial ontology in Part I are consistent with neither substantivalism nor relationism but do closely align with the property theory of space, a realization that will spur the development of the new taxonomy in Part III as well as inform the investigation of contemporary spatial ontologies in Part II.

In Part II, additional aspects of seventeenth-century spatial hypotheses will be explored, but the obstacles that face the application of the standard dichotomy to contemporary theories in physics will increasingly come to the forefront. In Chap. 5, for instance, both historical and present-day formulations of a property theory of space, as well as two other "third-way" conceptions, the definitional approach and structural realism, will be examined against the backdrop of Newtonian mechanics and general relativity—where, by "third-way," we signify alternative spatial hypotheses that reject the standard dichotomy. One of the goals of this chapter is to demonstrate that these alternative, and often property-oriented, spatial ontologies are as effective, if not more so than the standard dichotomy, despite the fact that alternative spatial ontologies have long been neglected in the literature. Chapter 5 will also introduce a number of issues that have dominated discussions in the philosophy of space and time over the past several decades, such as the hole argument and the array of substitute formulations of substantivalism and relationism that have been fabricated in its wake, issues that will continue to shape the discussion throughout the remainder of the investigation. While Chap. 6 will resume the examination of Newton's spatial ontology in order to address his immobility arguments, the lessons gathered from the seventeenth century's handling of these concepts will be shown to have relevance for the interpretation of spatial ontology in a modern setting. Specifically, Newton relies on deeper aspects of his overall ontology to ground the identity of the parts of space, an approach that runs counter to the emphasis on spatiotemporal symmetries and translations that one finds in contemporary assessments of similar problems in the philosophy of space and time. Chapter 7, in contrast, will examine the prospects for utilizing platonism and nominalism as an alternative means of assessing spatial ontologies in lieu of the traditional dichotomy. Platonism and nominalism, and fine-grained variants of these concepts, will play a major factor in the conclusions ultimately advanced in our investigation, both in the evaluation of seventeenth-century and contemporary spatial theories, but also as regards scientific realism more generally. In this chapter, a form of nominalism, dubbed truth-based nominalism, will be offered as the most defensible and versatile option for spatial ontologists, especially given the difficulties involved with the standard dichotomy. The structural realist approach to spatiotemporal theories, first examined in Chap. 5 and also discussed in Chaps. 6 and 7, will be the sole topic of Chap. 8, with emphasis placed on the contrasts between the ontic and epistemic branches of this structuralist philosophy and its underdetermination problems. As a new entrant in the evaluation of spatial ontologies, structural realism constitutes one of the most popular contemporary alternatives to the traditional dichotomy; thus, a more comprehensive examination of its potential is warranted. In these chapters, a version of epistemic structural realism will be defended on the grounds that it accommodates the competing demands of scientific realism and spatial ontology better than its various rivals.

Many of the separate themes in our investigation, in particular, the complex ontological structure of seventeenth-century spatial hypotheses and the platonism-nominalism scheme, will come together in Part III, which comprises Chaps. 9, 10, and 11. The troubles that plague the application of substantivalism and relationism in the context of quantum gravity theories will begin in Chap. 9, setting the stage for the introduction of a new two-tier system of classifying and evaluating spatial ontologies. The remainder of Chap. 9 will apply this new taxonomy to Early Modern spatial ontologies and address various competing interpretations and other related issues. Chapter 10, in contrast, will examine the general structure of quantum gravity hypotheses using the new taxonomy and, alongside comparisons with the similarly structured Early Modern spatial hypotheses from Chap. 9, conclude with an overall assessment and defense of this new approach to spatial ontology. This new system, which relies on a foundational and derived level of spatial ontology along with variants of platonism and nominalism at these different levels, will be shown to be more successful than the standard dichotomy on both historical and metaphysical grounds, i.e., points (1), (2), and (3) above. Likewise, in addition to the incorporation of nominalism, point (5), it will be demonstrated that the new taxonomy has more in common with the third-way spatial ontologies examined in Part II, especially the property theory, point (4), but also as regards supervenience and emergence, point (6). This new taxonomic scheme does not claim to be the final word on the assessment of spatial ontologies, but it will be argued that it constitutes a more successful and fruitful basis for conceptual analysis than the contemporary substantivalism/relationism distinction. One of the subordinate goals of our project, it should be noted, is to inject a wealth of overlooked historical and metaphysical concepts into the discussion of spatial theories, alternative ideas that have the capacity to reshape the current state of the philosophy of space if given the same level of attention that they received in the seventeenth century. If the course of that imagined evolution of spatial philosophizing were to diverge from the conclusions reached in this work, however, then our investigation will have at least assisted in the (long overdue) turn to a more accurate and successful system of assessing spatial ontology. Finally, Chap. 11 will briefly examine the development of the standard dichotomy in the eighteenth and later centuries, with the emphasis placed on several key thinkers who played a significant role in ushering in this new approach to spatial theorizing that followed in the wake of Newton and Leibniz.

Since the material in this investigation is almost evenly divided between two philosophical subdisciplines, Early Modern philosophy and contemporary philosophy of space and time, an outline of the chapter contents as it pertains to these two fields might prove useful to many readers. Hence, for those philosophers interested in the philosophy of space and motion within Early Modern philosophy, the following chapters and sections will be of interest: 1 (Sect. 3), 2, 3, 4, 5 (Sect. 1.1), 6

(Sects. 1 and 2), 7 (Sect. 3), 9 (Sects. 2 and 3), and 11. On the other hand, for philosophers of space and time and/or philosophers of physics: 1 (Sects. 1 and 2), 5
(Sects. 1.2 through 3), 6 (Sect. 3), 7, 8, 9 (Sects. 1, 2, 4, and 5), and 10. It should be
noted, however, that many portions of these chapters blend historical, metaphysical,
and contemporary philosophy of physics, and thus drawing a clear division of their
content along the lines just mentioned is a difficult task.

Part I
Substantivalism and Relationism Versus Newton and Leibniz

Chapter 1
A (Contrarian's) Reappraisal of the History and Current State of the Ontology Debate in the Philosophy of Space

If substantivalism (absolutism) holds that space is an independently existing entity, and relationism claims that space is only a relation between such entities, then what do alternative ontologies of space, which are allegedly neither substantivalist nor relationist, contend? This chapter will begin the discussion of this central question, although, as will become apparent throughout our investigation, there is much uncertainty regarding the content and scope of both substantivalism and relationism as well. In §1.1, a brief overview of the ideas and influences that have shaped the contemporary evolution and understanding of the standard dichotomy, i.e., substantivalism and relationism, will be provided, whereas §1.2 will offer a preliminary categorization of alternative spatial ontologies. In contrast, §1.3 will examine the background to the modern dichotomy at the beginning of the seventeenth century, that is, in the period just prior to the contributions of Newton and Leibniz, as well as provide a brief synopsis of the concepts that guided Descartes' natural philosophy of space and motion. This section of the chapter will begin to reveal the extraordinary obstacles that face any attempt to draw a clear line of conceptual lineage from the natural philosophy of space typically practiced in the seventeenth century to the presuppositions that inform the modern dichotomy.

1.1 The Standard Dichotomy: Substantivalism and Relationism

As noted in the Introduction, Newton and Leibniz are, without a doubt, the historical figures inevitably linked to, respectively, substantivalism and relationism.[1] The subsequent chapters in Part I, as well as many portions of Part II and Part III, will

---

[1] No references will be offered here since one cannot cite *every* work on the philosophy of space that postdates Newton and Leibniz's time (or, to be a bit more careful, *nearly every* work). However, Descartes is often paired with Leibniz as a chief early exponent of relationism.

© Springer International Publishing Switzerland 2016
E. Slowik, *The Deep Metaphysics of Space*, European Studies in Philosophy of Science, DOI 10.1007/978-3-319-44868-8_1

investigate the accuracy of this historical association, but this section will examine some of the central ideas implicated in the modern dichotomy, i.e., substantivalism and relationism, as well as alternative ontologies. The work of Sklar (1974), Friedman (1983), and Earman (1989) has significantly influenced the modern dichotomy and the general direction that the philosophy of space and time has taken over the past several decades, so much of our analysis will focus on their various contributions to defining the principle contenders in the current debate.

1.1.1 Substantivalism, Theoretical Entities, and the Spacetime Approach

Newton's absolute space is the commonly accepted forerunner to modern substantivalism, needless to say: "absolute space, of its own nature without reference to anything external, always remains homogeneous and immovable" (N 64). While Chap. 2 will critique the substantivalist interpretation of Newton, and §1.3 will explore possible sources of the absolute versus relational distinction itself, the origin of the term "substantivalism" merits some discussion. Huggett (2006a) has traced the use of the word "substantival" with respect to space (in Earman's sense, see below) back to W. E. Johnson (1964, [1922]), but the elevation of this particular label to canonical status is almost certainly a result of Sklar's text. Since a modern mathematically-informed audience might take Newton's original term "absolute" to signify a non-relational geometric structure alone (which is how we will employ that term henceforth), the empiricist direction of Sklar's overall philosophy may have contributed to his adoption of the term "substantivalism" as a means of highlighting space's status as an independently existing entity (1974, 161), with a possible subsidiary goal being to cast the substantivalist's commitment to space as the analogue of the scientific realist's belief in unobservable theoretical entities. Sklar ultimately develops an alternative property account of acceleration in lieu of the standard dichotomy (see Chap. 5), but his sympathies lie with relationism over substantivalism as regards space, probably due to the fact that relationism is more amenable to an empiricist conception of scientific theories. As for Friedman and Earman, the emphasis of their separate investigations of the standard dichotomy is largely placed on the mathematical structures of spacetime theories, but the realism/anti-realism division in the philosophy of science is a major factor as well, especially in Friedman's text: e.g., Friedman (1983, 242) appeals to the alleged "unifying power" of absolute space to uphold his realist account of spacetime structures against relationism; and one of Earman's most discussed criticisms of non-substantivalist theories (such as Sklar's) invokes the specter of instrumentalism, an anti-realist view that treats the unobservable theoretical entities in scientific theories as mere tools for saving the phenomena.

The main problem with conceiving the standard dichotomy as an instance of the larger scientific realism/anti-realism dispute, however, is that substantival space

lacks one of the principle characteristics of a theoretical entity, namely, causal powers (although general relativity complicates the issue; see Chaps. 5 and 7). At any
rate, given its unique status and function within the framework of any theory in
classical physics, treating space as just another unobservable theoretical entity—
when it is normally viewed as the backdrop to such entities—would likely strike
most philosophers as a questionable inference at best, or a category mistake at
worst, and this may explain why the more general realism/anti-realism controversy
concerning theoretical entities has not played a more openly formative role in shaping the course of the contemporary spatial ontology debate, despite its obvious
impact on that debate. Additionally, the fairly unique status of the spatial ontology
dispute vis-à-vis scientific realism/anti-realism may stem from fact that the philosophy of space and time straddles the larger disciplines of metaphysics and the philosophy of physics, and hence the older metaphysical approach to space still retains
a large measure of influence.

   In place of the scientific realism/anti-realism debate, Earman (1989), while still
echoing aspects of that outlook (as noted above), helped to usher in a new set of
conceptual distinctions that have shaped the investigation of spatial ontology over
the past quarter century: first, a codification of an array of spacetime structures to
serve as the basis for modeling proposed substantivalist and relationist interpretations of physical theories, an approach that drew inspiration from, and expanded on,
the pioneering article by Stein (1967), as well as Friedman's later effort (1983); and,
second, the elaboration of a specific challenge to substantivalism, the hole argument, put forward slightly earlier in Earman and Norton (1987), along with an
assessment of a range of substantivalist strategies that seek to neutralize its impact.

   As to the first, spacetime structures provide a means of testing the ontological
commitments of a proposed conception of space and time against the more specific
spatiotemporal requirements of a theory of physics. Ideally, and put simply, the
spacetime symmetries—the perspectives in space and time that uphold the spatial
geometry and the allowable states of motion (without consideration of force)—
should be identical to the dynamical symmetries of the proposed physical theory—
the perspectives in space and time that uphold the laws or principles of that physical
theory (especially conserved quantities that incorporate a measure of force).
Utilizing the spacetime structure approach, Newton and Leibniz' respective natural
philosophies can be seen to violate this identity requirement in different ways:
Newton's professed spacetime symmetries uphold "Newtonian spacetime", a structure that can discern whether a body is at rest or in motion with a determinate velocity (absolute velocity, which includes absolute position), whereas his dynamical
symmetries only require the weaker "neo-Newtonian (or Galilean) spacetime",
which can provide a determinate measure of absolute acceleration (change in velocity) but has no meaningful measure of absolute velocity and absolute position; for
Leibniz, his analysis of motion seems to equate his spacetime symmetries with
"Leibnizian spacetime", which can only meaningfully discern relative changes in
both velocity or acceleration among bodies, even though the dynamical symmetries
implicit in his conservation law necessitates a measure of absolute acceleration (i.e.,
whether a body moves inertially or accelerates), and hence at least neo-Newtonian

spacetime structure (which can calculate whether a body is, or is not, accelerating in an invariant or absolute sense).[2]

As the labels affixed to these different geometrical constructions indicate, Newton and Leibniz are typically associated with a distinct class of modern spacetime structures, but the accuracy of this method for gauging their spatiotemporal schemes depends crucially on the selection and interpretation of the various historical texts—and, as will be argued in Part I, a thoroughgoing investigation of the historical evidence casts serious doubt on many of the conclusions elicited using the spacetime structure approach, especially regarding Leibniz' alleged espousal of Leibnizian spacetime. This is not to suggest that the spacetime structure method is not a valuable tool for assessing the match between a theorist's spatiotemporal and physical hypotheses; rather, the problem resides in the tendency of the practitioners of this system to project modern conceptions of substantivalism and relationism into the complexities of seventeenth century spatial metaphysics. Furthermore, the spacetime structure approach, with its reliance on state-of-the-art techniques from differential geometry, can as easily obscure important features of Newton and Leibniz' spatial ontologies as expose them, e.g., by imposing conceptual distinctions and structures upon these seventeenth century spatial ontologies which the textual evidence does not support, such as a clear separation of the metrical and topological aspects of spatial geometry in conjunction with a spatial mapping (diffeomorphism) argument (see below, and §6.3). Conversely, since the spacetime method was developed within, and intended for, the domain of classical mechanics and classical gravitation theories (Newtonian gravitation and general relativity), it may lack the conceptual resources to characterize both the deep metaphysical elements contained in seventeenth century spatial ontologies as well as the microlevel physical components and geometrical structures employed in many contemporary quantum gravity hypotheses, microlevel entities that, assumedly, forsake the standard spatiotemporal background structures manifest at the marcrolevel of bodies/fields (see Chap. 10).

A second component of Earman (1989) that has influenced the spacetime ontology debate concerns substantivalism explicitly. Building on the delineation of spacetime structures mentioned above, and following Einstein's initial estimate of his own form of hole argument quandary, Earman singles out the point manifold as the geometric structure that defines, or should be equated with, substantivalism (1989, 155), a view he dubs "manifold substantivalism". Informally, the manifold in a theory like general relativity (hereafter, GR) furnishes the topology and continuity of spacetime, and hence presupposes that its basic elements, i.e., points, possess an identity apart from all of the higher level geometric structures that are mapped on to

---

[2] The analysis of symmetries in physics is itself a contentious issue, especially in modern field theories, but the simplified description provided above is adequate apropos the relationship between the metaphysical spatiotemporal hypotheses and the type of classical mechanical concepts implicit in Newton and Leibniz' natural philosophy. See, Earman (1989, chapter 2) for a detailed presentation of these spacetime structures, and, Slowik (2002, chapter 2), for an informal discussion that is similar to, but more elaborate than, the one offered above.

the manifold. While the details will be provided in Chap. 5, a host of substantivalist-inclined commentators have opted instead for metrical structure as the basis of substantivalism, where the metric in GR encapsulates the geometry of spacetime, such as distance, curvature, and volume, as well as causal structure. This formulation of substantivalism, called "metric field substantivalism" or "sophisticated substantivalism", has become the default position for most substantivalists in the context of GR, the theory that replaced Newtonian gravitation and which still offers the best description of the large scale structure of space and time (see, e.g., Hoefer 1996 on these substantivalisms). However, for the purposes of our investigation, the dispute between the manifold and metric field substantivalists lacks the historical and broad theoretical requirements to represent substantivalism. First, it is not clear that the successor theory to GR will continue to employ its topological and geometric structures as fundamental, non-reducible features, thus calling into question the identification of substantivalism with either the manifold or the metric. And, second, the relationship between spatial points (parts) and the metric, though undoubtedly relevant for the mathematics of GR, is not the principal issue that divides seventeenth century spatial ontologists into opposing factions. Rather, it is the independent existence of space in the absence of matter or any other entity that principally interests, and sometimes disturbs, seventeenth century thinkers like Newton and Leibniz. For these reasons, Sklar's more basic definition of substantivalism (1974, 161), which does seem to fit Newton and Leibniz' collective notion and usage, will be preferred throughout the ensuing investigation, i.e., space as an independently existing entity. Finally, substantivalists additionally hold that the spatial relations among bodies are derivative of the spatial relations among the parts of space that the bodies occupy.

1.1.2 Relationism

The problems intrinsic to defining substantivalism pale in comparison to the obstacles that face a viable formulation of relationism. Once again, Earman (1989) furnishes a set of useful conceptual distinctions that has moved the debate in a more positive direction, even if these distinctions rely upon the machinery of contemporary spacetime theories to the exclusion of a more historically neutral approach. In what follows, we will introduce several relationist themes that are modeled on Earman (1989) and Belot (2000, 2011), but which alter the emphasis and details in various ways.

The first type is "eliminative relationism", which conceives spatial relations as the distance relations among material bodies, or points/parts of physical fields, and thus limits the domain of spatial relations to the domain of material relations. More carefully, it is the relations that hold *directly* between bodies, or *directly* between the points or parts of a physical field, that constitute space, hence there is no space if there are no bodies, or the points/parts of a physical field, that stand as the relata of the relations. As will be mentioned below, Huygens may have been the first truly modern eliminative relationist, whereas the work of Barbour and Bertotti (1977,

1982) reflects a contemporary eliminative relationist outlook. An obvious upshot of eliminative relationism is that, provided a finite number of bodies, the scope of the spatial geometry is confined to these bodies, and hence the structure of space cannot be determined to be, say, infinite Euclidean as opposed to a different metrical structure that is only Euclidean in the materially occupied region. A closely related notion to eliminative relationism we will dub, "super-eliminative relationism", further confines the domain of spatial relations to the material relations *within* matter (or physical fields): i.e., unlike eliminative relationism, which allows spatial relations among or between bodies, super-eliminative relationism stipulates that all spatial relations must be coextensive with a body or field. Descartes' claim that there is only a conceptual distinction between space and matter, in conjunction with his denial of any possible void space, is consistent with super-eliminative relationism, but this retrospective assessment, like the supposed Newton-substantivalism link, risks running afoul of the textual evidence (on Descartes, see §1.3.2 below and §9.3.6).

The ideal geometrical setting for the eliminativist forms of relationism is Leibnizian spacetime structure, which can only meaningfully determine changes in position, velocity, and acceleration among bodies. A Leibnizian spacetime does not have the ability, as briefly noted above, to delineate the rectilinear paths required for the conservation principles and laws of motion that comprise Newton's mechanics and the other seventeenth century mechanical hypotheses: in Leibnizian spacetime, all trajectories through space are equal, so to speak, whether curvilinear or straight-line (via the group structure of the transformations that defines that spacetime). Accordingly, Leibnizian spacetime is the obvious choice for conjoining eliminative relationism with that other foundational principle of the modern relationist viewpoint, namely, relational motion. As defined by Earman, relational motion holds that "all motion is the relative motion of bodies, and consequently, spacetime does not have, and cannot have, structures that support absolute quantities of motion" (1989, 12). Whether this strong conception of relational motion was actually endorsed by Descartes and Leibniz is rather controversial, as will be discussed (see, also, Rynasiewicz 2000), but many modern exponents of eliminative relationism do accept these strictures, such as the work of Barbour and Bertotti cited above, and at least one seventeenth century natural philosopher, Huygens, straightforwardly embraced the notion, claiming that "[m]otion is only relative among bodies" (H 46). Furthermore, in an ironic historical twist, Newton's attack on the concept of relational motion (in the thought experiments involving rotation in the *Principia*'s scholium on space and time; N 68–70) likely elevated the relational motion concept in the minds of many contemporary and later philosophers, and through the simple expedient of merely standing as an alternative to Newton's contentious absolutism.[3]

---

[3] This is, in fact, the case with Huygens, who was originally content to side with an absolutist (or non-relational) understanding of rotation, but took the challenge of Newton's spinning bucket and rotating globes thought experiments as a motivation to develop an account consistent with relational motion; see, Stein (1977a), Bernstein (1984).

Unfortunately, the construction of a viable system of mechanics founded on eliminative relationism and relational motion has proven to be an especially difficult enterprise, with rotational motion still epitomizing the greatest obstacle for the relationist approach (see, Earman 1989, chapter 4). Although some researchers have persevered in their attempts to devise a strict relationist scheme (e.g., Barbour 1999), many relationists have adopted a number of more lenient forms that we will collectively label, "sophisticated relationism". The first sophisticated strain is modal relationism, which allows the utilization of "possible" bodies in the construction of spatial relations; i.e., rather than restrict the domain of spatial relations to existing bodies, modal relationist admit possible bodies as well, and hence one can now account for, e.g., the structure of an infinite spatial geometry even if only a few (finite) bodies actually exist (see, e.g., Manders 1982). Many sophisticated relationists also permit spacetime structures that allow absolute quantities of motion, in particular, neo-Newtonian spacetime (absolute acceleration), as opposed to the more natural Leibnizian spacetime (whose quantities are relational, as explained above). In contrast, "metric field relationism" conceives the metric field in GR as a physical entity, which thus renders the structure palatable for relationists (see, e.g., Rovelli 1997). What all of these sophisticated relationisms have in common is the rejection of substantivalism, i.e., space as a unique non-material (or non-physical) entity—rather, sophisticated relationists reckon space to be either a material/physical entity (metric field relationism) or the relations among material/physical entities (modal relationism).[4] Since sophisticated relationism has a natural affinity with alternative, third-way spatial ontologies, we will postpone an investigation of specific examples, as well as the substantivalist response, until Parts II and III. The criticisms raised by substantivalists against the sophisticated variety of relational hypotheses can be easily guessed, however: by allowing possible bodies, neo-Newtonian spacetime, etc., sophisticated relationism is a substantivalist (absolutist) conception masquerading as relationism.

1.2   The Third-Way in Spatial Ontology

As used in this investigation, "third-way" theories of spatial ontology reject the standard dichotomy and strive to construct an alternative conception. These third-way spatial ontologies cover an array of diverse approaches, but we will sort them according to their ontological orientation and ambition, with the emphasis placed exclusively on the property theory, spacetime structural realism, and the definitional interpretation of space. While Chap. 5 will provide a more detailed account, the rationale behind our selection follows from the increased attention and importance that these three strategies have received in the recent literature, and their close relationship with various philosophical disciplines or methods, specifically, traditional

---

[4]As used in our investigation, "physical" will pertain to both matter and fields, since, for some readers, "matter" and "material" may not be associated with the modern field notion in physics.

metaphysics/ontology (property theory), scientific realism (structural realism), and empiricism/positivism (definitional view). In effect, given the predominance of the standard dichotomy, there may be no further third-way conceptions of recent note, although one could presumably invoke the many modern hypotheses that analyze space from the perspective of human experience alone (whereby "modern", we are leaving aside the German Idealist and other phenomenalist schools prior to the second half of the twentieth century). These theories of "lived-space", which stem from contemporary psychology, sociology, geography, and other fields in the human sciences, do not attempt to explicate the nature and ontology of physical space, hence they will not form a part of our investigation—but, see Slowik 2010, for a skeptical overview of various modern lived-space notions that do wander into this terrain (pardon the pun), often treating physical space as a subjective or social construction.[5]

1.2.1   Space as Property

The property theory of space is the most forthrightly ontological of these third-way conceptions, and, since our investigation is devoted to alternative ontologies, it will receive the most attention and development as a direct result. Although not well-known, some of the principle players in the spatial ontology debate have discussed the property theory, if only briefly or in passing (Earman 1989, 13–15; Belot 2000, 576). On Earman's estimate, the property in question is spatiotemporal location, which can be thus seen as the non-substantivalist analogue of manifold substantivalism (where the manifold, as discussed above, is the topological structure that provides spatiotemporal location, but not distance or other metric functions). Although

---

[5] See, e.g., Casey 1997. One should mention Foster's (1982) spatial idealism, or anti-realism, in this context; see, Dainton (2010, chapter 10), for a detailed overview. In short, Foster's arguments are as plausible as Berkeley's, i.e., not very, despite their sophistication. Foster treats space as akin to an unobservable theoretical entity whose intrinsic character is forever unknowable, a conclusion that is part of his more general anti-realism about the entire physical world. This is different from geometric conventionalism, it should be noted, where only a finite and constrained class of possible geometries plus physical theories fits the evidence (see chapter 8). As noted above, some lived-space theorists seem to favor subjectivism or social constructivism, such that physical space and its properties are purely (or largely) relative to the individual or group—but this approach leads to a spatial form of solipsism, and the embarrassing scenario whereby the flat-earth believers live on a flat earth while the round-earth believers simultaneously live on a spherical earth. The fact that there are no substantive scientific disagreements on the earth's shape, or other empirically accessible objects, undercuts these spatial subjectivisms, as well as raises doubts for Foster's claim that the intrinsic nature of spatial geometric structure is unknowable—and, invoking the underdetermination of geometric structure across theory change (e.g., Newton's Euclidean geometry replaced by Einstein's non-Euclidean) does not help, since the history of these theoretical replacements reveals that the prior theory's geometry has been upheld as limiting cases of the successor theory's (e.g., Euclidean geometry is accurate except near the speed of light or near massive bodies like the sun).

Earman's analysis largely concerns why a spatiotemporal location property violates relationism, we will adapt his concept for the needs of a property theory of space.

P(loc): there are irreducible, monadic spatiotemporal properties, such as "is located at spacetime point p".

The limitations of P(loc) is that, as hinted above, it is wedded to a particular structure, the point manifold, which may not be a feature of all property theories, nor does it address the key ontological issues involved.

While the P(loc) concept will factor into the investigation of the property theory, the main emphasis will be placed on an adaptation of an idea that is often dubbed "ontological dependence" (see, e.g., Tahko and Lowe 2015; Rosen 2010), and which will be designated as:

P(O-dep): space is either an emergent, supervenient, or internal property of a grounding entity, and space cannot exist in the absence of that grounding entity.

Throughout our investigation, this property version of ontological dependence will be taken in a very broad sense, with the intention being to outline the numerous ways that space may originate from various types of entities. Informally, A supervenes on B if there cannot be a change in A without a change in B, although this approach does not delve into the ontological details of supervenience. Emergence, while similar to supervenience, is normally associated with a type of holism that arises from the combination of the more basic parts of the whole; i.e., there are novel phenomena or causal powers that arise from the whole that are not found in the individual parts that make up the whole (nor in a simple summation or arrangement of the parts). For instance, water has properties that are not found in the separate hydrogen and oxygen elements that constitute water. Reduction, on the other hand, would likely hold that the whole is nothing more than the sum of the parts. Leaving aside these definitional complexities, on our interpretation of ontological dependence, the emergent, supervenient, or internal property must retain a degree of ontological autonomy, and not comprise a mere logical or external relationship among the grounding entities which can be entirely eliminated by a shift in theoretical focus or logical classification. Accordingly, space as an internal feature of a body or field straightforwardly supports P(O-dep) under the venerable substance/accident (substance/property) distinction, but supervenience, emergence, or other relationships of that ilk would also satisfy P(O-dep) even though these latter concepts are much more difficult to pin down as regards their ontological import. Finally, and most importantly, P(O-dep) sanctions spatiotemporal structures that violate the tenets of eliminative and super-eliminative relationism, such as inertial structure (i.e., neo-Newtonian spacetime and stronger). This aspect of P(O-dep) thereby reveals one of the touted advantages of a property theory of space, namely, that it incorporates the best feature of substantivalism, inertial and other rich spatial structures, alongside the best feature of relationism, that space does not exist in the absence of all entities and physical processes (see Chap. 5).

Besides a metaphysical or ontological reading of the dependence relationship, supervenience and emergence are often conceived, as noted above, as theoretical relationships in the older deductive-nomological sense of reduction, whereby the higher level theory's entities and laws are really just effects or results of the lower level theory's entities and laws. Since later chapters will allow this form of reduction to be included in the discussion, along with the associated concepts that pertain to how one theory may approximate another as a limiting case or idealization, a property version of this type of relationship will also play a major role in our investigation, an idea that will be coined "theoretical level dependence", and defined as:

P(TL-dep): space is an element of a higher level theory that either emerges from, or can be reduced (possibly as an approximation or limiting case) to, elements of a lower level theory.

One might question whether a theoretical reduction of this sort counts as a type of property, but this worry can be easily alleviated if one examines some of the classic examples of this theoretical relationship, such as the reduction of chemistry to physics. The elements of chemistry (e.g., hydrogen, helium) do not exist in the absence of the elements of physics (i.e., electrons, protons, neutrons), and thus the former can be viewed as properties that result from different combinations of the latter. Put differently, different theories can pertain to the same physical (as opposed to theoretical) entities and their various properties at different phenomenal levels or scales. In fact, since a theory can be taken quite generally, including even theories of ontology, the ontological dependence relationship can be translated into a theoretical level dependence relationship as well; e.g., if one declares that material bodies are ontologically dependent on force, then the higher level theory that pertains to material bodies stands in an obvious relationship of dependence, in the P(TL-dep) sense, on the lower level theory that describes force. Like P(O-dep), P(TL-dep) also authorizes spatiotemporal structures (such as inertial structure) that contravene eliminative relationism and super-eliminative relationism. As noted above, P(TL-dep) will be taken up later in the investigation, especially Chaps. 9 and 10, but, given their close relationship, this version of a property theory is tacitly presumed in all discussions of P(O-dep) as well.

As will be demonstrated in Part I, Newton and Leibniz subscribe to P(O-dep), as have most spatial ontologies over the centuries. The non-eliminative sophisticated versions of relationism also fit P(O-dep), as well as P(TL-dep), since the spatial structures that arise from matter cannot be eliminated or reduced in the manner sanctioned by eliminative relationism: i.e., eliminative relationism holds that space is merely an external distance relation among actually existing bodies, whereas sophisticated relationism often posits inertial structure that transcend (emerge from, supervene on) bodies and their actual/possible external distance relations, thus there is close association between sophisticated relationism and P(O-dep) as it pertains to the nature of this emergence or supervenience feature of the spatial geometry required for physical theories (but §10.5.2 will air the property theorist's rebuttal).

Substantivalism, in contrast, holds that space is an independently existing thing, and so there is no deeper grounding entity that would constitute the basis of a P(O-dep) relationship. One might try to distinguish different aspects of this spatial substance that do stand in such a dependence relationship, yet, in order to qualify under the P(O-dep) classification, space would have to be seen as arising from a deeper layer of ontology—but then it would follow that this conception of space fits the property theory as opposed to substantivalism.

1.2.2   Spacetime Structuralism and the Definitional Approach

If the property theory joins the company of substantivalism and relationism by falling in line with traditional metaphysical categories—i.e., substance, property, relation—the other entrants in our survey of third-way theories represent less orthodox ontological approaches, either by way of scientific structural realism or a positivist-inspired substitution of spatial metaphysics for a conceptual system tied to material phenomena. Only a brief introduction will be presented at this point, since later chapters will take up these theories in more depth. The structural realist conception of space, and structural realism in general, is derived from scientific realism, but with the important proviso that the realist commitment is directed towards the essential structure of a theory (especially the structures that remain invariant across theory change), and not its theoretical entities. There are ontological and epistemological versions of structural realism that complicate any assessment of the content and function of structural realism, yet, because the attention is centered on invariant theoretical structures, and not entities, structural realism both marks a departure from the standard dichotomy on spatial ontology and more naturally compliments the spacetime structures approach discussed in §1.1 (see, e.g., Dorato 2000). Lastly, DiSalle (e.g., 1994, 2006), following the work of Stein (e.g., 1967, 1977b) has put forth a view of spatial geometry as the conceptual or definitional structures required for a system of laws of motion and dynamical interactions, a line of thought that perhaps can be traced back to Euler (or even Newton and Kant), and finds a twentieth century analogue in the empiricist's (or positivist's) constructivist perspective on scientific theories. This "definitional" or "dynamical" conception of space, as we will often call it, goes one step further than structural realism by championing a sort of deflationary attitude towards the typical metaphysical issues associated with space. That is, if structural realism rejects the standard dichotomy but takes a scientific realist stance on spatial structures, the definitional approach would seem to drop the scientific realist commitment as well: spacetime structures play a constitutive role in the formation of physical laws and principles, but we should restrict our philosophical speculation to just this constitutive role.[6] Given its deflationary

---

[6] DiSalle references Kant's "empirical realism" about space in his interpretation of the definitional school's metaphysical outlook, although it is unclear just what this entails as regards the standard dichotomy, or even scientific realism; see, DiSalle (2006, 67). Throughout the investigation, Stein

standpoint, the definitional conception of space may seem to fall outside the scope of our investigation of alternative spatial ontologies, yet, by addressing these ontological issues—even if obliquely via its definition-based resolution of the worries generated by the inadequacies of the standard dichotomy—it nonetheless satisfies the requirements of our study.

In what follows, these three third-way conceptions of space will be examined against the backdrop of the standard dichotomy, although other alternative approaches will occasionally be discussed. Once again, the justification for our selection stems from the increased interest in, and prevalence of, these three schemes in current philosophy of science and the philosophy of space and time. In §10.5.3, a final synopsis of the relative merits and the close relationship between these different third-way strategies will be offered.

1.3   The Complex Geneses of the Standard Dichotomy

While their persistent efforts to enlist Newton and Leibniz into the ranks of modern substantivalism and relationism exposes the importance of establishing an historical lineage, present-day spatial ontologists have shown a curious absence of interest in extending their historical research into the more general question of the very origins of the standard dichotomy itself. An in-depth investigation is beyond the bounds of our analysis of alternative spatial ontologies, but a brief survey of the likely late Medieval and Renaissance sources of the absolute/relational (substantival/relational) division will be provided. In particular, we will begin with an examination of Descartes' hypotheses on space and motion as means of disclosing the extreme difficulties involved in any attempt to read the modern dichotomy into seventeenth century natural philosophy, and finish with an overview of several precursors to the standard dichotomy in the period just prior to Newton's canonical presentation of these concepts in the *Principia*.

1.3.1   *The Relativity of Perceived Motion*

The late Medieval natural philosophers developed a number of concepts that would influence and inform those seventeenth century theories of motion and space that we now regard as relationist and absolutist. One of these concepts, the relativity of the perception of motion (or kinematic relativity), was a common theme in fourteenth century natural philosophy, although its origins can be traced back to the

---

will be listed as a member of the definitional approach, although it is unclear that he would accept this designation. But, since the conception that he advances, especially in the context of Newton (see chapter 2), seems quite close to DiSalle's, the designation seems justified.

ancients. As explained in Clagett (1959), both John Buridan and Nicole Oresme "illustrated the doctrine by positing observers on ships which are moving relatively to each other, noting that the observer would not detect his own motion (assuming a calm sea) but would describe the motion of another ship in terms of his own ship being at rest" (1959, 585). "[T]he doctrine of the relativity of the detection of motion" adds Clagett, "can be juxtaposed with the view of William Ockham and his nominalist successors that movement (*motus*) is not an entity having existence independent of the moving body, but is simply a shorthand way of saying that a body occupies first this position, and then this position, [etc.]" (589). Ockham's nominalist reduction of local motion to a mere change of place, a view with strong relationist overtones, should be contrasted with the latent absolutism in Buridan's property-oriented account of motion (which is similar to, but distinct from, his better known impetus concept).[7] To quote Anneliese Maier, "local motion [for Buridan] is by no means merely a name [as it is under Ockham's nominalist account], but rather something real that inheres in the moving object and is distinct from it. It is a pure successiveness (*res pure successiva*) or flux that is not reducible to other [Scholastic] categories and must simply be affirmed in its uniqueness" (Maier 1982, 35; she also describes Buridan's *res succesiva* as "a flux that to a certain extent inheres in the object as an accident"; 86).[8] For Descartes and Leibniz, the motion of bodies *per se*, i.e., as opposed to the cause of their motion, would seem to accord with Ockham's nominalist hypothesis more closely than any account, like Buridan's, which conceives motion as akin to a property (see Chap. 4 on Leibnizian motion).

On the other hand, given the dominance in the later Medieval period of the Aristotelian view that extension is a bodily property (i.e., substantial form), it is therefore not surprising that the development of the various Scholastic ideas associated with space often bear more resemblance with themes in modern relationism than with absolutism/substantivalism. One school of thought, based on Aquinas' work, made a distinction between formal place, which is place as determined relative to the universe as a whole, and material place, which is the standard Aristotelian-based idea that place is the boundary between the contained body and the surrounding containing bodies (in a material plenum). By this means, one could uphold the widely accepted notion that place is immobile, via formal place, whereas material

---

[7] Impetus functions much like an impressed force for Buridan, and is the cause of local motion, although it is ontologically distinct from local motion (see, Maier 1982, 86–87). Impetus was a commonly accepted notion in the fourteenth century, it should be noted.

[8] Ockham's nominalist conception of local motion and Buridan's *res pure successiva* hypothesis can be seen, respectively, as variations on the concepts, *forma fluens* (that motion is the terminus or form acquired) and *fluxus formae* (that motion is an additional flux over and above the terminus or form acquired), put forth in previous centuries by Averroes and Albertus Magnus. In addition, Buridan's idea was prompted by the concern over the rotation of the universe, where no new places (termini) are acquired, and thus a problem for Ockham's nominalist account and, more generally, the *forma fluens* view (see, Murdoch and Sylla 1978, 215–219). Grant refers to Buridan's *res pure successiva* account as establishing "the absolute nature of motion", and as an accidental form possessed by the body (Grant 2010, 79).

place is not (since the bodies that make up the boundary of material place might themselves be in motion). This approach evokes relationism in that it defines place, both formal and material, relative to physical entities (respectively, the whole universe and the boundary between bodies), as well as denies the motion of the whole universe (which explains the immobility of formal place). The alternative account, attributed to Scotus and developed up to the end of the sixteenth century by many others (e.g., Eustachius), rejects formal place but strives to recapture that useful notion by means of "equivalent places", or "equivalent relations" among the contained and containing bodies, and thereby retains the ability to refer to a fixed place that, say, has been vacated by one body and occupied by another, etc. (The problem is, in short, that any change to the material boundary results in a new place, so a motionless body in a fluxing plenum is constantly, and counter-intuitively, changing its place.) The Scotist approach resembles modern relationism in that it defines place via an equivalence relation among bodies, or the boundaries of bodies, but Scotus (and many Scotists) also sanctioned the possible motion of the (finite) universe as a whole through the imaginary void space outside the universe, a view that appears to flout relationism for an absolutism of motion (since it is a motion that is not relative to matter).[9] Imaginary space, by the seventeenth century, was usually conceived as a non-dimensional capacity to receive extended three-dimensional bodies, with God's immensity serving as the ontological grounds of this capacity (and not matter; see, Grant 1981, chaps. 6 and 7). However, the development of these imaginary space theories in the late Medieval and Renaissance periods did not factor into the construction of Descartes' own theory of space, or at least not much. His rejection of the possibility of a void (Pr II 16) is a likely factor in his refusal to take up the imaginary space concept, but one can still employ variants of this type of space to account for the immobility of place, as will be explained below.

---

[9] In the discussion above, the problem of the motion of the outermost sphere of the universe has not been discussed, although it was one of the motivating factors behind the debates on the status of place: since the final sphere does not have an outer boundary, and hence is not in a place, many ideas were put forward to resolve this dilemma (such as the relationship of the parts of the final sphere to the inner parts, the latter providing a roundabout way of securing the place of the former). On the influence of late Scholastic ideas of place on Descartes, see also, Des Chene (1996), Ariew (1999, chapter 2). Ariew provides a nice summary of these conceptual developments: "In sum, while later Scholastics agreed in rejecting the independence of space from body, they disagreed about other important issues. Hidden within the debate between Thomists and Scotists on the question of the mobility/immobility of place and the place of the ultimate sphere were questions about the relativity of motion or reference of motion. Some thinkers supported a Thomist doctrine in which the motion of a body is referred to place, conceived as its relation to the universe as a whole, a universe which is necessarily immobile; others supported a Scotist doctrine in which the motion of an object is referred to its place, conceived as a purely relational property of bodies" (1999, 53). By the sixteenth century, various natural philosophers, such as J. C. Scaliger, resurrected the ancient atomist's concept of a three-dimensional void space as a means of overcoming the inherent limitations involved with Aristotle's conception of place; see, Grant (1981, chapters 7 and 8), and Leijenhorst and Lüthy (2002, 387).

1.3.2   Cartesian Spacetime Revisited

Given the influence of the many different Scholastic theories of space and motion on Descartes' own thought, any attempt to discern original relationist themes or concepts in his work is thus rendered a quite problematic exercise. In essence, Descartes' hypotheses in this area of natural philosophy represent a mere extension of the debate that he inherited from his illustrious Scholastic predecessors, so, just as with the earlier Scholastics, one can locate a number of ideas that lean, somewhat haphazardly, either relationist or absolutist, but more often relationist given the Aristotelian framework and its substance/property scheme (i.e., spatial extension is a property, probably an essential property, of Scholastic material substance). His chief contributions to the development of our modern understanding of motion and space are, in fact, either neutral with regard to relationism or more absolutist in orientation; namely, the laws of motion (which include the collision rules and the conservation law) and his mechanical method of explaining physical phenomena. Like the other members of the mechanical school of natural philosophy in the seventeenth century (excluding Leibniz), Descartes abandons the Scholastic's substantial forms and accidents, claiming instead that material phenomena "can be explained without having to assume anything else…in their matter but motion, size, shape, and the arrangement of their parts" (CSMK III 89). Likewise, he deems the multi-faceted Aristotelian definition of motion ("as the actualization of a potential in so far as it is a potential") to be hopelessly obscure (CSM I 93–94), and insists that all movement is "local movement, because I can conceive no other kind" (Pr II 24).

However, in working out the details of his theory of motion and space, Descartes adapts a host of Scholastic ideas for his own purposes, with the nominalist conception of motion assuming a leading role. In the *Principles*, he defines motion, in the proper sense of the term, as "the transfer of one piece of matter or of one body, from the neighborhood of those bodies immediately contiguous to it and considered at rest, into the neighborhood of others" (Pr II 25). The two-dimensional inner surface of the containing bodies that border the contained body, also named the "external place" of the contained body in an earlier section, is thus involved in the account of proper motion, whereas "internal place" is the abstraction of the three-dimensional extension of bodies "considered in a general way" (Pr II 12, and Pr II 15; the internal and external place dichotomy is itself a scholastic development, and can be found in such thinkers as Toletus, among many others; see, Garber 1992, 148–155).[10] Internal and external place function much like the Scholastic equivalent places idea described above, since he argues that the extension of a place can be common to the many different bodies that come to occupy that place "provided only

---

[10] More carefully, Descartes defines proper motion with respect to the "immediately contiguous" bodies, and does not mention the common surface between contained and containing bodies. However, as noted above, external place would seem to be implicated in this process. See, also, Zepeda (2014).

that it is of the same size and shape and maintains the same situation among the external bodies which determine that space" (Pr II 12). Descartes' proper motion is next contrasted with a common understanding of motion, which he describes as "the action by which a body travels from one place to another" (Pr II 24), and which can admit a host of different perspectives for the measurement of the motion. Hence, Descartes' common conception of motion embraces the relativity of the perception of motion outlined above, for a body can simultaneously partake in many motions relative to different reference bodies or viewpoints, as when an object on board a ship is viewed at rest relative to the other parts of the ship, but moving relative to the shore; and, the principle of Galilean relativity, that physics is the same in all inertial frames, is also upheld since passengers in the ship's cabin would not be able to determine if they were at rest or in uniform motion (Pr II 24). Descartes strives to resolve this relativity of common motion by utilizing the contiguous neighborhood of bodies: if a body's motion is judged by way of its contiguous neighborhood, then a body is either at rest, or in translation away from, its contiguous neighborhood, thus dispelling the relativity of the judgment of motion by means of this invariant fact (i.e., translation or no translation; Pr II 25).

As often noted by commentators, however, Descartes' definition of proper motion courts relationism by declaring that the contiguous bodies that constitute the reference frame for determining motion are "considered at rest", a description that seems to imply that the decision as to which bodies are at rest or in motion, either the containing or the contained, is arbitrary. Therefore, despite the facts pertaining to a transfer/non-transfer from the contiguous neighborhood, it would seem that his concept of proper motion upholds the relativity of perceived motion as does his definition of common motion. Indeed, he states that "we cannot conceive of the body AB being transported from the vicinity of the body CD without also understanding that the body CD is transported from the vicinity of the body AB" (Pr II 29), and he concludes by adding that "all the real and positive properties which are in moving bodies, and by virtue of which we say they move, are also found in those [bodies] contiguous to them, even though we consider the second group to be at rest" (Pr II 30). For many of Descartes' contemporaries and later natural philosophers (a famous early example being Newton's analysis in *De gravitatione*), these last assertions about the reciprocal nature of the transfer between contained/containing bodies would be construed as supporting the relational theory of motion. Nevertheless, these statements may be simply intended do deny that motion is a property that resides in a body and which is causally responsible for its change of place. In opposition to any seventeenth century version of an impetus concept, perhaps even Buridan's *res successiva* account of local motion, Descartes' goal may have been only to support the nominalist reduction of motion to change of place, and this would explain why he consistently points out that there is nothing more "in" the moving body than in the contiguous neighborhood of bodies. He explains that motion "is only a mode, and not a substance, just as shape is a mode of the thing shaped, and rest, of the thing which is at rest" (Pr II 25).[11]

[11] On Descartes' conception, a "mode" of extension is a way that extension manifests itself, or as a property of extension (Pr I 53; for instance, shape is mentioned as a mode of extension). Presumably,

Overall, while it is unclear that any definitive explanation can be given for the relationist-like features in Cartesian motion, the nominalist interpretation just offered has the advantage that it is consistent with the many aspects of his natural philosophy that undermine relational motion. A notable instance involves the fourth and fifth collision rules, which are specific cases of bodily interactions that fall under the *Principles'* third law of nature: "The third law: that a body, upon coming in contact with a stronger one, loses none of its motion; but that, upon coming in contact with a weaker one, it loses as much as it transfers to that weaker body" (Pr II 40). In the fourth collision rule, Descartes stipulates that a large stationary body remains at rest after impact with a smaller moving body, while the smaller body reverses its path after the collision (Pr II 49). In the fifth rule, conversely, a large moving body will move a smaller stationary one, "transferring to [the smaller body] as much of its motion as would permit the two to travel subsequently at the same speed" (Pr II 50). According to the relational theory of motion, however, rules four and five constitute the very same collision, since both involve the interaction of a small and large body with the same relative motion prior to impact (once again, there are only differences in motion among bodies on the relational motion hypothesis), and so the outcomes should be identical. The divergent outcomes of these two collision rules thus implies that either Descartes did not accept a thoroughgoing relational theory of motion, or, less plausibly, that he was unaware of the relationist contradiction exemplified through his handling of these collision rules.[12]

Further evidence in support of the idea that Descartes' account of the reciprocity of transfer is actually intended to deny that motion is a property or substance, rather than support relational motion, can be found in the late correspondence with Henry More. If the transfer of a body is merely reciprocal, More asks if the wind blowing through a tower window falls under a reciprocal transfer as well, so that the air can be viewed at rest and the tower in motion (5 March 1649, AT V 312). Rather than answer More's question directly, Descartes attempts to sidestep the problem by invoking an example that involves the force (or strength) of two men attempting to free a grounded boat, with one on board pushing against the shore, and the one on shore pushing against the boat:

> I cannot better explain the reciprocal force [*vires*] in the separation of two bodies one compared to the other than by putting under your eyes a situation wherein a boat is stranded against the shore of a river and there are two men, one of which is on the shore and pushes the boat with his hands away from the shore, and of which the other being in the boat, in the same way, pushes the shore with his hands to also draw the boat away from the shore. If the

---

claiming that motion is a mode of extension eliminates the possibility that motion is a feature of a body that is separate from extension, i.e., a "substance" *in* extension, the latter view seemingly implicit in Buridan's concept.

[12] Schmaltz (2008, 89) denies the relationist interpretation of Descartes theory of motion on these very grounds, i.e., that it is inconsistent with the details of his laws of motion. Others, such as Blackwell (1966, 227), have argued that the relational aspects of his theory of motion are not intended to be taken seriously, and are merely an attempt appease the Church's potential condemnation of his Copernicanism (since Descartes' puts the earth at rest in a vortex circling the sun, and so the earth does not move in the proper sense, as defined above).

force of the men is identical, the effort of the man on the shore, who is thus connected to the land, contributes no less to the boat's motion that the effort of the man on the boat, who is transported along with it. Therefore it is obvious that the action by which the boat recedes from the shore is equally in the shore as in the boat (15 April 1649, AT V 346).

In other words, the reciprocity of transfer between the boat and shore, a purely kinematical event, is reinterpreted as a sort of reciprocity of force (or action) between two possible (dynamic) sources of the motion, i.e., the push of each man. As noted in Slowik (2002, 126–128), this account would seem to admit that it is actually the boat (or wind, in More's example) that is in motion, and not the shore (tower); likewise, the kinematical and dynamical aspects of motion are conflated (as is also argued by Shea 1991, 323). Yet, if the reciprocity hypothesis was originally formulated to deny that motion is a sort of property that a moving object obtains from other bodies, then Descartes' reply to More adds valuable information about his conception of motion, and is not simply an evasion of an embarrassing consequence pertaining to relative motion. In effect, Descartes is invoking an early version of what would become Newton's third law of motion (N 71–72), that every action has an equal and opposite reaction, as an alternative to the idea that motion is brought about by the transference of a (substantive) force among bodies, i.e., motion "is a *transference*, not the force or action which transfers" (Pr II 25). Put simply, if the force that moves the boat "is equally in the shore as in the boat", then the hypothesis that the boat moves due to a transference of force (from shore to boat) is obviously refuted. This interpretation remains consistent with relational motion, it should be noted, but it does suggest that the goal of the reciprocal transfer hypothesis is not relational motion—instead, Descartes' real aim is to undermine the concept that motion is a bodily property, with an account of motion that strongly resembles modern relationism forming a byproduct of that strategy.[13]

In fact, given a body's translation from its neighborhood, the possibility that one could not establish whether it is the body or the neighborhood that really moved apparently posed a problem for Descartes, for he includes a discussion of a potential method for breaking the symmetry of reciprocal transfer via two smaller bodies that move in opposite directions on the surface of a larger body. In brief, since positing motion to the larger body leads to the contradictory conclusion that it moves in two directions at the same time relative to the resting smaller bodies, it must be the two smaller bodies that move in opposite directions on the resting larger body (Pr II 30; see Slowik 2002, chap. 7). This system for determining a body's individual state of motion may be acceptable to a relationist, but it is clearly limited in scope: e.g., what happens when the motion is between continuous bodies, such as two nested spheres? More problematic is Descartes' claim that rest and motion are opposite states (Pr II 37), a stance that runs afoul of a truly relationist hypothesis, needless to say. Accordingly, once all of the non-relational aspects of Descartes' system are

---

[13] The rejection of the idea that motion is a substance/property possessed by the moving body is also a feature of various Cartesian texts in the late seventeenth century, such as Rohault's popular text (1969, 42–43). Likewise, various modern commentators have interpreted Descartes' reciprocity of transfer hypothesis along these lines, e.g., Miller and Miller, in Descartes (1983, 51, n. 13).

taken into account, it becomes quite difficult to defend the relationist reading of Cartesian motion that would seem to have gained hold by the end of the seventeenth century.

On the whole, an upshot of Descartes' method of determining place is that there would seem to be no empirical means of securing a fixed location within the material world, a conclusion that nicely accords with relationist doctrine. Indeed, he openly admits that it is probable that there are "no truly motionless points" in the universe, and "from that, we shall conclude that nothing has an enduring fixed and determinate place, except insofar as its place is determined in our minds" (Pr II 13). An epistemology that renders the determination of motion a relative feature is not sufficient in itself to count as a theory of relational motion, however. One additionally requires an ontology that denies that there are any frameworks or methods by which to determine the true (i.e., absolute) states of bodily motions as opposed to mere relative motions—and, besides the issues raised above (motion and rest as opposite states, etc.), there are additional reasons to believe that Descartes' overall ontology falls short of this demand. In several discussions, he appears to acknowledge that there is an absolute fact about bodily states of motion and rest from God's perspective, even though his comments about the lack of a fixed place suggest that these facts cannot be obtained from within the material world. In the *Principles*, it is argued that God placed a determinate amount of motion into the world at the moment of creation, a conserved quantity (dubbed "quantity of motion", measured by the product of size and speed) that God still maintains (Pr II 36; Pr III 48). From God's perspective, consequently, there is a determinate, absolute fact about the state of motion of each body in the plenum, as opposed to a mere relative motion, since God is the ultimate and sustaining cause of that motion.[14] In the late correspondence with More, this point is addressed more clearly, and he even suggests that all bodies are naturally at rest absent God's intervention: "I agree that 'if matter is left to itself and receives no impulse from anywhere' it will remain entirely still. But it receives an impulse from God, who preserves the same amount of motion or transfer in it as he placed in it at the beginning" (CSMK 381).

One might respond to this last inference, i.e., God's role as the arbiter of the true states of Cartesian motion, by insisting that it is an irrelevant feature of Descartes' system: the solution to the relativity of perceived motion—the transfer/non-transfer of a body from its neighborhood—is presented as an objective fact *within* the material world, and so Descartes' God does not directly effect the details and function of his theory of motion. This form of reply, while correct in the strict sense that any human application of Descartes' theory must rely solely on the invariant transfer/non-transfer to gauge states of motion, fails nonetheless to render Cartesian motion

---

[14] A God-based conception of the Cartesian conservation law has also been suggested by Hübner (1983, 130), whereas Dugas (1958, 196) favors the material world at the initial moment of creation as the preferred frame (which, of course, only God can know). It should be added, furthermore, that Slowik (2002) does not actually reject a relationist interpretation of Descartes' natural philosophy, since the main goal is to examine the prospects for a consistent relationist reconstruction. This chapter, in contrast, develops a case against relationism.

acceptable to the relationist since there still remains an ontological foundation upon which to single out true motions from mere perceived motion. Indeed, employing God to resolve the problem of the mobility of place, whether material place or another matter-based surrogate, had a long pedigree in Scholastic and non-Scholastic thought by Descartes' time. The rise to prominence of the imaginary space concept in the sixteenth and seventeenth centuries was largely prompted by this very concern, with an incorporeal and (usually) non-dimensional imaginary space serving as a foundation upon which to fix the immobility of place (and thus refute the counter-intuitive consequences associated with the mobility of matter-based versions of place). Toletus, whose internal versus external place distinction served as the model for Descartes, used imaginary space in this fashion (Toletus 1589, 122–123), along with the Coimbra Jesuits and many others: as Leijenhorst explains, "when place is conceived as a real surface-limit inhering in a physical body, it cannot be immobile. However, if we consider it to be that part of imaginary space that coincides with the surface-limit of a given body, it clearly fulfills Aristotle's requirement that place should be immobile" (Leijenhorst 2002, 112). The immobility of imaginary space was, in turn, a consequence of God's immensity, although the exact ontological details were a topic of much debate. For example, another seventeenth century Scholastic who greatly influenced Descartes' work, Suárez (1965, 95–111), described imaginary space as the way we understand God's immensity, an interpretation that he hoped would circumvent the difficulties involved with associating this peculiar form of non-entity with God: i.e., imaginary space was often regarded as a pure privation, or a mere capacity to receive bodies, lacking corporeal dimension but somehow linked to, and congruent with, God's immensity. Descartes does not resort to a God-grounded conception of imaginary space to save the immobility of place, preferring his invariant transfer/non-transfer hypothesis instead, but his statement to More that matter would remain at rest in the absence of God's generation or conservation of the world's quantity of motion would seem to attest to the continuing influence of the late Scholastic ontology of place and motion on his natural philosophy—to be specific, if matter has a natural rest state relative to God, then the immobility of place is guaranteed despite our epistemological inability to access the facts pertaining to fixed places from within the world.

Hence, when Descartes declares that "nothing has an enduring fixed and determinate place, except insofar as its place is determined in our minds" (Pr II 13), one way to read this passage is as an endorsement of the abstractionist or conceptualist processes involved in nominalism: a "fixed place" does not refer to the Scholastic's imaginary space, but is simply a nominalist designation for a feature of the material world that, although not itself a substance or accident, can be grasped by our minds via the concepts of internal and external place. We can only gain a rough or approximate notion of a fixed place in the material world employing his place concepts, of course, since all bodies are in motion, but there are nonetheless determinate facts about immovable places grounded in God's perspective of the material world. As will be argued in later chapters, both Leibniz and Newton embrace various aspects of this interpretation of space and motion, since a God-grounded, and nominalist, approach is central to their respective spatial ontologies as well, albeit in quite different ways.

1.3.3   Huygens' Crucial Role in the Evolution of the Standard Dichotomy

It may come as a surprise to many philosophers and historians of science, but Newton did not invent the standard dichotomy, i.e., the concepts of absolute and relational space and motion. In fact, these ideas probably originated, at least in their more familiar guise, during the Medieval period, and they were likely prompted by the relativity of perceived motion (see §1.3.1), in particular, the need to delineate perceived motions from actual motions. Ockham, for example, employed the absolute and relational distinction with respect to "effects" while putting forth his nominalist interpretation of motion: "A local motion is not a new effect, neither an absolute nor a relative [*respectivus*] one. I maintain this because I deny that position (*ubi*) is something. For local motion is nothing more than this; the moveable body coexists with different parts of space" (Ockham 1967, 140).[15] Regardless of its pre-seventeenth century origins, the most likely source of influence on Newton's espousal of the absolute/relational dichotomy can be traced to the numerous works on impact mechanics that preceded the *Principia* in the 1650–1687 period. For instance, absolute and relative space and motion are part of the conceptual apparatus in Giovanni Borelli's *De Vi Percussionis* (1667): "local motion occurs either from one place [*locum*] of world space to another or in the relative space of some container; the former shall be called real and physical motion [*motus realis & physicus*], the latter we will call relative motion [*motus relativus*], although oftentimes it does not involve a change of region [*situs*] in the place or the space of the world" (Borelli 1667, 3). Borelli uses the description "world space" instead of "absolute", but another treatise on collisions and bodily interactions from the same period, Ignace Pardies' *Discours du mouvement local* (1670), does employ the latter label for motion, albeit "respective motion" replaces "relative motion": "I call...*absolute* velocity, which is consider'd in a Body compared with the Space wherein it moveth; and *respective*, that which is considered in two Bodies compared together, by which velocity these two Bodies mutually approach to, or recede from, one another" (DLM 28–29).

---

[15] Given Ockham's complex analysis of language, which employs "absolute" and "respective" in a more general sense, it is unclear how to interpret this passage as regards absolute and relative motion *per se*. However, the use of these terms was likely common by the seventeenth century, since one can find various natural philosophers in the first quarter of that century that refer to space/place as *ens absolutum* or *ens relativum*, such as Balthasar Meisner (see, Leijenhorst and Lüthy 2002, 392). From the perspective of our investigation, the main question concerns the specific historical source and inspiration behind Newton's adoption of those terms and concepts, hence our treatment of this (immense) topic is fairly limited.

In short, there were several natural philosophers who accepted an absolute/relational motion (and space) distinction prior to Newton's *Principia*, but the person who likely elevated this distinction in the minds of late seventeenth century natural philosophers, namely, Huygens, did not. Like Galileo, Huygens' brand of mechanical philosophy is pragmatic and decidedly mathematical in orientation, a feature that probably explains his somewhat ambiguous relationship to orthodox Cartesian natural philosophy.[16] Nevertheless, like Newton, Huygens interpreted Descartes' analysis of motion as favoring relationism, and relational motion ultimately became a central doctrine of his oeuvre. Huygens specifically affirms several hypotheses that are synonymous with modern relationism. First, he reasons that a single body in an otherwise empty universe cannot move, or that the possibility of its motion is a contradiction (since motion is a relation among bodies). In response to the standard Scholastic counter-argument, that God could move a lone body, he states:

> [S]ince I claim that motion is nothing unless in relation to other bodies, without any offense to God we shall say that He cannot make it thus that there is a relation to something that does not exist, i.e. that a body be several bodies. Similarly, I maintain that God also cannot create [*constitui*] a single body at rest, for rest, just as motion, is relative to something else, and neither can be predicated of a single body.[17] (H 27)

This forthright avowal of a central tenet of relational motion is absent in Descartes' work, it is important to note; and, while Descartes offers the fairly neutral suggestion that there may be no motionless points in the material world (Pr II 13), Huygens takes the strict relationist stance that the very question concerning an immovable place/space is meaningless:

> Those who know that the Earth moves will say perhaps that the fixed Stars are at rest... [H]owever, when asked what is it to be at rest, [they] have nothing else to answer but that rest is when a body and each of its parts maintain the same place in the space of the World. But, as this space is infinitely extended on all sides, without any limits or a centre — which is all too obvious to need any proof — they must confess that there is nothing whereby one could determine a place in this space....For when they say that this space is immovable, in order that its parts are likewise immovable, I don't know what their idea thereof is. But they don't realize that knowing what it is to be immovable is what we are still searching for, and thus they fall into a vicious circle. They thought, perhaps, that it would be absurd to say that space is in motion; hence they concluded that it is therefore immovable. Instead, they ought to have inferred that neither motion nor rest can be properly ascribed to this space, but only to bodies — or only improperly to space insofar as it is occupied by or enclosed in a body. (H 54–55)

Huygens' second conclusion, that an immovable space is a confused notion, follows from his first, that motion is exclusively a relation among bodies. Descartes, as

---

[16] Huygens' acceptance of atoms and an intercorpuscular void stand out in this regard, i.e., as particularly anti-Cartesian: see, Snelders (1980, 120), Westman (1980); and, for more on Huygens' physics and the center-of-mass frame, see, Barbour (1989, 473–478), and Slowik (2002, chap. 8).

[17] As an interesting admission of his lack of interest in theological matters, as well as a sign of the impending decline of the theological component of natural philosophy in the following century, Huygens adds in the margin: "Perhaps there is no need to involve God in this matter, after all" (H 27).

discussed in §1.3.2, admits that a given body's rest or motion can be established from God's perspective, and his laws of nature posit individual states of bodily motion in direct violation of strict eliminative relationism. Provided this evidence, it is therefore not surprising that some commentators have singled out Huygens as the first truly modern relationist; e.g., Huygens "was the first physicist who believed in the exclusive validity of a principle of kinematic as well as dynamic relativity" (Jammer 1993, 126).

The most important achievement of Huygens' commitment to relational motion is his use of the center-of-mass reference frame to generalize Descartes' first collision rule to cover the impact of all bodies.[18] From the perspective of an observer situated at the origin of the center-of-mass reference frame (which is often inertially moving itself), two bodies that approach one another from opposite directions along a straight line, irrespective of their size and speed, will rebound after their collision while retaining their initial speeds; or, more formally, the center-of-mass frame is the point where ratio of their (non-accelerating) speeds is reciprocal to the ratio of their sizes. The center-of-mass frame not only upholds the Cartesian conservation law for the quantity of motion (product of size and speed) via Descartes' first collision rule (which is the only correct one out of the set of seven, although confined to equally sized bodies), but, since the colliding bodies move relative to one another, the center-of-mass perspective also upholds relational motion:

> After having shown that the speed [*vitesse*] of rebound, or separation of two elastic or hard bodies depends on the relative speed with which they collide, I shall postulate that there are perfectly elastic bodies that rebound with the same speed with which they approached one another. From this I demonstrate that when they come into collision with velocities inversely proportional to their weights [*poids*] or quantities of matter [i.e., size], they will each rebound with the same speed they had before. And from this I subsequently determine all cases. (H 13)

Overall, one would be hard pressed to find a more faithful exponent of strict eliminativist relationism, whether in the seventeenth or any other century: "We ought not to say that bodies change their relative distance [*inter se*] and position [*situs*] through motion; rather, motion itself is the change in that distance, and is not anything different from it" (H 24).[19] Barbour, moreover, contends that Huygens' analysis of impact anticipates the modern concept of a transformation law among inertial frames (1989, 464).

---

[18] Throughout our investigation, "center-of-mass" will be used interchangeably with "center-of-gravity", although the two are only identical if the gravitational field is uniform in the region that the bodies occupy.

[19] Huygens also discovered the laws for the conservation of momentum (mv) and, roughly, kinetic energy (mv^2), the latter eventually embraced by Leibniz, of course. However, as a devotee of Descartes' concept of speed over the concept of velocity (speed in a given direction), Huygens continued to favor the Cartesian conservation law for the quantity of motion (ms, where s is the scalar quantity, speed), via his revision of Descartes' first collision rule and his use of the center-of-mass frame. Westfall concludes that "the concept [of velocity] did not please him greatly, however, and he continued to speak formally of quantity of motion in the Cartesian sense, a scalar quantity which is always positive in value" (Westfall 1971, 156).

Unfortunately for Huygens, many of his contemporaries interpreted his relationist-based work on impact mechanics in a quite contrary manner. Rather than accept his center-of-mass reconstruction as the proper method for understanding Descartes' collision rules and conservation law—a reconstruction that is firmly rooted in strict eliminative relationism—several natural philosophers drew the conclusion that the Cartesian conservation law is erroneous, since it does not hold true with respect to absolute (or world) space, but only works with respect to the relative space and motion of the colliding bodies. Pardies, whose treatise on impact we have examined above, advances this precise claim, arguing that Descartes' quantity of motion is not conserved in impact: "'tis not true that there is always as much *absolute* Motion after the percussion, as there was before. But 'tis easie to demonstrate, that the *respective* Motion is always the same; so that the Bodies recede one another *after* the percussion, as fast as they approached *before* it" (DLM 48). As is clear from the context, Huygens' center-of-mass method forms the basis of Pardies' interpretation, and Descartes' conservation law and collision rules serve as its target: Pardies insists that he has shown "that *Monsieur Des-Cartes* hath been deceived in Six Rules of the Seven, which he hath delivered about Motion" (DLM 71), and that Pardies' own conclusion, quoted above, "is against *Monsieur Des-Cartes*, who hath not distinguished the Motion which is here called *absolute* from that, which is called *respective*. And when he saith, that there is always an equal quantity of Motion *before* or *after* the percussion, he means it of this *absolute* Motion; or it is very apparent, that he hath mistaken" (DLM 71–72). Similarly, Mariotte's treatise on mechanics, first published in 1673, stipulates as one of its three basic definitions that "[t]he respective velocity of two bodies is that with which they approach each other, whatever may be their own velocities" (*Traité de la Percussion ou Choq des Corps*, 3rd ed., 1684, 1–2; trans. in Dugas 1958, 290). Huygens provides the following rebuttal: "In his Definition 3, Mariotte distinguishes the relative celerity of two bodies from their 'proper velocities'. I contend that there is no proper celerity. Instead of saying, 'no matter what their proper velocities may be', he ought to say, 'no matter what their velocities relative to some other body'" (H 51).

Huygens' pioneering investigation of collisions in the 1652–1656 period provided the catalyst for the interpretation of Pardies, Mariotte, and several others, such as Déchales, hence the espousal of absolute space and motion by Pardies et al., along with the allegation that his work was not duly cited by these authors, must have been a source of much aggravation to Huygens.[20] He states: "The dispute on absolute and relative motion. It is commonly believed that there is some true motion, as opposed to relative motion. Borelli, Mariotte. Maybe Pardies, too? Newton thinks so. Wallis, perhaps" (H 21)? While the reference to Newton reveals that this passage postdates 1687, the year of the *Principia*'s appearance, the essays that sanction absolute space by Borelli, Pardies, Mariotte, Déchales, and likely others, were pub-

---

[20] As Elzinga explains, "Ignace Gaston Pardies (1636–1673) seems to have been the first (or at least one of the first) to have applied Huygens' relativity principle [i.e., relational motion and the center-of-mass method] to the study of impact after Huygens himself—in a work, *Discours du mouvement local* (1670). Edne Mariotte (1620–1684) also made use of it in, *Traité de la Percussion ou Choq des Corps* (1673, 1674, 1676). Huygens was suspicious of Mariotte for borrowing his ideas without mentioning Huygens [see, H 51, above]. Another who soon used Huygens' principle and method was C. F. M. Déchales, *Cursus seu Mundus Mathematicus* (1674)" (Elzinga 1972, 134).

lished in the 1660s and 1670s. Consequently, given the evidence that Newton was familiar with the work of some of these authors, the direct inspiration for Newton's own use of the absolute/relational distinction almost certainly stems their earlier efforts.[21] Just as Borelli, Pardies, and Mariotte's analyses of mechanical phenomena rely on the distinction between absolute space/motion and relative space/motion (see the passages at the beginning of this section), Newton employs this distinction in the exact same manner: "Absolute motion is the change of position of a body from one absolute place to another; relative motion is change of position from one relative place to another" (N 65). However, while Borelli, Pardies, and Mariotte attempt to discern the true motions of bodies from their merely relative motion during impact (with Huygens' center-of-mass frame providing this distinction in Pardies and Mariotte's treatment), Newton must have realized that the inertial motion of a colliding pair of bodies could not reveal the true rest frame of the material world—and for the same reasons that Huygens' recounts above, i.e., Newton's Corollary 5 (N 78). Rotational motion, on the other hand, is not subject to the same limitations, for the non-inertial effects of the rotation do not align with the presence or absence of a relative rotation, a point that Newton vaguely grasped in his critique of the Cartesian vortex theory in the earlier *De gravitatione*, but only developed in full in the scholium on space and time in the *Principia*.

In subsequent chapters (2 and 6), our examination of the attempts by recent commentators to reinterpret Newton's implementation of the standard dichotomy will draw upon these findings, especially those reconstructions that claim that Newton's absolute space only signifies an inertial/non-inertial frame distinction. Yet, shifting to the larger historical context, it is fairly safe to conclude that previous investigations of the origins of the absolute/relational debate have hitherto failed to take into account the upsurge in the utilization of that dichotomy in the essays on mechanics in the decades prior to its celebrated appearance in the *Principia*, mechanical treatises that, moreover, appear to have been directly inspired by Huygens' study of impact in the 1650s. On the whole, and leaving aside the catalyst of Descartes' rather equivocal conception of motion, the absolute/relational distinction underwent a remarkable transformation in the period 1650–1687, evolving from a largely metaphysical treatment of the relativity of perceived motion within a larger Scholastic and theological setting to a conceptual system within mechanics that is designed to gauge the scope and correct application of various conservation laws and rules of impact. In other words, much like natural philosophy as a whole, the absolute/relational distinction was itself swept up and transformed by the increasing mathematization and mechanization of the late seventeenth century's scientific revolution.

---

[21] Westfall (1983, 242–245) recounts Newton's correspondence with Pardies on optics in 1672. Furthermore, Marius Stan, who provided the translation of Huygens' codex 7A, comments that the term "absolute" (as regards place/space or motion) does not appear often in the fragments that comprise the codex. He adds that "[t]his suggests that [Huygens] was familiar with this notion (true motion as translation in world space) before 1687, and rejected it before the first edition of Newton's *Principia* came to light", and that Huygens "speaks of 'they' who explicate true motion as change of absolute place", which "may be Borelli, Pardies, and Mariotte" (H 45). That is, Huygens had objected to the use of absolute space (or world space) prior to Newton's more famous treatment of the concept.

Chapter 2
Newton's Neoplatonic Ontology of Space: Substantivalism or Third-Way?

Among philosophers of space and time, two aspects of Newton's ontology of space have seldom been questioned: first, that Newton qualifies as a substantivalist, since he reckons space to be an independently existing substance or entity (see §1.1.1); and second, that Newton's views were deeply influenced by his seventeenth century Neoplatonic predecessors, especially Henry More, whose ontology grounds the existence of space upon an incorporeal being, i.e., God or World Spirit. While the majority of the interpretations of Newton's spatiotemporal ontology in the twentieth century supported these conclusions, a number of important investigations over the past several decades have nonetheless begun to challenge even these ostensibly safe assumptions. Among the most important of these reappraisals can be found in the work of Howard Stein (e.g., 1967, 2002) and Robert DiSalle (e.g., 2002, 2006), who both conclude that the content and function of Newton's concept of absolute space should be kept separate from the question of Newton's alleged commitment to substantivalism. More controversially, Stein (2002) further contends that Newton's natural philosophy treats space as akin to a basic fact or consequence of any existing thing, a view categorized as one of the more epistemologically-oriented, third-way alternatives in Chap. 1, i.e., the definitional conception of space, and therefore non-substantivalist. A related, albeit much more nuanced, interpretation that parts company with traditional substantivalism may also be evident in an influential article by J. E. McGuire (1978a), who argues that space for Newton is "the general condition required for the existence of any individual substance" (1978a, 15). As regards the second of our traditional assumptions associated with Newton's spatial ontology, Stein (2002, 269) forthrightly rejects any Neoplatonic content, whereas McGuire's (1983) essay conjectures that, though "Platonic in character", the primary influence on Newton's ontology is "Descartes' *Meditations,* rather than the eclecticism of Renaissance Neo-Platonism, of which we find little evidence in *De gravitatione*" (1983, 152).[1]

---

[1] In contrast, we will argue that Cambridge Neoplatonism and elements of the (often similar) Gassendi-Charleton philosophy are the primary influence on Newton's ontology of space. McGuire

© Springer International Publishing Switzerland 2016
E. Slowik, *The Deep Metaphysics of Space,* European Studies in Philosophy of Science, DOI 10.1007/978-3-319-44868-8_2

This chapter will examine the ontology of Newton's spatial theory in order to determine the accuracy of these novel historical reassessments; namely, that Newton's concept of space is (i) non-substantivalist, (ii) anti-Neoplatonist, and (iii) is in line with the particular third-way interpretation supported by Stein and DiSalle. Many of the themes that will inform our overall investigation of the ontology of space, such as the different ways that an entity can relate to space, will quickly take center stage, thus securing a basis for comparison and further analysis with other spatial hypotheses, whether from the seventeenth century or in contemporary debates, throughout the remaining chapters. While §2.1 will introduce the basic outline of the non-substantivalist and third-way components of these alternative interpretations, §2.2, §2.3, and §2.4 will be devoted to a lengthy critical analysis of the type of third-way case presented in Stein (2002), an investigation that will require a close examination of many of Newton's works, especially the unpublished tract, *De gravitatione*. As will be demonstrated, Newton's spatial theory is not only deeply imbued in Neoplatonism, contra the revisionist trend, but these Neoplatonic elements likewise compromise Stein's definitional, third-way interpretation. Nevertheless, Newton's spatial ontology does in fact accord with another third-way approach, the property theory of space, or P(O-dep), introduced in Chap. 1—and hence Stein's non-substantivalist reading and third-way classification are, on the whole, correct, although for drastically different reasons, and with respect to a different third-way hypothesis. Throughout our investigation, furthermore, the specific details and subtleties of Newton's particular brand of Neoplatonism will be contrasted with the ontologies of his contemporaries and predecessors, and by this means a more adequate grasp of the content of his theory of space will be obtained.

2.1 Two Third-Way Conceptions of Newton's Absolute Space

Before launching into an investigation of the historical details associated with their arguments, it would be helpful at this point to delineate the strategies employed by the principle proponents of a non-substantivalist, third-way interpretation. As

---

also comments on "the question of Henry More's influence on Newton's doctrine of extended space", concluding "that it is minimal in the period from 1664 to 1668" (1983, 152; where the four year span, 1664–1668, covers the then accepted period for the composition of Newton's major treatise on the ontology of space, *De gravitatione*—see footnote 10 on the recent dating of this work). However, McGuire later conceded "that a possible influence" on Newton's concept of emanation (see §2.2) "is Henry More" (1990, 105); and, in his most recent work (2000, 2007; McGuire and Slowik 2012), he successfully pursues a number of Neoplatonic threads in Newton's natural philosophy. Nevertheless, these post-1990 reappraisals fall short of openly retracting McGuire's earlier demotion of the Cambridge Neoplatonist influence, and thus the justified authority of McGuire's pre-1990 work is likely to give a misleading impression of his evolving conception of these issues (if examined in isolation from the later output). In private discussion, McGuire has indeed confirmed this potential mischaracterization of his overall Newton scholarship, adding that the new dating of *De gravitatione* provided a crucial stimulus to the evolution of his views. Finally, unless otherwise noted, all references to Neoplatonism refer to the seventeenth century varieties then popular in England.

described in Chap. 1, third-way theories of space reject the standard dichotomy between substantivalism (that space is an entity of some sort) and relationism (that space is a mere relation among entities) for an alternative ontology that aims to combine the best aspects of the traditional pair while avoiding their major weaknesses. While ascertaining the details of these theories, and how they differ from the more sophisticated forms of substantivalism and relationism, is a daunting task, Newton's spatial ontology may present a viable candidate for a third-way interpretation, since Newton's spatial theory is difficult to place using the standard dichotomy.

The first of these third-way interpretations stipulates that, apart from his metaphysics, Newton's concepts of absolute space and time in the *Principia* (N 64–70) are best regarded as definitions, mathematical concepts, or structures required for the successful application of his physics, namely, for the three laws of motion and the theory of gravity (and including the mathematical apparatus associated with these hypotheses). That is, Newton may have engaged in the speculation on the nature of space common among seventeenth century natural philosophers, but the crucially important feature of his overall theory is the realization that "a spatio-temporal concept belongs in physics just in case it is defined by physical laws that explain how it is to be applied, and how the associated quantity is to be measured" (DiSalle 2002, 51). This approach to Newton, as mentioned in Chap. 1, can also be described as the definitional or dynamical interpretation, the latter indicating the role that force plays in singling out inertial and non-inertial motions. In contrast to the advocates of relational motion (Descartes, Huygens, Leibniz), Newton understood that the motions and dynamical interactions of material bodies could not be adequately treated by recourse to their relative motions alone (e.g., the famous bucket experiment in the *Principia*, N 68–70). Yet, while he was correct as regards absolute time and absolute acceleration (and, hence, rotation), absolute spatial position and absolute velocity would eventually be recognized as overly rigid and unnecessary structures for an adequate treatment of acceleration within the context of Newtonian physics. As DiSalle comments, a four-dimensional spacetime structure equipped with an affine connection (neo-Newtonian spacetime) would have sufficed for Newton's purposes (DiSalle 2002, 35).

We will label this definition-based strategy the "weak third-way" interpretation of Newton's spatial ontology *if* it also accommodates other approaches that *do* take into account the metaphysical disputes common in that era. In reflecting on the question, "What concepts of time, space, and motion [in the *Principia*] are required by a dynamical theory of motion?", DiSalle offers what is possibly his most forthright endorsement of the weak third-way line:

> Asking this question about Newton's theory *does not deny its connection with his profound metaphysical convictions*—not only about space and time, but about God and his relationship to the natural world. On the contrary, it illuminates the nature of those convictions and their relationship to Newton's physics. For Newton, God and physical things alike were located in space and time. But space and time *also* formed a framework within which things act on one another, and their causal relations become intelligible through their spatio-temporal relations—above all, through their effects on each other's state of motion." (DiSalle 2002, 38; emphasis added).

While acknowledging Newton's "profound metaphysical convictions", which include space, time, and "God and his relationship to the natural world", DiSalle adds that space and time "also" formed a framework for understanding the dynamical relationships among bodies. The implication of this assessment, at least at face value, is that the content or role of space and time may not be exhausted by their constitutive function in Newton's dynamics. Although the evidence is open to interpretation (see footnote 2), the fact that DiSalle does not seriously engage the details of Newton's metaphysics in order to counter the traditional substantivalist position thus lends some support for a weak third-way reading. He later adds that "Newton was not a 'substantivalist', *at least not in the now-standard use of the term*" (emphasis added), since Newton was critical of substance ontologies, and he did not regard the parts of space as possessing an intrinsic individuality, whereas the modern substantivalist (often) views spacetime points as irreducibly basic existents (DiSalle 2006, 37; see also Chaps. 1 and 6). Once again, this appraisal leaves open the possibility that other notions of substance, "not in the now-standard use of the term", might apply in Newton's case.[2]

The weak third-way reading gains credibility in the first edition of the *Principia* (1687), which contains little metaphysics treating those all-important Neoplatonist concepts, substance and God—yet, later editions of the *Principia* (the General Scholium of the second edition, 1713), the later Queries to *The Opticks*, and various non-published writings (to be discussed below) do indeed pick up these ontological themes, thus trying to infer Newton's overall commitment to the weak thesis remains difficult to gauge. One of the most meticulous investigations of Newton's concepts of space and motion, namely, Rynasiewicz (1995), would seem to be consistent with the weak third-way conception, however. Among twentieth century commentators, one of the earliest cases for a weak third-way interpretation of Newton is Toulmin (1959), although Stein's landmark (1967) is better known today (the latter also hinting towards a strong third-way reading, see below).

Stein (2002), on the other hand, openly sanctions a much stronger position that goes well beyond the weak third-way view: Newton's *metaphysics* does not, in fact, advocate a form of substantivalism—call this the "strong third-way" interpretation of Newton. Whereas the weak interpretation is confined to Newton's handling of the concepts of space and time as they appear in his *physics*, the strong thesis actually engages Newton's metaphysical writings in an attempt to counter the prevailing consensus that Newton endorsed substantivalism or a seventeenth century equivalent. Stein claims that "Newton's 'metaphysics of space' is…that space is (some

---

[2] To be specific, private correspondence with DiSalle (Princeton, Spring 2009) seemed to support the weak third-way interpretation, but a presentation at the 2012 meeting of the International Society for the History of Philosophy of Science (King's College, Halifax) by DiSalle, entitled, "Transcendental Philosophy from a Newtonian Perspective", appeared to favor the strong third-way reading (to be discussed below). Huggett's analysis of DiSalle (2008, 404–405) would seem to side with the strong interpretation as well, since DiSalle claims that Newton's concept of absolute space does not allow the material world to possess different, but uniform, positions or velocities (see Chap. 6 for these discussions). Consequently, while DiSalle's work (2002, 2006) is open to a weak third-way construal, it is unclear if that is the position he accepts.

kind of) effect of the existence of anything, and therefore of the first-existing thing" (2002, 268). In essence, Stein interprets Newton's metaphysics as sanctioning a conception of space that fits neither substantivalism nor its chief rival, relationism. Rather, the view that Stein attributes to Newton is very much like Stein's *own* metaphysical interpretation of space (spacetime), as a passage from an earlier essay makes clear: Stein claims that spacetime structures are "an 'emanative effect' of the existence of anything" (Stein 1977b, 397), where the phrase in quotation marks, "emanative effect", is an obvious allusion to Newton's spatial hypotheses (as will be explained shortly). If space is conceived as an "effect of the existence of anything", as Stein regards both his own theory and Newton's, then it is quite difficult to pin an ontology to this thesis, let alone a commitment to substantivalism. That is, space is not an independently existing substance/entity because it depends (in some manner) on the existence of "anything", presumably, physical bodies or fields, thus violating the independence clause for substances. But, neither is it a mere relation, since the domain (as the set of possible values) of the spatial relations in a given universe at any instant is not limited to the actual spatial relations among the material existents at that instant, as it is under strict eliminative relationism (see Chap. 1). In short, Stein's metaphysics-avoiding and physics-centered interpretation of Newton, like his own hypothesis, would seem to favor the third-way, definitional conception of space as opposed to the prevailing substantivalist and relationist ontologies.[3] Although an in-depth examination of several third-way interpretations of spatial ontology will be confined to Chap. 5, we will ultimately judge the merits of a proposed strong third-way classification for Newton's spatial theory in subsequent sections of this chapter.

2.2   The Case for a Strong Third-Way Interpretation

At a first glance, the strong third-way analysis of Newton's spatial concepts appears quite promising. In his unpublished work, *De gravitatione* (henceforth, *De grav*), which most likely predates the *Principia* by roughly six to eight years (i.e., c. 1680), Newton insists that space "has its own manner of existing which is proper to it and

---

[3] Substantivalism, as noted in §1.1.1, is a complex topic, with various interpretations spotlighting different aspects of the concept. However, if forced to give a quick synopsis of the conclusions of this chapter, then Newton's ontology of space is substantivalist *if* one defines substantivalism as an entity that can exist in the absence of all matter; but, Newton's space is not substantivalist *if* that concept denotes an entity that is independent of all other entities—since, as the traditional accounts have correctly insisted, Newton holds that space necessarily depends on God. Additional discrepancies with the modern approaches to substantivalism will be discussed in Parts II and III. Furthermore, while a strong third-way, non-substantivalism fails (since Newton's space is deeply metaphysical and theological), a weak third-way, non-substantivalism is nonetheless a consistent interpretation (although difficult to corroborate). Finally, as noted in Chap. 1, the pairing of Stein and DiSalle under the "definitional" label is conjectural, since Stein's views on space are open to interpretation; but, they are clearly quite similar in their deflationist approach to spatial ontology, hence the common pairing seems warranted.

which fits neither substance nor accident [i.e., property]" (N 21). Space "is not a substance…because it is not absolute in itself, but is as it were an emanative effect of God and an affection of every kind of being; on the other hand, because it is not among the proper affections that denote substance, namely actions, such as thoughts in the mind and motions in body" (21). He adds that space is not a substance since it cannot "act upon things, yet everyone tacitly understands this of substance" (21), but neither is it an accident of body, "since we can clearly conceive extension existing without any subject, as when we imagine spaces outside the world or places empty of any body whatsoever" (22). The substance/accident dichotomy holds that all existents are either self-dependent substances, or the properties (accidents) that can only exist "within", or "inhere in", a substance (see, e.g., Bolton 1998, 179 on the substance/property dichotomy in this period). In contrast, Newton consistently refers to space as an "affection" (*affectio*) or "attribute" (*attributa*)[4]:

> Space is an affection of a being just as a being (*Spatium est entis quatenus ens affectio*). No being exists or can exist which is not related to space in some way. God is everywhere, created minds are somewhere, and body is in the space that it occupies; and whatever is neither everywhere nor anywhere does not exist. And hence it follows that space is an emanative effect (*effectus emanativus*) of the first existing being, for if any being whatsoever is posited, space is posited. (25)

Much of the ensuing investigation in this chapter will strive to unravel the complexities of this quite enigmatic passage, but we will first investigate Stein's strong third-way, non-substantivalist interpretation in more detail.

2.2.1   Space as a Necessary Consequence or Result

Based largely on the evidence of the above quote, Stein concludes that "Newton does *not* derive his 'Idea' of space—its ontological status included—*from* his theology (as has often been claimed); for he tells us that if *anything* is posited, space is posited" (Stein 2002, 268). Because God is the first existing thing, "space (in some sense) 'results from' the existence of God" (268), but this does not detract from Newton's general hypothesis that "space (in some sense) 'results from' *the existence of anything*" (268). He adds:

> But this sense of the word—simply *a necessary consequence,* with no connotation of "causal efficacy" or "action"—*exactly* fits the rest of what Newton says; indeed, this meaning might have been inferred directly from Newton's words: "[S]pace is an emanative effect

---

[4] Overall, Newton's concept of substance is difficult to accurately fix relative to his contemporaries and predecessors, largely because he seldom provides any details when employing this term. The same is true as regards his employment of "affection" and "attribute", which seem to denote a property that is necessary for a being's existence, whereas an "accident" (such as red, triangular, etc.) is not. Newton refers to space as an attribute/affection of all being, while denying that it is an accident, thus (apparently) demonstrating its necessity for all being (see also footnote 20). For these metaphysical categories, see, once again, Bolton (1998), as well as Carriero (1990) for more on Newton's use of the term "affection".

of the first-existing being, for *if I posit any being whatever I posit space*": the second clause tells us precisely what the first clause *means*. (269)

Stein's strong third-way interpretation of Newton stands out rather clearly in this passage, for, stripped of its ontological connotations, "space as an emanative effect" becomes simply "space as a necessary consequence of the existence of anything"—and, of course, it is just this type of ontologically deflationary reading that Stein counsels, i.e., space as a basic fact, neither causally generated nor possessing causal powers, and hence quite difficult to read as the product of Neoplatonist, or substantivalist, dogma. On the whole, only the definitional conception of space would appear to meet the requirements of this stridently non-ontological assessment (but see also §10.5.3).

How plausible is Stein's case for a strong third-way construal? While some of the objections will have to await subsequent sections of the chapter, wherein the ontology of the Cambridge Neoplatonists will be discussed, there are a few difficulties that can be raised directly. First of all, Newton never explicitly states that space is a *necessary* consequence or result, which is a description that, as noted above, seemingly equates space with a form of logical or conceptual fact, as opposed to an ontological feature of entities.[5] Presumably Newton would have emphasized this "necessary consequence" notion of space in a more lucid manner, since his application of the relevant terms, especially "emanative effect", often parallels the decidedly ontological meaning given to these very same terms in earlier Neoplatonist tomes. Moreover, other passages would seem to support the traditional ontological picture of Newton's spatial theory, such as his comment that space "is something more than an accident, and approaches more nearly to the nature (*naturam*) of substance" (N 22). If Newton's concept of space, as Stein contends, is more logical than ontological, it would seem to follow that Newton should reject any application of the substance/accident dichotomy to space—one would not expect that he would try to place the concept somewhere between these traditional ontological positions.

Overall, the best evidence for Stein's interpretation appears in the quotation examined at length above, where Newton claims that "space is an emanative effect of the first existing being, for if any being whatsoever is posited, space is posited" (25), whereupon Stein reasons that "the second clause tells us precisely what the first clause *means*" (Stein 2002, 269). In *De grav*, however, the term "emanative effect" is *not* used with reference to "any being whatsoever", but *only* to God or the "first existing being". To avoid the obvious theological implications, Stein takes the phrase, "first existing being", to pertain to *any* first existing being, presumably even

---

[5] In a later writing, Newton does refer to infinite space and time as "modes of existence in all beings, & unbounded *modes* & consequences of the existence of a substance that which is really necessary & substantially Omnipresent & Eternal" (Koyré and Cohen 1962, 96–97; see, also, §2.4.1). The use of the term "consequences" in this passage might be taken to support the strong third-way, non-substantivalist interpretation—yet, it is used in conjunction with the basic ontological term, "modes", which denotes the specific way in which a being manifests a general property (e.g., circular is a mode of shape). Consequently, it is not clear whether this passage actually assists or harms Stein's third-way reading.

a mere corporeal being—but this interpretation strains credibility. On Newton's ontology, only God (or possibly a world soul) can qualify as the first existing being, as the context of *De grav* makes clear. Once again, the evidence for space's incorporeal ontological foundation will emerge in more detail in the ensuing sections, where the distinction between emanation and space as an attribute of "being qua being" will be explained.

Second, if Newton's emanation concept is merely the claim that "if any being whatsoever is posited, space is posited", then one would expect a much more general application of this concept to other beings, especially corporeal being. The fact that Newton never entertains the possibility that space could emanate from a material body, or anything other than God (or a world soul), strongly suggests that Stein's readings of "emanative effect" and "first existing being" are much too broad.[6]

Third, concerning Stein's attempt to equate "first existing being" with "any first existing being", a serious difficulty resides in the historical fact that there were clear precedents among the earlier Cambridge Neoplatonists for employing such phrases, like "first existing being", with reference to God alone. In More's *Enchiridium Metaphysicum* (1679), there are several instances of such terms in his well-known comparison of the metaphysical titles ascribed to both God and spatial extension:

> For this infinite and immobile extension will be seen to be not something merely real...but something divine after we shall have enumerated those divine names or titles which suit it exactly,...Of which kind are those which follow, which metaphysicians attribute to First Being. Such as one, simple, immobile, eternal, complete, independent, existing from itself, subsisting by itself, incorruptible, necessary, immense, uncreated, uncircumscribed, incomprehensible, omnipresent, incorporeal, permeating and encompassing everything, being by essence, being by Act, pure Act. (EM 57)

As is evident given the references to "being by essence", "being by Act", etc., More's discussion of "First Being" relies heavily on concepts that originate from Aristotle's *Metaphysics*; in particular, the existence of an eternal, immovable "first" substance required to ground the world's lesser, finite, and mutable substances.[7] These traits of First Being, therefore, are only applicable to God or a world soul, although More's contention is that most also apply to space: "That which, however, is the first Being and *receives all others,* without doubt exists by itself, since nothing is prior to that which sustains itself" (59; emphasis added). Accordingly, the historical context of the terms and phrases used in Newton's work would seem to fatally

---

[6] The term "emanative effect" only appears three times in *De grav*. Besides Stein's favorite of these three quotations (i.e., "space is an emanative effect of the first existing being"), there are: "[space] is as it were an emanative effect of God and an affection of every kind of being" (N 21); and, "space is eternal in duration and immutable in nature because it is the emanative effect of an eternal and immutable being" (26).

[7] In his depiction of the traits of infinite extension, More adds that "it is necessary that it be immobile. Which is celebrated as the most excellent attribute of First Being in Aristotle" (1995, 58). In the *Metaphysics* (CWA 1071b 1-1071b 10), Aristotle concludes that "it is necessary that there should be an eternal unmovable substance. For substances are the first of existing things, and if they are all destructible, all things are destructible".

undercut Stein's somewhat peculiar interpretation of the phrase, "first existing being". In short, the intended meaning of Newton's "first existing being" almost certainly follows More's usage, which, in turn, is based on a long line of Aristotelian/ Scholastic argument. More importantly, as the subsequent investigation of *De grav* will demonstrate, Newton likewise demands an infinite, immobile "first existing being" to ground the existence and extension of all lesser, mobile entities.

2.2.2   *Efficient Causation and Cambridge Neoplatonism*

As Stein admits (2002, 271), his interpretation runs counter to the prevailing consensus among Early Modern and Newton scholars that Newton's spatial ontology is thoroughly imbued with Cambridge Neoplatonic natural philosophy: see the commentaries by Burtt (1952, 261), Jammer (1993, 110), Koyré (1965, 89), Funkenstein (1986, 96), Hall (2002), to name only a few.[8] Edward Grant's assessment is fairly representative: "if space is God's attribute, does that not imply it is somehow an accident or property of God" (Grant 1981, 243)? A notable exception to this line of reasoning, however, is presented in an influential early article by J. E. McGuire (1978a), which presents a view of Newton's concept of space that can be interpreted, albeit only superficially, as similar in content to Stein's assessment. McGuire argues that space for Newton is "the general condition required for the existence of any individual substance including its characteristics", and adds that:

> The relation between the existence of a being and that of space is not causal, but one of ontic dependence. Newton is defining one condition which must be satisfied so that any being can be said to exist. In short, the phrase, 'when any being is posited, space is posited' denotes an ontic relation between the existence of any kind of being and the condition of its existence. (McGuire 1978a, 15)

Possibly in response to Carriero's (1990) criticisms, McGuire later qualified this account of Newton's spatial theory, concluding that the relation between the divine being and the infinity of space "can be seen (in a curious sense) as a causal dependency, and, moreover, one that has a legacy in theological and philosophical thought" (McGuire 1990, 105). It will be useful to explore these issues in greater

---

[8] While providing a brief synopsis of the natural philosophy of the Cambridge Neoplatonists is difficult, the central feature is probably the rejection of a purely mechanical account of the material world (i.e., that all material phenomena can be completely explained through the interactions and impact of inert matter in motion). Rather, the Neoplatonists appealed to God, or spirit, as an active agent, or foundational basis, for all natural phenomena (see, e.g., MacKinnon's summary in More 1925, 315). Concerning the details of Charleton's natural philosophy, which is decidedly Gassendian at least as regards space, an incorporeal basis for space is posited, and thus it is strikingly similar to Cambridge Neoplatonism on this particular issue, although there are important differences on many other issues.

depth, for they shed light on a likely source of Newton's descriptive phrase, "emanative effect".

McGuire contends that there is a Scholastic precedent for construing the relationship between God and space as "under the rubric of *efficient* causation", yet, "since the notion of an eternal and efficient cause does not involve any activity, production, or active efficacy between it and its effect, it is difficult to distinguish natural or ontic dependence in these contexts from the notion of causal dependence between eternal things" (105).[9] McGuire offers the example of "Augustine's foot eternally embedded in dust, and thus eternally causing its footprint" (105), to characterize this special form of efficient causation that links these eternal "things", namely, God, space, and time, which are not temporally prior to one another. Henry More, in his *The Immortality of the Soul* (1659), had employed the concept of an emanative cause in just this manner in explicating the extension of incorporeal substance. More contends that there exists a spatially extended, immaterial "Secondary Substance" that is coextensive with the extension of material substance: we have a "rationall apprehension of that part of a Spirit which we call the *Secondary Substance*. Whose Extension arising by graduall Emanation from the First and primest Essence [of the immaterial being], which we call the *Center* of the Spirit" (IS 35). More adds that "*an Emanative Effect is coexistent with the very Substance of that which is said to be the Cause thereof*", and explains that this "Cause" is "the adequate and immediate Cause", and that the "Effect" exists "so long as that Substance does exist" (33). And, while relationship between space and God is not openly addressed in this work, it would seem to be tacitly implied since God is included in his definition of spirit (IS 20–24; "Divine Amplitude" is also referenced at IS 23) and in his discussion of the "Spirit of Nature" (IS 449–450), the latter comprising one of the first emanations from God in standard Neoplatonic thought. To sum up, although there remain significant differences between McGuire's and Carriero's understanding of Newton's use of emanation, they nonetheless concur that traditional ontological, and specifically causal, issues are at play, and that there are a number of Scholastic and Neoplatonist precedents for Newton's handling of emanative causation as a unique type of efficient causation.

Stein criticizes the view that Neoplatonism underlies Newton's spatial hypotheses, however. Commenting that "the grounds for thinking that Newton's theory of emanation is neo-Platonic, or 'Cambridge Platonic', are very weak" (2002, 269), Stein asserts that emanation is distinct from creation for the Neoplatonists, and, since all being (except God) is created, thus space is not created, and hence not a being. Yet, as just disclosed, there is a Scholastic-influenced form of causation in the work of More that does fit Newton's use of emanation. It is true that Newton lists

---

[9] McGuire (2007), following (1990, 105), likewise connects his earlier "ontic dependence" hypothesis with efficient causation: "It seems evident that emanative causation, as Newton understands it, reflects this relationship between God's necessary existence and space's uncreated nature: space exists always because God exists necessarily. Moreover, since the notion of an eternal and efficient cause does not necessarily involve activity, production, creation, or active efficacy between it and its effect, the distinction between ontic and causal dependence essentially collapses" (2007, 123–124).

"uncreated" (*increata*) among the characteristics of space (N 33); but, as Carriero (1990, 113–115) explains, this use of "uncreated" is almost certainly due to the fact that, for Newton, the cause of a created being is prior in time, whereas an emanative cause is co-existent with its effect (i.e., not temporally prior to its effect). This interpretation of Newton's use of creation is corroborated in *De grav* through his claim that "extension is not created (*creatura*) but has existed eternally" (since extension is an emanative effect of an eternal being; N 31). Equally important, Newton's hypothesis closely follows More's line of reasoning, since More both defines an emanative cause as co-existent with its effect, as well as lists "uncreated" among the attributes of space, in the passage quoted in §2.2.1 (EM 57).[10] As Carriero also observes (1990, 114), while Newton states that space is uncreated, he never states that space is uncaused.

Consequently, despite Stein's best effort to argue for his anti-Neoplatonist, "necessary consequence" reading of Newton's term "emanative effect", the historical context renders such an interpretation extremely implausible. In short, since More also defines the spatial extension of incorporeal substance as an emanative effect, and given that More's hypothesis stands as a clear instance of his Neoplatonist ontology, there can be little doubt as to the direct inspiration, and thus intended meaning, of Newton's use of the identical phrase "emanative effect".[11]

---

[10] More is less forthcoming on the uncreated status of space in his earlier *The Immortality of the Soul*, although it is strongly implied in his discussion of emanative causation: "By an *Emanative Cause* is understood such a Cause as merely by Being, no other activity or causality interposed, produces an Effect" (IS 32). Newton's list of the characteristic of space versus matter in *De grav* thus reveals a knowledge of More's later *Enchiridion*, first published in 1671, as do many of the other features detailed in our investigation (namely, the "being as being" hypothesis, in §2.4). Indeed, it is highly unlikely that Newton was not familiar with this quite important work of More's later years. The arguments for a later dating of *De grav* (in Dobbs 1991, 130–146), i.e., after 1680, thereby gains support, since Newton's treatise exhibits the influence of several of More's major works, including the *Enchiridion*. McGuire (1978a, 41, n. 27) had earlier remained a bit circumspect about the influence of More's *Enchiridion* based on the earlier date supplied by Hall and Hall for *De grav* (Newton 1962a, 90), i.e., circa 1666; but McGuire has since advocated the later date (2007, 112).

[11] In her collection of More extracts, MacKinnon summarizes the emanation concept as follows: "The universe of Neo-Platonism is formed by emanation from the One, through the descending stages of intelligence, the soul, and the world, with formless matter, or unreality, as the ultimate limit of the emanative power" (More 1925, 315). Needless to say, much in *De grav* discloses a penchant for a Neoplatonist, emanationist ontology, such as his comments on the possibility of a world soul: "the world should not be called the creature of that soul but of God alone, who creates it by constituting the soul of such a nature that the world necessarily emanates [from it]" (N 31). Throughout *De grav*, as will be explained, Newton places incorporeal beings (spirits, souls) at the foundation of his hierarchy, with the lesser, corporeal world emanating from these incorporeal beings. Another instance of the use of emanation that parallels Newton's, although with respect to time, is employed by J. B. van Helmont, a natural philosopher in the early half of the seventeenth century who Newton had studied (see, Ducheyne 2008). Finally, it should be noted that this investigation does not take sides on the complex issues associated with causation in Newton's natural philosophy, e.g., whether emanative causation more closely resembles an efficient or formal cause, or something else. The main purpose of the discussion of causation is to refute Stein's (2002) strong third-way interpretation by disclosing Newton's Neoplatonism. However, if forced on this

Besides emanation, a veritable host of sixteenth and seventeenth century natural philosophers proposed similar hypotheses on the nature of space that closely parallel Newton's, including many thinkers in England with whom he was directly acquainted (most importantly, in addition to More, Isaac Barrow and Joseph Raphson). For example, the idea that space lies outside the Scholastic substance/accident categories was almost commonplace in this period and in the sixteenth century: Fonseca, Amicus, Bruno, Telesio, Patrizi, and Gassendi, to name a few, all favored this notion; and, more significantly, theological concerns are heavily implicated in their respective views. These last two, Francesco Patrizi and Pierre Gassendi, in addition to More, probably comprised the main source of influence on Newton's developing conception of space, although their influence was likely obtained indirectly through More and Walter Charleton, Gassendi's foremost English advocate. One can find a surprising number of close similarities between the individual hypotheses of these natural philosophers and Newton's spatial hypotheses: for instance, an atomistic or stoic cosmology, with a finite material world set within an infinite, three-dimensional void space, is common to all (except Patrizi, who fills all of space with light). Patrizi, like Newton, also emphasizes the mathematical aspect of space, which can receive all geometric shapes, and further argues that, while neither substance nor accident *per se*, space is nevertheless a type of entity not covered by the Scholastic categories; i.e., space is closer to the traditional concept of substance (PS 241), a view that Newton also holds (see §2.2.1, and §9.3.3). Gassendi, whose conception of space is similar to Patrizi's, would also influence later English natural philosophers: Gassendi's theory of space, according to Grant, "is an absolutely immobile, homogenous, inactive (resistenceless), and even indifferent three-dimensional void that exists by itself whether or not bodies occupy all or part of it and whether or not minds perceive it" (1981, 210). More importantly, Gassendi holds that space is both uncreated and co-eternal with God, although, like many Scholastic predecessors, he also believes that God is in every place while not actually extended in the same manner as body (see §2.4 below).[12] It is against this historical backdrop that any assessment of the import of Newton's concept of space must begin, especially the relationship between God and space.

---

issue, then we might hazard the conjecture that emanative causation is a sort of hybrid of both formal and efficient causation, inspired by More's secondary substance concept of a being's spatial extension in the earlier *Immortality*, and a property view of space that draws on More's later view from the *Enchiridion* (i.e., that space is God's attribute). McGuire's (2007) inference that ontic and causal dependence is hard to separate in the God-space context is accurate as well (see footnote 9). See, also, Gorham (2011), for a more Cartesian interpretation of these issues.

[12] Translations of Patrizi's spatial hypotheses are provided in Brickman (1943). For Gassendi's philosophy of space, see his *Syntagma philosophicum* (in his *Opera Omnia,* 1658), parts of which are translated in Brush (1972), and Capek (1976). Gassendi's ideas deeply influenced the content of Charleton's discussion of space in his, *Physiologia Epicuro-Gassendo-Charletoniana* (1654), a work known to Newton (see footnote 17). As noted, Gassendi and Patrizi's respective spatial ontologies will be examined in more depth in Chap. 9.

2.3 Neoplatonism and the Determined Quantities of Extension Hypothesis

In order to more accurately assess both Newton's conception of space and its third-way interpretations, this section will examine a host of Neoplatonist-leaning hypotheses in *De grav*, especially the space-body relationship. Once again, the fruits of our analysis will serve as a crucial backdrop to many discussions throughout the remainder of our investigation (especially Chaps. 6, 7, and 9).

2.3.1 Newton Against Ontological Dualisms

As described previously, the strong third-way reading insists that the ontological grounding of space need not be specifically theological in kind. In other words, matter, or any being other than God, is sufficient to provide space's ontological foundation, a possibility that is allegedly reflected in Newton's claim that space is "an affection of every kind of being" (N 21). Stein, for example, argues that "on the objective or ontological side,..., Newton's doctrine about space and time, in the light of his explicit statements, did not teach that space and time *per se,* or their attributes, depend upon the nature of God" (2002, 297).[13] Concerning the question, "Can we conceive space without God?" (271), Stein cites a passage from Newton's critical assessment of Descartes' hypothesis that equates spatial extension with matter:

> If we say with Descartes that extension is body, do we not manifestly offer a path to atheism, both because extension is not created but has existed eternally, and because we have an idea of it without any relation to God, and so [in some circumstances] it would be possible for us to conceive of extension while supposing God not to exist? (N 31)[14]

In commenting on this passage, Stein concludes that, for Newton, "extension does not require a subject in which it 'inheres', as a property; and it can be conceived as existent without presupposing any *particular* thing, God included" (2002, 271).

---

[13] A similar interpretation of Stein (2002) has been put forward by Andrew Janiak: "Stein (forthcoming)...notes...that Newton's view is not first and foremost a theological one, for its first premise is that space is an affection of all entities. The fact that God's infinite and eternal existence makes it the case that space is infinite an eternal is logically parasitic on this first premise. That is, given the logical structure of Newton's view, space would emanate from the first existent, whatever that first existent happened to be, because for Newton once we posit an entity we posit space. This just means that spatiality is what we might call—following Galileo's discussion of the primary qualities of objects—a *necessary accompaniment* of the existence of entities" (Janiak 2000, 222, fn.67).

[14] The phrase in brackets, "in some circumstances", is excluded from Stein's translation. The differences are not relevant to the above arguments against his overall position, however, so it will not be discussed.

The problem with this rendering of Newton's statement, put simply, is that Newton's ability to conceive space without God does not entail that he believed that space can exist without God. Indeed, a significant part of Newton's argument against Descartes' ontology stems from precisely this worry, namely, that it allows a conception of spatial extension, as the essential property of corporeal substance, without any apparent connection to, or need of, the concept of God. In its place, Newton tentatively advances a Neoplatonic ontology in which *both* spatial extension and body depend upon God.

In *De grav*, Newton's contemplates a world wherein God directly endows spatial extension with bodily properties, such as impenetrability or color, without requiring an underlying corporeal substance to house these accidents: "If [God] should exercise this power, and cause some space projecting above the earth, like a mountain or any other body, to be impervious to bodies and thus stop or reflect light and all impinging things, it seems impossible that we should not consider this space really to be a body from the evidence of our senses" (N 27–28). If we accept this hypothesis, then Newton contends that "we can define bodies as *determined quantities of extension which omnipresent God endows with certain conditions*" (28); the "conditions" being, first, that these determined quantities are mobile, second, that they can bring about perceptions in minds, and third, that two or more cannot coincide. By this process, Newton argues that these bundles of quantities can exactly replicate our everyday experience of material bodies without need of Descartes' material substance, or the Scholastic notion of prime matter (27–31). These determined quantities, moreover, are sustained and moved through the exercise of the divine will alone, and Newton makes repeated references to the relationship between the human mind and human body, on the one hand, and God's will and the determined quantities, on the other, to make this point: "God, by the sole action of thinking and willing, can prevent a body from penetrating any space defined by certain limits" (27). This conception foreshadows Newton's later description of space as God's "sensorium" (in the Queries to the *Opticks*, N 127–140), since the omnipresence of the divine will is directly analogous to the omnipresence of human thought and sensation in the human body—e.g., just as humans can move their limbs at will, God can likewise move bodily quantities through space at will.

Therefore, with respect to the citation provided by Stein above, i.e., (N 31), Newton's point is that any theory, like Descartes', that links bodily extension to corporeal substance alone, so that mental properties are excluded, is apt to mistakenly infer that extended corporeal substance can exist independently of God— Why?: because the divine will is erroneously presumed to be more akin to a mental property on the Cartesian scheme, so that it has little or no relationship with the extension of corporeal substance. Newton's "determined quantities of extension" hypothesis (hereafter, DQE) is, in fact, the means by which he circumvents the supposedly atheistic implications of both Descartes' dualism of mental and material substance, as well as the Scholastic's dualism of primary matter and substantial form. The passage cited by Stein is preceded by an explanation that clearly shows that Newton rejects any theory, such as Descartes', that ties body, and thus bodily extension, exclusively to corporeal substance: "For we cannot posit bodies of this kind [i.e., on the DQE hypothesis] without at the same time positing that God exists, and has created bodies in empty space out of nothing" (31). For the Cartesians and Scholastics, however:

> They attribute no less reality in concept (though less in words) to this corporeal substance regarded as being without qualities and forms, than they do the substance of God, abstracted from his attributes....And hence it is not surprising that atheists arise ascribing to corporeal substance that which *solely belongs to the divine*. Indeed, however we cast about we find almost no other reason for atheism than this notion of bodies having, as it were, a complete, absolute, and independent reality in themselves. (32; emphasis added)

Accordingly, leaving aside issues of conceivability, the quote provided by Stein is not evidence that Newton actually accepts an ontology that allows bodily extension to be "as it were, a complete, absolute, and independent reality in themselves"—i.e., apart from God—rather, Newton argues at length that any theory that allows such an autonomous conception of bodily extension is completely misguided.

Newton's Neoplatonism is evident throughout his assault on these Cartesian and Scholastic dualisms. Directly *after* the quote provided by Stein, he offers a number of additional criticisms against strictly demarcating the incorporeal and the corporeal via Descartes' distinction in substances:

> Nor is the distinction between mind and body in [Descartes'] philosophy intelligible, unless at the same time we say that mind has no extension at all,...; which seems the same as if we were to say that it does not exist, or at least renders its union with body thoroughly unintelligible and impossible. Moreover, if the distinction of substances between thinking and extended is legitimate and complete, God does not eminently contain extension within himself and therefore cannot create it; but God and extension would be two separate, complete, absolute substances, and in the same sense. But on the contrary if extension is eminently contained in God, or the highest thinking being, certainly the idea of extension will be contained within the idea of thinking, and hence the distinction between these ideas will be such that both may fit the same created substance, that is, but that a body may think, and a thinking being be extended. (31; modified translation)

One of the remarkable facets of Newton's assessment is that it anticipates the Empiricist's skeptical analysis of substance, but, for our purposes, the important question pertains to the relationship between eminent containment and emanative causation, two separate, but similarly named, metaphysical hypotheses. While the evidence is sketchy, it is possible that Newton may regard the emanative causation of various attributes, such as extension, as an ontological consequence of their eminent containment in a foundational incorporeal being.[15] If Newton does accept this

---

[15] The distinction between eminent containment and emanative causation is somewhat vague in the literature, but presumably they are distinct hypotheses. For example, one can hold that God eminently contains the reality manifest in, say, a stone, but that God's creation of a stone does not employ the emanationist model favored by many Neoplatonists. The emanationist form of explanation, such as the type encountered in More's oeuvre (e.g., EM 135), marshals an assortment of light metaphors to describe the causal process whereby the foundational level entity (the light source) brings about the existence of lesser entities (the light itself, or the shadow), the latter additionally characterized as an image of, or a radiation from, the foundational level being. In the above quotation, Newton may be simply contending that the Cartesian dualism of mind and body undermines Descartes' own eminent containment hypothesis, so the reference to eminent containment in

type of metaphysical relationship, then his claim that "extension is eminently contained in God" (in the above passage), raises insurmountable obstacles for the strong third-way position, needless to say, as we will elaborate in the next section.

2.3.2   The Ontological Foundation of Newton's Spatial Ontology

In previous research devoted to *De grav*, careful attention has seldom been devoted to the aspects of the DQE hypothesis that specifically concern the nature of space. Part of the explanation for this oversight might be due to the context in which the DQE hypothesis is introduced, namely, as an account of the nature of body, and not—explicitly, at least—on the nature of space. Yet, if one desires to understand the ontological presuppositions of Newton's overall spatial theory, then the DQE hypothesis is of crucial importance. In addition, although Newton states that his DQE hypothesis is "uncertain", and that he is "reluctant to say positively what the nature of bodies is" (N 27), these declarations of uncertainty do not detract from its significance as the only hypothesis that he does, in fact, present and develop. Not only is a large portion of *De grav* allotted to the DQE hypothesis, but (as is evident above) Newton makes repeated claims as to the superiority of this hypothesis in comparison with the Cartesian and Scholastic alternatives: e.g., "the usefulness of the idea of body that I have described [the DQE hypothesis] is brought out by the fact that it clearly involves the principal truths of metaphysics and thoroughly confirms and explains them" (31). In summarizing its importance, he adds: "[s]o much for the nature of bodies, which in explicating I judge that I have sufficiently proved that such a creation as I have expounded [the DQE hypothesis] is most clearly the work of God, and if this world were not constituted from that creation, at least another very like it could be constituted" (33).

Admittedly, since the endorsement of the DQE hypothesis in *De grav* remains tentative, it is possible that other God-based conceptions of corporeal existents may have been amenable to Newton. Yet, there is both direct and indirect evidence that supports the contention that Newton accepted the DQE hypothesis, or a close analogue, throughout his later years. In an unpublished tract from the 1690s brought to light in McGuire (1978b), dubbed "Tempus et Locus" (henceforth, TeL), Newton concludes:

---

this passage need not imply that Newton actually accepts this view. On the other hand, a bit earlier in *De grav*, Newton remarks that "created minds (since it is the image of God) is of a far more noble nature than body, so that perhaps it may eminently contain [body] in itself" (N 30)—and, importantly, the context strongly favors the view that Newton is elaborating his own view here. However, Newton's statements employing just "emanative effect" and "emanate" are alone sufficient to demonstrate his Cambridge Neoplatonist stance (and thus uphold the argument of this chapter), regardless of whether or not he accepted eminent containment. On the vexed issue of eminent containment in Descartes, see, Gorham (2003).

The most perfect idea of God is that he be one substance, simple, indivisible, live and making live, necessarily existing everywhere and always, understanding everything to the utmost, freely willing good things, by his will effecting all possible things, *and containing all other substances in Him as their underlying principle and place;* a substance which by his own presence discerns and rules all things, just as the cognitive part of a man perceives the forms of things brought into his brain, and thereby governs his own body" (TeL 123; emphasis added).

As for the indirect evidence, in a footnote to Pierre Coste's French translation of Locke's *Essay Concerning Human Understanding* (third edition), Coste reports that Newton provided an account of the creation of matter, in 1710, that correlates with the DQE hypothesis in *De grav* (Koyré 1965, 92). Less specific, but also important, is David Gregory's summary of his 1705 conversations with Newton: "He believes God to be omnipresent in the literal sense…for he supposes that as God is present in space where there is no body, he is present in space where a body is also present" (Hiscock 1937, 29).

In short, what the DQE hypothesis reveals about Newton's spatial ontology is that God, or some spiritual entity at (or near) the level of God, is the emanative cause of corporeal being, and perhaps eminently contains corporeal being. All of Newton's examples of emanative causation and eminent containment in *De grav*, as revealed above, involve a mental/spiritual entity as the source — i.e., God, the world soul, created minds — and either matter or space as the emanative effect or the eminently contained entity. Thus, leaving aside the question of God's attribute of extension, since body is at (or near) the lowest rung in the hierarchy of being, and thereby *depends* for its existence on these incorporeal beings, space *cannot* be the emanative effect of matter/body. Put differently, it would be highly unorthodox for Newton to have conceived spatial extension as the emanative effect of a material being or beings, especially given the deep disparity in Newton's characterization of space and matter: "extension is eternal, infinite, uncreated, uniform throughout, not in the least mobile, nor capable of inducing changes of motion in bodies or change of thought in the mind; whereas body is opposite in every respect" (N 33). Newton's final use of "emanative effect" in *De grav* makes this point quite clearly, i.e., that space's qualities could only originate from God: "space is eternal in duration and immutable in nature because it is the emanative effect of an eternal and immutable being" (26). Indeed, the hierarchical relationship between God and body, mediated via God's attribute of spatial extension, is very likely the motivation behind Newton's DQE hypothesis, since he constantly criticizes the opposition (i.e., Cartesians, Scholastics) for "ascribing to corporeal substance that which solely belongs to the divine" (31; see §2.3.1) — and, once again, what the Cartesians and Scholastics have erroneously ascribed to corporeal substance is extension. Therefore, the strong third-way theorist's contention that, for Newton, "space (in some sense) 'results from' *the existence of anything*" (Stein 2002, 268), is inconsistent with the DQE hypothesis — apparently, only God or an incorporeal being akin to a world soul can be the emanative cause of space.[16]

---

[16] McGuire (1978a, 15) explores a hypothetical interpretation that would allow beings other than God to ground the existence of space; yet, as disclosed in personal discussion, McGuire's purpose was only to explore the implications of an emanationist ontology, and not to put forward the view

Apart from the issue of incorporeal being, any interpretation that would posit matter as the emanative origin of space is likewise unacceptable given the basic ontological relationship between body and space on the DQE hypothesis. Material bodies are, in effect, portions of space that exhibit certain empirical properties, such as impenetrability or color, hence body *presupposes* spatial extension—body cannot, therefore, be the emanative cause of space. Newton emphasizes body's dependence on space in describing the DQE hypothesis, furthermore: "extension takes the place of the substantial subject in which the form of the body [i.e., the determined quantities] is conserved by the divine will" (N 29). It would be quite odd, consequently, if Newton additionally held that the determined quantities, as the forms or properties, were the ontological foundation of their own, as it were, substantial subject.

At this point, we should return to the topic of attributes and their dependence on God. Throughout Newton's analysis in *De grav*, God as the foundation of all possible substances, attributes, or accidents is constantly acknowledged, and this includes space:

> For certainly whatever cannot exist independently of God cannot be truly understood independently of the idea of God. God does not sustain his creatures any less than they sustain their accidents, so that created substance, whether you consider its degree of dependence or its degree of reality, is of an intermediate nature between God and accident. And hence the idea of it no less involves the concept of God, than the idea of accident involves the concept of created substance. And so it ought to embrace no other reality in itself than a derivative and incomplete reality. Thus the prejudice just mentioned must be laid aside, and *substantial reality is to be ascribed to these kinds of attributes* [i.e., extension], which are real and intelligible things in themselves and do not need to be inherent in a subject [i.e., an accident inherent in corporeal substance], *rather than to the subject* [i.e., corporeal substance] *which we cannot conceive as dependent* [upon God], much less form any idea of it. And this we can manage without difficulty if (besides the idea of body expounded above) we reflect that we can conceive of space existing without any subject when we think of a vacuum. And hence some substantial reality fits this. (32–33; emphasis added)

Since this passage clarifies to some degree the relationship between God and the attribute of extension, it is worth examining in more detail. As described in §2.3.1, Newton rejects the Cartesian and Scholastic accounts on the grounds that they foster a mistaken conception of corporeal substance that is independent of God (as well as incoherent). In its place, Newton champions a view that regards extension as an affection or attribute of God, which naturally implies that the concept of extension, unlike corporeal substance, "cannot be truly understood independently of the idea of God". Indeed, having rejected corporeal substance, Newton then argues that we should ascribe "some substantial reality" to extension as opposed to corporeal substance, with the possibility of a vacuum (matterless void) offered as further evidence against assigning extension to corporeal substance. One should not, accordingly, construe the term "subject" (*subjecto*) as referring to *any* subject, whether God or a lesser substance; rather, "subject" consistently refers to corporeal substance in the

---

that Newton actually accepted this hypothetical scenario. Unlike Stein, McGuire has always accepted that Newton's theology is central to understanding his theory of space (see, e.g., 1978a, 38–39).

above passage. It is only on this interpretation that Newton's overall argument is rendered coherent: it would be inconsistent for Newton to criticize the Cartesians and Scholastics for positing a conception of corporeal substance that is independent of God, *and then* put forward his own preferred thesis that makes spatial extension independent of all "subjects", taken broadly, and thus God. With respect to the pivotal sentence italicized in this quotation, we can give a more accurate rendering as follows: "substantial reality is to be ascribed to the attribute of spatial extension, which is a real and intelligible thing-in-itself and does not need to be an accident inhering in corporeal substance, rather than ascribe substantial reality to corporeal substance, which we cannot conceive as dependent upon God, much less form any coherent idea of it." Implicit in this statement, not surprisingly, is the belief that the substantial reality of the attribute space is dependent upon God, an idea that provides the basis of Newton's preference for the DQE hypothesis.

Newton's DQE hypothesis, therefore, quite clearly assigns to space a form of substantial reality, an admission that may help to elucidate Newton's earlier claim that space "approaches more nearly to the nature of substance" (22). Yet, while space is declared to have "some substantial reality", it is also an attribute of God, and is neither a substance nor an accident (i.e., given his rejection of the substance/ accident dichotomy regarding space), a point nicely encapsulated in the passage quoted earlier: space "is not a substance…because it is not absolute in itself, but is as it were an emanative effect of God" (21). The strong third-way interpretation of Newton is, as a result, quite correct in claiming that Newton's absolute space is not a substance. Nonetheless, given that space is an emanative effect of an incorporeal being (God), the substantial reality that Newton does bestow upon extension makes it practically equivalent—"approaches more nearly"—to the traditional substance concept: not only can space exist absent all corporeal existents, but, on the DQE hypothesis, spatial extension replaces corporeal substance as the container of his mobile, determined bodily quantities (see, also, footnote 21). Consequently, the more radical anti-Neoplatonist and non-theological reading of Newton's spatial theory championed by the strong third-way theorists is simply not upheld under a close scrutiny of the relevant texts.

2.4   Space as an Affection of Being

In order to more adequately diagnose the defects of the strong third-way interpretation, two important features of Newton's spatial theory must be addressed: first, why did Newton utilize affections/attributes in place of the more familiar accidents, as manifest in his well-known claim that "space is an affection of a being just as a being" (N 25)?; and second, why is space associated with "being just as a being" (*ens/entis quatenus ens*)? In this section, a more detailed comparison of Newton, More, and Charleton, will help to shed light on these aspects of Newton's spatial ontology. As will be argued, the second question discloses a predominate feature of the spatial ontologies of late seventeenth century English natural philosophy, a feature that Newton shares with both More and Charleton, whereas the first question is indicative of Newton's general discontent with the substance/accident distinction, and in this manner marks a point of departure from More's ontology towards the line favored by Gassendi-Charleton. The rationale for focusing on the these two philosophers, More and Charleton, is that, first, their influence on Newton is well-documented, and second, they represent the two leading positions during Newton's time on the relevance of the substance/accident dichotomy for space (with Charleton sponsoring Gassendi's popular solution).[17]

2.4.1   Extension and Accidents

If one seeks a rationale for Newton's characterization of space as an attribute or affection, a likely explanation is the metaphysical difficulties associated with classifying space as an accident. In his later, *Enchiridion Metaphysicum* (1671), More offers an ontology that treats space as an internal feature of God that has much in common with the seventeenth century's traditional sense of accident, although he uses the term "attribute", and sometimes "affection", instead of "accident" to describe space's metaphysical status:

> The real attribute of some real subject can be found nowhere else except where in the same place there is some real subject under it. And, indeed, extension is the real attribute of a real subject.... Indeed, we cannot not conceive a certain immobile extension pervading everything to have existed from eternity...and really distinct, finally, from mobile matter. Therefore, it is necessary that some real subject be under this extension, since it is a real attribute. (EM 56–57)

More thereby concludes that spatial extension must be the attribute of an incorporeal substance, and, while the details are not explicit, More seems to embrace the notion that attributes inhere in substances in the traditional way that an accident inheres in a substance: "extension indeed is in the real subject" (EM 68).[18] In

---

[17] See, Westfall (1962), and McGuire (1978a), on the references within Newton's work to More and Charleton. Newton's early notebook, *Quaestiones quaedum Philosophicae*, contains evidence that he read, at the least, both Charleton's *Physiologia,* as well as More's, *The Immortality of the Soul.*

[18] More tends to complicate his hypothesis that space is God's attribute by often referring to space as an incorporeal substance; e.g., in the ensuing section of the *Enchiridion*, he reasons that his theory utilizes "*the very same way of demonstration which Descartes applies to proving space to be a substance, although it be false in that he would conclude it to be corporeal*" (EM 57). More rejects Descartes' theory of space for many of the very same reasons that Newton provides in *De grav*, for instance, that Descartes cannot account for possibility of a vacuum (which is a conceivable state-of-affairs). An early formulation of this argument appears in *An Antidote against Atheism* (1655): "If after the removal of corporeal matter out of the world, there will be still Space and distance, in which this very matter, while it was there, was also conceived to lye, and this distant Space cannot but be something, and yet not corporeal, because neither impenetrable nor tangible, it must of necessity be a substance Incorporeal, necessarily and eternally existent" (AAA 338). More's penchant for conflating "space as God's attribute" and "space as identical to God's substance" may have prompted Newton's more careful attempts to deny the latter (see below).

contrast, Walter Charleton's popular work, although quite similar to More's views in many ways, parts company with More by declaring that space is neither substance nor accident, since it is *"more general than those two"* (Charleton 1654, 66), an opinion earlier adopted by Gassendi (SWG 384).[19]

While rejecting More's conception of space as God's accident, Newton's DQE thesis does resemble More's theory in that all extended things, whether body or spirit, necessitate the infinite spatial extension grounded in God's existence. In *De grav*, Newton repeatedly claims that extension "does not exist as an accident inhering in some subject" (N 22), and this argument also surfaces much later (1719–1720) in a paragraph he intended for the Des Maizeaux edition of the Leibniz-Clarke correspondence:

> The Reader is desired to observe, that wherever in the following papers through unavoidable narrowness of language, infinite space or Immensity & endless duration or Eternity, are spoken of as *Qualities* or *Properties* of the substance which is Immense or Eternal, the terms *Quality & Property* are not taken in that sense wherein they are vulgarly, by the writers of *Logick & Metaphysics* applied to *matter*; but in such a sense as only implies them to be modes of existence in all beings, & unbounded *modes* & consequences of the existence of a substance which is really necessarily & substantially Omnipresent & Eternal; Which existence is neither a substance nor a quality, but the existence of a substance with all its attributes properties & qualities (Koyré and Cohen 1962, 96–97).

In the correspondence with Leibniz, Clarke had suggested that space is a property of God (C.III.3, C.IV.10), and this problematic notion may have prompted Newton, in the above passage, to qualify and correct Clarke's argument so that it does not appear to sanction the view that space inheres in God in the same way that bodily accidents, properties, or qualities inhere in matter, or that space is a merely contingent feature of God.[20]

On this last point, Carriero notes that, "it is a standard theological position that there are no accidents in God" (1990, 123), yet, a further difficulty with viewing spatial extension as God's accident is that it might encourage the view that God inherits all of the consequences normally associated with extension, e.g., divisibil-

---

[19] Besides Newton's contemporary, Joseph Raphson (see, Koyré 1957, chap. 8), another Cambridge Neoplatonist who held that space is an attribute/accident of God is Ralph Cudworth (see, Grant 1981, 230). However, the Patrizi-Gassendi solution, that space is neither accident nor substance, was quite popular in England: besides Charleton and Barrow, one should add the earlier Neoplatonic philosophies of Warner and Hill (see, Garber et al. 1998, 558–561). For additional assessments of More's spatial theory, see, Boylan (1980), Copenhaver (1980), and, for the theological aspects of Newton's theory, Snobelen (2001).

[20] That is, an "accident", as the name implies, was often taken to be an unessential feature of a being, but space cannot be unessential given his view that all beings manifest spatial extension. Overall, it is unclear if Newton draws a principled distinction among the terms, "accident", "property", and "quality", but it is highly unlikely. Furthermore, while Newton is careful to designate space an attribute and affection in *De grav*, and not as an accident, the Des Maizeaux draft mentions "attribute" alongside "property" and "quality", as does the 1713 General Scholium (N 91), so it is possible that he may have abandoned his special use of "attribute" at a later date. Indeed, "mode" and "consequence" now seem to take over the special role that he had earlier accorded to "attribute", at least in the Des Maizeaux draft (see also footnotes 4 and 5).

ity, location, three-dimensionality. In the *Enchiridion*, More strives to circumvent this dilemma by ascribing to space some of the same incorporeal features that belong to God, for example, that God and space are both "simple", i.e., indivisible, such that they lack separable parts (EM 58; see, also, Chap. 6). It is not surprising, therefore, that More ultimately concludes on the basis of these similarities (among God and space) that there are two types of extension, namely, the divisible extension of corporeal matter and the indivisible extension of incorporeal spirit (with infinite spatial extension being an attribute of the latter; EM 118).[21]

2.4.2   Nullibism and Holenmerism

More's conclusion that all being is spatially extended is likewise supported by his rejection of two popular hypotheses on the relationship between God and space: first, he rejects the "nullibist" view favored by the Cartesians, that God is not in space; and, second, he rejects "holenmerism" (or "holenmerianism"), a belief common among the Scholastics, that God is whole in every part of space (which thereby guarantees that God is not divisible even if matter and space are divisible; EM 98–148). Given the rejection of these two hypotheses, the inevitable outcome is that incorporeal spirit is extended, a conclusion also adopted by Newton's Neoplatonist contemporary, Joseph Raphson (see, Koyré 1957, chap. 8).

In *De grav*, Newton's position parallels More's anti-nullibism. As first disclosed in §2.2, Newton also reckons that both corporeal and incorporeal beings are extended: after declaring that, "Space is an affection of a being just as a being", he explains that, "No being exists or can exist which is not related to space in some way. God is everywhere, created minds are somewhere, and body is in the space that it occupies; and whatever is neither everywhere nor anywhere does not exist" (N 25). A bit further on, he adds: "If ever space had not existed, God at that time would have been nowhere; and hence he either created space later (where he was not present himself), or else, which is no less repugnant to reason, he created his own ubiquity" (26). As for Gassendi and Charleton, both reject nullibism for similar reasons as More, and thus their natural philosophy may have also been a source for Newton's anti-nullibism: e.g., "no substance can be conceived existent without Place and

---

[21] Both More and Charleton believe that space is incorporeal, and this belief is based largely on the idea that the dimensions of space, like spirit, penetrate the dimensions of corporeal substance (EM 123–124; Charleton 1654, 68). Newton's DQE hypothesis nicely captures this aspect of their philosophy, since bodies are just parts of space endowed with material properties—consequently, Newton's reference to the extension (diffusion) of mind throughout infinite space (see §2.4.2) also follows these earlier philosophies by closely associating space with a spiritual entity. Yet, while both More and Charleton incorporate two types of extension, i.e., an incorporeal extension that penetrates corporeal extension, Newton's DQE hypothesis is more parsimonious in that it employs only one, namely, the divine attribute of extension. Indeed, Newton never (to the best of our knowledge) refers to space as "incorporeal" (or "immaterial"), a quite significant fact that is noted by McGuire as well (1978a, 42, n.38).

Time" (Charleton 1654, 66). On the other hand, Gassendi accepts holenmerism ("the divine substance is supremely indivisible and whole at any time and any place"; RIV 94); as does (presumably) Charleton (1654, 70).

If Newton clearly articulates his anti-nullibism, his opinions on holenmerism are more difficult to discern, and may comprise one of the more enigmatic elements of his spatial metaphysics. Overall, numerous passages in the *De grav*, as well as some later works, support a close analogy between the extension of material beings and God's extension. He begins by explaining that "[space and time] are affections or attributes of a being according to which the quantity of any thing's existence is individuated to the degree that the size of its presence and persistence is specified" (N 25). He then compares the "quantity of existence" among God and created being: "So the quantity of the existence of God is…infinite in relation to the space in which he is present; and the quantity of the existence of a created thing…in relation to the size of its presence, it is as great as the space in which it is present" (25–26). This explanation suggests that God and created beings do not differ as regards extension, contra holenmerism, since the same, as it were, metric—quantity of existence—applies equally to both, but with the important qualification that God possesses an infinite quantity of existence and created beings do not (or need not). Yet, since quantity of existence is an undefined notion in *De grav*, it is difficult to draw a specific conclusion based on this use of terminology.

In the ensuing passage, however, a better case can be made that Newton does side with More's anti-holenmerism: "lest anyone should for this reason imagine God to be like a body, extended and made of divisible parts, it should be known that spaces themselves are not actually divisible" (26). So, Newton not only *fails* to reject the claim that God is extended, but he also claims, like More, that space is not actually divisible (although space is conceptually divisible; see Holden 2004, on the different forms of divisibility in the Early Modern period). It would seem, therefore, that Newton posits an indivisible space to resolve the controversy concerning God's potential divisibility (see also, Janiak 2000, 224). Newton then draws an interesting analogy between the extension of both God's being and a temporal moment: "And just as we understand any moment of duration to be diffused (*diffundi*) throughout all spaces, according to its kind, without any concept of its parts, so it is no more contradictory that *mind* also, according to its kind, can be diffused through space without any concept of its parts" (N 26; emphasis added). That is, just as "a moment of duration is the same…throughout all the heavens" (26), Newton maintains that God, conceived as a mind-like spiritual being, is likewise the same part-less being throughout all space. A similar claim is made in the 1690s manuscript examined previously: "[t]he most perfect idea of God is that he be one substance, simple, indivisible" (TeL 123). Newton's characterization of God as simple, i.e., without parts, thus matches his own description of space in the same work: "[S]pace itself has no parts which can be separated from one another,.... For it is a single being, most simple, and most perfect in its kind" (TeL 117). Like More, Newton characterizes both space and God as simple, single beings (see §2.2.1 and §2.4.1).

In Chap. 6, the indivisibility of space for both More and Newton, as well as the holenmerism issue, will be revealed to have additional far-reaching implications for various components of their respective theories of space, and it will inform aspects of the alternative conceptual system for analyzing spatial ontologies developed in Chap. 9.[22]

2.4.3   Ens Quatenus Ens

Having exposed the Neoplatonist undercurrent in *De grav*'s rejection of nullibism and holenmerism, we are finally in a position to grasp the import of his various claims that space is an attribute/affection of "a being just as a being" (*ens quatenus ens*). The purpose of this explanation, in brief, is *not* to offer a unique anti-Neoplatonist proposal on the relationship between space and existents, i.e., that space is a logical or conceptual presupposition associated with any existent, as the strong third-way theorists counsel; rather, Newton's intention, replicating More's earlier maneuver, is to put forward an *ontology* that counters nullibism (that God is nowhere in space) and holenmerism (that God is in complete in every part of space). Or, put differently, Newton's use of *ens quatenus ens* does not amount to a repudiation of Cambridge Neoplatonist hypothesizing about the nature of space: *ens quatenus ens* is, in fact, an instance of such ontological speculation. In addition, Newton's *ens quatenus ens* hypothesis may have been motivated by similar discussions in More's *Enchiridion*, where a metaphysics of "being just as a being" forms that backdrop of More's thought, including his spatial hypotheses: e.g., "the essence of any being insofar as it is a being is constituted of amplitude [extension] and differentia [form], which distinguishes amplitude from amplitude" (EM 9). Overall, the strategy of both More and Newton is to link spatial extension, in some form at least, to all being, even God, rather than to just a sub-class of being (as, for example, in the Cartesian identification of extension with corporeal being).

Returning to Newton's well-known quote, that "space is an emanative effect of the first existing being, for if any being whatsoever is posited, space is posited" (N 25), Stein contends that "the second clause tells us precisely what the first clause *means*" (Stein 2002, 269). But, as first argued in §2.2, this explanation hinges on a questionable interpretation of the phrase "first existing being"; likewise, there are numerous precedents in the earlier Cambridge Neoplatonist literature for employing "emanative effect" to signify a unique form of God-based ontological dependency. Given the discussion above, we are now in a better position to grasp that the phrase, "for if any being whatsoever is posited, space is posited", is not intended to explicate the meaning of "emanative effect", but is instead another instance of

---

[22] Some commentators (e.g., Pasnau 2011, 338, n.21; Reid 2007) defend holenmerism as Newton's preferred ontology of space, but this is dubious given the paltry evidence for holenmerism and the powerful evidence in favor of God's actual extension (for other anti-holenmerist interpretations of Newton, see Grant 1981, 253; McGuire and Slowik 2012). As argued above, since Newton, following More, makes a parallel case for the simplicity of both God and space (i.e., as beings without parts), whereas holenmerism is predicated on spatial parthood, it is extraordinarily difficult to ascribe holenmerism to Newton.

Newton's *ens quatenus ens* thesis that space is an attribute/affection of "a being just as a being". Stein's error, in short, is that he conflates two distinct hypotheses, namely, emanative causation and *ens quatenus ens*. Additional evidence for the separation of these hypotheses is contained in Newton's first reference to emanation in *De grav*, where he claims that "[space] is as it were an emanative effect of God *and* an affection of every kind of being" (N 21, emphasis added). What is important about this passage is that it does not run together the construal of space as an "emanative effect of the first existing being" and space as "an affection of every being" (= "if any being is posited, space is posited") in the manner advocated by Stein— rather, these two hypotheses are clearly distinguished in this quotation, thus raising an obstacle for Stein's attempt to use the latter concept to explain the meaning of the former.

To better grasp the intent of Newton's much-debated claim, it will be useful to quote the broader context of the full paragraph:

> Space is an affection of a being just as a being. No being exists or can exist which is not related to space in some way. God is everywhere, created minds are somewhere, and body is in the space that it occupies; and whatever is neither everywhere nor anywhere does not exist. And hence it follows that space is an emanative effect of the first existing being, for if any being whatsoever is posited, space is posited. And the same may be asserted of duration: for certainly both are affections or attributes of a being according to which the quantity of any thing's existence is individuated to the degree that the size of its presence and persistence is specified. So the quantity of the existence of God is eternal in relation to duration, and infinite in relation to the space in which he is present; and the quantity of the existence of a created thing is as great in relation to duration as the duration since the beginning of its existence, and in relation to the size of its presence, it is as great as the space in which it is present. (N 25–26)

The first sentence begins with the exposition of the *ens quatenus ens* hypothesis, and Newton explains, in the second sentence, that extension also pertains to God in some manner ("No being exists"). The third sentence posits an omnipresent God in infinite space, "God is everywhere", and Newton adds that created minds and bodies are located in, and occupy, this same space (and cannot be nowhere)—i.e., minds are "somewhere", and body "is in the space that it occupies", but, since God is "everywhere", these lesser beings must also partake of God's extension. This last point is presented quite explicitly a bit later in the paragraph: "the quantity of the existence of God is…infinite in relation to *the space in which he is present*" (emphasis added).[23] The fourth sentence is Stein's favored quote ("space is an emanative effect"), but it begins with the phrase, "And hence it follows that", which is an important qualification since it relates the subsequent content to the previous three sentences. Put simply, Newton is arguing that, *since* God is omnipresent, and *since* the other beings occupy finite portions of the infinite space brought about by God, "hence it follows that space is an emanative effect of the first existing being, for if any being whatsoever is posited, space is posited". That is, space must be the

---

[23] The infinity of space is presented as akin to an a priori certainty in *De grav*, as McGuire also concludes, "in Newton's view the presence of matter presupposes ontologically the infinitude of spatial extension" (1983, 184). See §2.5 below as well.

emanative effect of an infinite, omnipresent being ("the first existing being") because all being manifests extension in some fashion ("for if any being whatsoever is posited, space is posited" = "Space is an affection of a being just as a being"), and thus the remaining (finite) beings *require* an omnipresent being to ground the existence of the infinite space in which they reside; likewise for time, as disclosed in the two remaining sentences of the paragraph. Here, it is important to recall an identical line of reasoning in More's *Enchiridion*, where an infinitely extended "first Being", God, "receives all others" (EM 59; see §2.2), as well as Newton's assertion, in his 1690s work, TeL, that God contains "all other substances in Him as their underlying principle and place" (TeL 123). To conclude, space is not, as Stein contends, a necessary consequence of the existence of any being, rather, the entailment goes in the other direction: the actual existence of any being necessarily presupposes an infinite, immutable space, and only God can secure that precondition.[24]

2.5 Newton's Spatial Theory and Substance/Property Ontologies

Thus far, we have examined Newton's spatial ontology largely from the perspective of a non-substantivalist, strong third-way standpoint, in particular, Stein's (2002) conception. As revealed in previous sections, the abundant Neoplatonist elements in Newton's spatial theory (emanative causation, the primacy of incorporeal being over corporeal being, etc.) raise insurmountable obstacles for any strong brand of third-way interpretation, but in this section we will focus on how the substance/property (or substance/accident) dichotomy factors into Newton's theory, and how this dichotomy, in turn, effects a potential substantivalist or third-way classification more generally.

To recap our earlier discussion, although Newton denies that space is an accident and that space "inheres" in God (see §2.2), this does not change the fact that space is, so to speak, God's predicate—and there is nothing in *De grav*, or in any of Newton's other works, for that matter, that would suggest that any lesser being can play the role of space's subject (besides the hypothetical Neoplatonist world soul). Indeed, this more limited subject-predicate relationship between God and space remains an undeniably pervasive feature of Newton's natural philosophy, as we

---

[24] As noted in footnote 13, a strong third-way interpretation, similar to Stein (2002), is also adopted by Janiak, although he strives to distance his reading from some aspects of Stein's interpretation (2008, 155–163). Janiak, however, follows Stein in running together "being as being" and emanative causation, which are two distinct hypotheses, as argued above. After claiming that "the affection thesis entails the claim that space is an 'emanative effect' of the first existing being" (142), Janiak concludes that "space emanates from whatever entity is the first to exist" (146), thereby sanctioning a major aspect of Stein's strong third-way case (since space is no longer dependent on God). Yet, as we have seen, Newton's *De grav* links the emanation of space to a higher, infinite incorporeal/spiritual being alone, and thus the claim that space would emanate from any type of being is simply unsupportable.

have seen. But what does this relationship between God and space—namely, that space is God's attribute, without any sense of inherence—imply for his overall spatial theory? Is the relationship between God and space, moreover, analogous to a logical or conceptual association, as Stein's version of a strong third-way reading supports, or is this relationship quite similar to the notion of inherence, and thus provides little support for the strong third-way interpretation?

Newton's endorsement of the infinity of space helps to shed light on these questions. Although we cannot imagine the infinity of space, he claims that "we can understand it" (N 23), and he mentions a rough geometric proof involving the intersection of two lines that slowly approach a parallel configuration: "therefore there is always such an actual point where the produced [lines] would meet, although it may be imagined to fall outside the limits of the physical universe", or, as he also puts it, "the line traced by all these points will be real, though it extends beyond all distance" (23). Despite human limitations, "God at least understands that there are no limits, not merely indefinitely but certainly and positively, and because although we negatively imagine [extension] to transcend all limits, yet we positively and most certainly understand that it does so" (24–25). Consequently, our examination of Newton's *ens quatenus ens* hypothesis gains an important qualification, namely, if anything is posited, *infinite* space is posited, a point first revealed in §2.4.3.

Newton's God-grounded reification of an infinite Euclidean space has important implications for understanding the role of the substance/property distinction in his spatial ontology. Despite rejecting the notion of inherence, Newton's "space is an attribute of *ens quatenus ens*" hypothesis nonetheless entails that the domain of space is closely tied or restricted to God's domain; in other words, space is infinite *because* God is infinite, an hypothesis that is stated clearly in several later works, such as in the 1713 edition of the *Principia*: "[h]e endures forever, and is everywhere present; and by existing always and everywhere, he constitutes duration and space" (N 91); and *De grav*: "space is eternal in duration and immutable in nature because it is the emanative effect of an eternal and immutable being" (N 26). This "congruence", as we may call it, between the ontological domains of both God and space is also relevant to his quantity of existence concept examined earlier, where he claims that "the quantity of the existence of God is eternal in relation to duration, and infinite in relation to the space in which he is present" (25–26) Therefore, notwithstanding his repudiation of the substance/property dichotomy, Newton presumes a metaphysics of space that closely mimics that dichotomy, save for the notion of inherence. Not only does Newton require an entity to ground the existence of space, but the domain or extent of the former determines the domain of the latter: specifically, space can only be infinite if the entity that provides the foundation for space is infinite (i.e., omnipresent), and, as revealed above, the infinity of space is tantamount to a certain (a priori?) truth on Newton's scheme. This interpretation of Newton's spatial theory is partially confirmed in a passage from the 1690s TeL manuscript where Newton denies the possibility "that a dwarf-god should fill only a tiny part of infinite space with this visible world created by him" (TeL 123). In other words, God must be infinite to ground the existence of his (necessarily) infinite space. Applying these lessons to the quantity of existence quotation above (N

25), it would thus seem to follow that a being's spatial attribute must match, or be congruent with, its quantity of existence.

Newton's tacit utilization of a sort of surrogate substance/property concept may not at first appear to be a setback for a non-substantivalist interpretation of his spatial theory; yet, if the non-substantivalist's interpretation additionally strives to depict Newton's theory as comparable to the Stein-DiSalle definitional version of a third-way ontology of space discussed previously (§2.1), then Newton's latent form of substance/property metaphysics does indeed constitute a serious problem. The definitional approach to space favored by Stein and DiSalle departs from a standard substance/property approach by admitting the existence of spatial structures that transcend the actual relations among the world's actual inhabitants. For example, third-way theories of this sort are consistent with a hypothetical island universe wherein the structure of space is infinite Euclidean despite a finite material distribution. A definitional third-way theory can meaningfully entertain, for example, whether space possesses a flat (infinite, unbounded) Euclidean structure, or a spherical (finite, unbounded) non-Euclidean structure, since the structure of space, whether finite or infinite, can be obtained through our experience of the behavior of a finite number of objects or entities, and perhaps only one. That is, the definitional wing of the third-way approach does not confine the domain of spatial structures so as to precisely match the domain of actually existing things—on the definitional approach to space, the inertial structure manifest in Newton's rotating bucket or spinning globes thought experiments, via the non-inertial force effects of rotation, is sufficient to reveal the inertial structure of the whole of space, whether space is finite or infinite. A similar possibility confronts other ontological interpretations, such as the modal variant of sophisticated relationism. Given the existence of a body, many modal relationists would declare that the mere *possibility* of that body's motion throughout the universe is enough to classify the space as, say, infinite Euclidean or spherical non-Euclidean, since the potential return of the moving body (from the opposite direction that it departed) is enough to rule out the flat Euclidean case in favor of a non-Euclidean structure. Substantivalists, strict eliminative relationists, and super-eliminative relationists will, of course, spurn these definitional third-way and sophisticated relationist ontologies, since they accept the substance/property dichotomy: for a substantivalist, space is a substance and so the domain of space coincides with that substance; and, for a strict non-modal eliminative relationist, the behavior of a few finite bodies in a local region cannot determine the structure of an infinite space (i.e., these bodies only reveal the structure of the part of space they occupy, hence the domain of spatial structures coincides, or is reduced to, the domain of these bodies and their interrelationships). But, as revealed in our investigation, Newton would likely concur in their disparagement of these kinds of spatial ontologies, for he advocates a surrogate form of the substance/property dichotomy. Like the substantivalists and strict relationists, Newton's spatial ontology necessitates a congruence between the domain of actual existing entities, either corporeal or incorporeal, and the domain of the attribute space. In the larger scheme of the ontological classification of spatial ontologies, consequently, Newton's rebuff of the notion of inherence does not appreciably effect the classification of his own theory, which runs counter to the third-way definitional hypotheses of space or sophisticated modal relationism.[25]

For these reasons, Newton's ontology of space is consistent with the ontological dependence version of the property theory of space, P(O-dep), surveyed in Chap. 1, since his theory fits a subclass of non-standard approaches that mimic a substance/ accident metaphysics. A property theory of space is not substantivalism, moreover, and thus the following interpretation proposed by Pooley is not justifiable: after citing the passage from *De grav*, where Newton states that the motion of bodies "be referred to some motionless being such as…space in so far as it is truly distinct from bodies" (N 20–21), Pooley draws the conclusion from Newton's use of the term "being" in this case that it "involves a variety of substantivalism" (2013, 526; see, also, Earman 1989, 11, which draws the same conclusion). Given Newton's claim, discussed above, that space "approaches more nearly to the nature of substance" (N 22), Pooley reasons that, "of the three categories—substance, accident, or nothing—Newton states that space is closest in nature to substance" (2013, 526). Yet, as we have seen, Newton denies that space is an accident (property) *of body*, "since we can clearly conceive extension existing without any subject, as when we imagine spaces outside the world or places empty of any body whatsoever" (N 22). That is, space is closer to the nature of substance if we confine our attention to the material or bodily level of ontology alone, since space is independent of body—but, as argued above, space is much closer to a property of God if his overall ontology is examined. Despite his qualms about the concept of inherence and a variety of associated theological concerns (the contingency of accidents vis-à-vis God, divisibility, etc.), the evidence of the texts overwhelmingly supports the dependent nature of space in accordance with a God-based P(O-dep) conception, a position nicely encapsulated in a passage that we have often quoted above: space "is not a substance…because it is not absolute in itself, but is as it were an emanative effect of God" (21).

One might strive to reclaim a substantivalist interpretation by appealing to the metaphysics of emergence or supervenience, a tactic that would equate Newton's

---

[25] There are a class of third-way theories that do not significantly part from the substance/accident dichotomy in certain contexts, however, such as the sophisticated relationist interpretation of general relativistic spacetimes in the work of, e.g., Dorato (2000) or Dieks (2001a). As will be explained in later chapters, if the metric field is conceived in the manner of a physical field, it thus follows trivially that the domain of spacetime, i.e., metric field, is congruent with the domain of physical fields (and thus the substance/property distinction can be claimed to have been upheld via their congruent domains). Could Newton's theory obtain a third-way classification by association with these modern third-way conceptions? Unfortunately, if Newton's theory were to acquire a third-way designation by this means, then the plethora of earlier theories that also posit a God-infused space, from Plotinus to More, would also obtain this same third-way label, as would any theory that links the domain of material phenomena with the domain of spatial extension (like Descartes')—and this, of course, would trivialize the third-way classification since nearly all spatial hypotheses would now count as third-way. In short, the theories of Dorato, Dieks, et al. are merely consistent with the substance/property distinction, but it is not a necessary requirement. For Newton, the congruence of God and space, i.e., his surrogate substance/property dichotomy, is indeed a necessary component of his natural philosophy.

"space as emanative effect" with space as a type of emergent or supervenient entity. Although Part II and III will concern these issues in detail, the difficulty with this potential substantivalist strategy is that, if forced to choose from the standard dichotomy in a modern setting (and thus leaving aside the property theory and Newton's God-infused world), then the emergence of space from a deeper level of ontology seems much more amenable to a relationist interpretation than a substantivalist one. Suppose, for example, that a theory's base entity is either a material object or a physical field: given this base ontology, it would be natural to infer that the emergent or supervenient entity is also material/physical, whereas it would be quite odd to categorize that entity as entirely different in kind, i.e., as a unique non-material spatiotemporal thing alone. An instance of this reasoning can be found in Norton's critique of Brown's interpretation of the spacetime structure of special relativity. After reciting Brown's claim that these types of spacetime structures are "a codification of certain key aspects of the behavior of particles and fields" (Brown 2005, 100), Norton labels this a "spacetime supervenes on matter" view, and classifies it "as a form of relationism" (2008, 822). As will be argued in later chapters, however, an emergent or supervenient conception of space resembles neither substantivalism nor relationism, but does qualify as a property theory of space.

2.6   Conclusion

Notwithstanding the rhetoric from the definitional wing of the third-way approach to spatial ontology, there is not much in Newton's treatment of spatial ontology that stands out as unique or groundbreaking, and what is original can be seen as a natural extension of, or variation on, the work of his older contemporaries. As revealed above, Newton's *ens quatenus ens* doctrine and his view that space emanates from God were almost certainly derived from More's earlier application of these concept; likewise, the belief that space does not conform to the traditional substance-property dichotomy was common among many of Newton's predecessors and contemporaries, such as Charleton. If forced to make a choice, the chief novelty would probably lie in Newton's combination of these spatial themes, namely, the pairing of More's idea of a spatially extended God (contra nullibism and holenmerism) with the Gassendi-Charleton rejection of the substance/property scheme for space. Yet, given his refutation of the corporeal/incorporeal divide (as it pertains to both substances and properties), and given his deeply rooted suspicion of substance ontologies in general, it is not surprising that Newton resorts to an alternative approach that eschews altogether the substance/property dichotomy for spatial extension.[26] Nevertheless, other components of the Gassendi-Charleton philosophy run counter

---

[26] It unclear to what degree Newton's rejection of an incorporeal/corporeal distinction in *De grav* extends to his later published works. In Query 29 to *The Opticks*, Newton does describe God as "a being incorporeal" (N 130), and, in the correspondence with Leibniz, Clarke likewise deems God an "incorporeal substance" (C.IV.8). Yet, in both *De grav* (N 33) and the General Scholium of the

to Newton's outlook. For instance, Newton's rejection of a corporeal/incorporeal distinction may also explain his apparent reluctance to embrace their holenmerism, since that view implies a sharp distinction between corporeal and incorporeal extension, and thus tacitly reintroduces the type of a mind-body division that he reckons offers "a path to atheism" (N 31). Yet, like both More and Charleton (among many others), Newton is similarly compelled to associate space with incorporeal existents, despite his reluctance to employ the incorporeal label: i.e., since space must be the affection of a being, there is only one being, God, that can secure the infinity, indivisibility, etc., of space.

Of course, Newton's skepticism regarding substance could be interpreted as supporting Stein's minimal claim (2002, 281) that Newton advanced the debate on spatial ontology in a more modern, third-way direction. But, as should be readily apparent by now, the basis for Newton's forward-looking, skeptical treatment of substance is deeply rooted in his Neoplatonic metaphysics. Newton's primary criticism of the Cartesians and Scholastics, as disclosed above, stems largely from his belief that their ideas of substance entail a troubling dualism of mind and body, as well as from a general worry "that atheists [may] arise ascribing to corporeal substance that which solely belongs to the divine" (N 32). In essence, it would appear to be specifically Neoplatonist concerns, related to the overlap of the Western conception of God and the mind/body problem in particular, that prompt Newton's unique approach to spatial ontology, and not any sort metaphysically-deflationary, strong third-way insight into the nature of space.

The foregoing analysis does not contest the merits of the weak third-way interpretation, however. The concept of absolute space espoused in the *Principia,* in particular, the 1687 edition with its notable absence of ontological speculation concerning God and substance, is an undeniable breakthrough—but it is a breakthrough for physics, and not metaphysics. That is, the virtues of the *Principia*'s notion of absolute space are methodological and epistemological, and not ontological, although it would take the shrewd assessment of an Euler or Kant to fully appreciate this point. Accordingly, since the weak third-way reading allows both the definitional conception of space and time required for his physics *and* the type of Neoplatonist ontology of space founded on God that we have detailed above, it constitutes a more convincing interpretation of Newton's spatial hypotheses. The weak third-way reading, in conjunction with P(O-dep), also accounts for Newton's endorsement of absolute position and absolute velocity. Although a more thorough discussion will be postponed until Chap. 6, Newton repeatedly refers to "the immobility of space" (N 25), or, as he famously declares in the *Principia,* "absolute space, of its own nature without reference to anything external, always remains homogeneous and immovable" (64). Unlike More, Gassendi, and the vast majority of other seventeenth century natural philosophers before him, Newton is reluctant to use the terms "immobile" or "motionless" with respect to God, but it is clear that he derives the immutability of space from the immutability of God, and, of course,

---

1713 *Principia* (91), Newton expresses skepticism regarding the concept of God's substance, but not his attributes; and, as observed in footnote 21, Newton never refers to space as incorporeal.

motion is a form of change: "space is eternal in duration and immutable in nature because it is the emanative effect of an eternal and immutable being" (26). Hence, it is the P(O-dep) conception that best accounts for some of the distinctive, if problematic, features of Newton's metaphysics and physics: specifically, since space functions much like a property of God, Newton apparently concludes that the immobility (immutability) of the latter grounds the immobility of the former, and thus the material world can assume any number of different positions and velocities with respect to immobile absolute space, even if the those different positions and velocities cannot be detected since they preserve their relative configurations.

Chapter 3
Leibniz' Ontology of Space: Whither Relationism?

One of the nagging puzzles that besets both the spacetime and Early Modern communities is the problematic fit between Leibniz' conception of space and relationism. C. D. Broad long ago hinted at the unsuitability of a spatial relationist interpretation of Leibniz (1981, 171–173), but the treatment of his spatial hypotheses in many of the canonical texts in the philosophy of space and time has continued to portray Leibniz as having sanctioned a straightforwardly contemporary version of relationism (see, e.g., Sklar 1974, 169, and, Friedman 1983, 219, all nicely recounted in Auyang 1995, 247). Sophisticated (or modal) relationism—that space is a mere relation among bodies, but that these relations may include within their scope possibilia or non-actual bodies—is still often defended (e.g., Khamara 1993, 478; Belot 2011, 184), while others promote the super-eliminative variety that insists that all spatial relations can be eliminated in favor of material relationships, so that a vacuum is impossible (see, e.g., Futch 2008, 48; and, for relationism in general, Hooker 1971, 111). Yet, in this chapter, not only will the majority of these relationist interpretations of Leibniz' theory be revealed as inadequate to the task, but the viability of any relationist interpretation of Leibnizian space will itself be called into question. As will be demonstrated, the underlying metaphysics of Leibniz' theory requires a different set of conceptual resources, despite the obvious fact that the aftermath of his rejection of the absolutism of his day, especially in the correspondence with Clarke, set in stone an idea of Leibnizian space that continues to mislead historians and philosophers of science. While the conclusions of this chapter may strike the reader as rather controversial, the preponderance of the evidence that will be presented has played a major role in the metaphysical investigations of Leibniz for the past several decades. Unfortunately, the lessons to be gathered from this research have not been sufficiently assimilated by spacetime philosophers in their analysis of the foundations of Leibnizian space, but neither have the subtleties of Leibniz' concepts been properly factored into various metaphysical and historical appraisals. Part of the goal of this chapter, in short, is to remedy this unfortunate oversight.

© Springer International Publishing Switzerland 2016
E. Slowik, *The Deep Metaphysics of Space*, European Studies in Philosophy of Science, DOI 10.1007/978-3-319-44868-8_3

In §3.1, the various brands of relationism will be compared and contrasted with Leibniz' spatial hypotheses, with the surprising result that most are either entirely inadequate or, at best, only tangentially relevant to his deeper metaphysical design. In §3.2, God's foundational role as the ontological basis of space will be examined in detail, along with an analysis on the substance/accident dichotomy, quantity and the order of situations, and the holism or monism of both geometry and the material world's interconnections. All of these elements of Leibniz' natural philosophy, including the role that nominalism plays in Leibniz' spatial ontology, will be brought together for a final synthesis in §3.3. Overall, a case will be made that the most plausible reading of Leibniz' spatial hypotheses resembles the unique form of property theory, P(O-dep), first introduced in Chap. 1, albeit this conception does resemble relationism on several important points.

Furthermore, throughout this chapter, the analysis will be largely confined to the metaphysical level of material bodies and a substance/accident/relation metaphysics, since a thorough treatment of the intricacies of the monadic component of Leibniz' theory will be postponed until Chap. 4. Yet, various hints as to how the monadic realm connects with the material realm will be briefly considered in §3.3.3. This choice, to specifically focus upon the level of bodies, is in keeping with both the traditional substantival/relational dispute in ontology as well as the Leibniz-Clarke correspondence, the latter being his most significant and detailed contribution to the philosophy of space. In short, the late correspondence with Clarke does not bring into play the underlying monadic foundation of Leibniz' philosophy, thus partly justifying an exclusive investigation of the more commonplace ontological themes associated with material substances, accidents, and God. Nevertheless, a complete account of Leibniz' views on space must ultimately explicate the monadic basis, and thus the ensuing analysis is but the first half of a larger story that will be picked up in the next chapter.

3.1   Relationism and Leibnizian Space

In this section, Leibniz' theory of space will be weighed against various forms of relationism while simultaneously engaging the commentary of a number of important contemporary studies (Arthur, Futch, Belot, De Risi, etc.). The Leibnizian corpus relevant to this investigation will largely be drawn from 1700 onward, with special attention dedicated to the *New Essays* and the Leibniz-Clarke correspondence, although works from earlier periods will also figure prominently in the discussion.

3.1.1   *Relationism Versus Universal Place*

As mentioned in Chap. 1, providing a precise definition of substantivalism and relationism is a daunting task in its own right, but, summarizing the earlier analysis, substantivalism (or absolutism) will be defined as the view that space is an independently existing entity of some sort, whereby the geometry of space is independent of bodies (i.e., distance relations are between the parts of space, with the distance relations among bodies supervening on these independent geometric facts). A central tenet of relationism is the rejection of substantivalism, hence, at least on this point, Leibniz' philosophical inclinations side with relationism. In various writings, Leibniz rejects the view that space is an entity that exists independently of material things, yet, unlike Descartes, he rejects the thesis that space is identical with matter: "I do not say that matter and space are the same thing. I only say that there is no space where there is no matter and that space in itself is not an absolute reality" (L.V.62).

Nevertheless, as also explained in §1.2, other approaches to space and time also deny substantivalism, such as the property theory, and so the rejection of substantivalism does not automatically equate with relationism. Additionally, relationism comes in many flavors, and so the question remains as to which type Leibniz subscribes, if any. One form is super-eliminative relationism, whereby all spatial relations must coincide with material relations, hence a vacuum is impossible; another is eliminative relationism, which also confines space to the relations among existing bodies but allows empty spaces; and the last is sophisticated relationism, which holds that the domain of spatial relations can include possible bodies.

Turning to the first of these relationisms, the super-eliminative type is not supported by the textual evidence, for on several occasions Leibniz insists that a vacuum is a possible, although not actual, state of affairs: "I don't say that the vacuum, the atom, and other things of this sort are impossible, but only that they are not in agreement with divine wisdom" (AG 170).[1] Likewise, the hypothetical scenarios envisaged in the *New Essays*, which discuss indirect methods of measuring a vacuum within the material world (NE II.xv.11), would seem to refute super-eliminative relationism. Yet, assuming the structure of space is Euclidean, and hence infinite, as Leibniz presumably does (see footnote 7), then his admission that God could have brought about a finite material world also undermines eliminative relationism, since the relations among the finite number bodies in this scenario are incapable of determining the geometry of an infinite Euclidean space: e.g., "[a]bsolutely speaking, it appears that God can make the material universe finite in extension" (L.V.30).

---

[1] In denying the possibility of a vacuum, Leibniz provides other rationales besides divine wisdom to secure the fullness of space, the most important being that a vacuum would violate the principle of the identity of the indiscernibles (PII) because the empty parts of space would be intrinsically identical: "Space being uniform, there can be neither any external nor internal reason by which to distinguish its parts and to make any choice among them" (L.IV.17). Futch relies on these sorts of auxiliary metaphysical arguments to support super-eliminative relationism, and thereby defend a plenum (matter-filled world) against a possible vacuum (2008, 47–57). In contrast, our rejection of the eliminativist and super-eliminativist interpretations rests straightforwardly on Leibniz' assertion that these hypothetical vacuum scenarios are indeed possible, hence his conception of space is, by itself, neither eliminativist nor super-eliminativist. Yet, if these auxiliary metaphysical arguments are included, then Leibniz' natural philosophy does not admit a vacuum, just as Futch contends. On the related issue of "compossibility", which pertains to the possible existents in a world (e.g., Lm 662), see Rutherford (1995, 181–188).

Given the evidence of these texts, it is not surprising that the type of relationism often associated with Leibniz takes a modal form, thereby equipping an infinite space with a full panoply of relations despite the (possible) presence of matter-less regions, whether internal or external to a finite material world. Nevertheless, a close inspection of Leibniz' writings casts doubt on the viability of even this more lenient construal of relationism, i.e., sophisticated relationism.

Consider the following account of "place" in the *New Essays* (1703), delivered by Leibniz' spokesman, Theophilus:

> [(a)] 'Place' is either *particular,* as considered in relation to this or that body, or *universal;* the latter is related to everything, and in terms of it all changes of every body whatsoever are taken into account. If there were nothing fixed in the universe, the place of each thing would still be determined by reasoning, if there were a means of keeping a record of all the changes or if the memory of a created being were adequate to retain them—as the Arabs are said to play chess on horseback by memory. However, what we cannot grasp is nevertheless determinate in the truth of things. (NE II.xiii.8)

While it may seem innocent enough at first glance, passage (a) undercuts the prospects for any *body-centered* interpretation of Leibniz' concept of space, i.e., modern relationism (where "*space* is that which results from places taken together", L.V.47). In the discussion that directly precedes (a), Philalethes, expressing Locke's conception, contends that "same place" is relative to different contexts, and can thus be applied, for instance, to a chess-board in a ship: "The chess-board, we also say, is in the *same place*…if it remains in the same part of the cabin, though, perhaps, the ship which it is in [has set sail]" (NE II.xiii.8). Leibniz responds, in quote (a), by referring to Philalethes' genuinely body-based relational conception as "particular" place, "in relation to this or that body", and he then goes on to contrast this idea with a "universal" notion of place which "is related to everything, and in terms of it all changes of every body whatsoever are taken into account". Leibniz' claim that "if there were nothing fixed in the universe, the place of each thing would still be determined by reasoning" is deeply antithetical to relationism, needless to say, and mimics Newton's use of absolute place in the *Principia* (N 66). Put simply, a relationist must define the notion of, for instance, "same place" by means of a material reference frame: they cannot, as does Leibniz, countenance the possibility that there may be no fixed material frames at all while simultaneously insisting that "same place" is still "determinate in the truth of things"—determinate with respect to what? Leibniz' claim implies that there is something else besides material existents, i.e., his universal place, that records all bodily changes of place, a conclusion that runs counter to all versions of relationism (see §1.1, and §3.1.2 and §3.3.1 below). Of course, Leibniz would deny that universal place is an independent entity that can exist apart from matter; rather, it is simply an internal feature of some sort in bodies (and, ultimately, monads) that, presumably, allows a reconstruction of the prior places that bodies had occupied. While this last inference would seem correct, it nonetheless still falls afoul of relationist doctrine since any record or memory of a universal place within matter—a record by means of which "all changes of every body whatsoever are taken into account"—is akin to absolute place or space, with the only difference being the reinterpretation of absolute place as an internal feature of each body.

Before proceeding further, it is important to note that the more elaborate and better-known arguments concerning space in the Leibniz-Clarke correspondence do not challenge the concept of universal place put forward in the *New Essays*. In contrast, his analysis of "fixed existents", as well as other passages, would seem to confirm the continuing applicability of universal place. After defining place via the relation of a body to "other existents which are supposed to continue fixed" over a period of time, Leibniz states that "*fixed existents* are those in which there has been no cause of any change of the order of their existence with others, or (which is the same thing) in which has there has been no motion" (L.V.47). Leibniz' assorted references to "suppose to continue fixed" could, as a result, be taken as signifying a body-centered framework for determining place that allows Galilean transformations to all equivalent, inertially-related frameworks (i.e., "suppose" means that the ensemble may itself be moving uniformly or at rest relative to other ensembles).[2] Yet, Leibniz elucidates the meaning of this supposition in his fifth letter: "And supposing or feigning that among those coexistents there is a sufficient number of them which have undergone no change, then we may say that those which have such a relation to those fixed existents as others had to them before, have now the *same place* which those others had" (L.V.46). In other words, the supposition is that there are any fixed existents at all, an interpretation that is verified by his reference to "a sufficient number of them [unchanged bodies]" needed to establish "same place", as well as by his later defense of the validity of rest: "It is true that, exactly speaking, there is not any one body that is perfectly and entirely at rest, but we frame an abstract notion of rest by considering the thing mathematically" (L.V.49). This last explanation parallels the assertion from the *New Essays,* examined previously, that upholds universal place: "if there were nothing fixed in the universe, the place of each thing would still be determined by reasoning." Finally, and most importantly, the explication of particular place in the *New Essays* brings into play inertially related frames, e.g., the stationary chess board on the moving ship, but he proceeds to contrast this bodily-centered notion of place with universal place—in summary, there is no evidence in the correspondence with Clarke to read "fixed existents" as denoting a (Galilean) system of inertial reference frames, nor to overturn the concept of universal place.

3.1.2 Universal Place and the Property Theory

One way to grasp how relationism conflicts with Leibniz' theory is to focus on the fact that universal place allows bodies to occupy different positions/places even though these bodies could bear the same relations of co-existence among existing

---

[2] The following characterization of "same place" in Arthur (1994), on the other hand, seems entirely appropriate, and does not bring into play Galilean transformations: "Thus the hypothesis of fixed existents allows us to define *place* in terms of an equivalence: it is the equivalence class of all things that bear the same situation to our (fictitious) fixed existents. And when we take all possible situations relative to these fixed existents, we have a manifold of places, or abstract space" (1994, 237).

and potentially existing bodies (with "co-existence" being Leibniz' preferred manner of describing bodies that exist at the same time; e.g., L.V.47). In other words, the
"body-centeredness" of the spatial relations mandated by relationism is violated,
even granting the modality involved with potentially co-existing bodies. For
instance, suppose body A stands one meter to the right of bodies C, E, F, G, and that
these five bodies alone comprise the material universe. If A_0, E_0, C_0, F_0, G_0 symbol-
ize this initial configuration, and A is moved one meter further to the right, to new
position A_1, after which E_0, C_0, F_0, and G_0 are moved one meter to the right as well,
obtaining new positions E_1, C_1, F_1, G_1, then the initial and final states of this relative
configuration are identical for all traditional body-centered relationists: that is, A_0,
E_0, C_0, F_0, $G_0 = A_1$, E_1, C_1, F_1, G_1; and this identity also holds for all of the potential
bodily positions defined relative to A, E, C, F and G. But, Leibniz' insistence in (a),
"that the place of each thing would still be determined by reasoning, if there were a
means of keeping a record of all the changes", signifies that the initial and final relative configurations *do* indeed occupy different positions in universal place, since
these changes are capable of being recorded via the distinct motions/forces applied
to the bodies (first to A, and then to the others). Thus, A_0, E_0, C_0, F_0, $G_0 \neq A_1$, E_1, C_1,
F_1, G_1, an outcome that reveals that there is an aspect of spatiality — "determinate in
the truth of things", as noted in (a) — that is *not* a direct relation among bodies (and
potential bodies) due to the fact that a body's position bears a truth value, so to
speak, *relative* to universal place ("in terms of it all changes of every body whatsoever are taken into account"). On the other hand, since Leibniz appeals to the "truth
of things" and also rejects absolute space, the truths of universal place are apparently retained by the bodies themselves (or, once again, monads), thus it would not
be correct to infer that universal place amounts to an independent entity that exists
apart from material existents.

The ramifications of universal place, passage (a), for Leibniz' overall theory of
space are quite profound, although little attention has been devoted to this notion
among Leibniz scholars. In short, the conclusion that A_0, E_0, C_0, F_0, $G_0 \neq A_1$, E_1, C_1,
F_1, G_1 entails that a version of the property theory of space and time, P(loc), first
surveyed in §1.1, must be accepted, even though it runs counter to relationism:

P(loc): There are irreducible, individual bodily spatiotemporal properties, like "is
located at spatial point *p*".

Because $A_0 \neq A_1$, etc., irreducible spatial locations *do* factor into Leibniz'
account — hence, Leibniz actually endorses P(loc), contra relationism and Earman's
interpretation of Leibniz (1989, 14). As a last ditch maneuver, one could accept a
severely minimalist interpretation of sophisticated relationism, which can be
dubbed, "super-sophisticated relationism", that rejects the independent existence of
space apart from all matter, but which holds that the spatial framework is itself independent of the bodies that instantiate space. Thus, according to super-sophisticated
relationism, the existence of, say, one body, *x*, is enough to ground all spatial relations, *S*, even if infinite in scope, and *x* and *S* are independent of each other. Put
differently, the body *x* that instantiates *S* can now occupy different locations relative

to S, in accordance with P(loc).[3] Yet, this strategy prompts the question whether super-sophisticated relationism is truly a relational hypothesis, since it *is* consistent with P(loc), and thereby seems to be a mere restatement of the type of property theory represented by P(loc).

Nonetheless, it is important to add that Leibniz' alleged endorsement of P(loc) is limited in various ways, and so Leibniz' sanction of P(loc) is both partial and apparently confined to certain cases. In what follows, we will employ Maudlin's "Leibniz shift" terminology (1993, 188) to characterize these limitations: a new position in space that preserves the relative configuration of all bodies is called a "static shift", whereas a "kinematic shift" adds a uniform velocity to all bodies. Hence, a true property theory of space would side with the absolutist in claiming that the consequence of a static or kinematic shift is a different state of affairs, namely, each body now occupies a different place. Yet, Leibniz denies this inference (L.III.5), and, furthermore, uses this hypothetical scenario as a means of attacking absolute space. Consequently, Leibniz' sanction of P(loc) fails to include the shift scenarios, and so his conception is only partially consistent with a property theory of space that upholds P(loc). In §3.3.2, the possible reasons for this limitation of Leibniz' endorsement P(loc) will be investigated.

Finally, if an assortment of Leibnizian ideas about space appear hostile to relationism, a natural defense might be to invoke his allegedly pure theory of relational motion, and thereby indirectly defend relational space via relational motion (since motion is change of place, and so relational motion requires relational place). Given the complexity of Leibniz' theory of motion, however, an in-depth investigation of this issue will be postponed until Chap. 4, although a few comments can be provided at this point. Leibniz contends that "[i]f we consider only what motion contains precisely and formally, that is, change of place, motion is not something entirely real, and when several bodies change position among themselves, it is not possible to determine, merely from a consideration of these changes, to which body we should attribute motion or rest" (AG 51). Accordingly, a central tenet of Leibniz' conception of motion is consistent with relationism. Yet, Leibniz' metaphysics of space is quite different from his metaphysics of motion, therefore, even if one were to grant that Leibniz espoused a relational theory of motion, it does not automatically follow that Leibnizian space is also relational.

More importantly, Leibniz' puzzling conviction that force, or the cause of motion, breaks the symmetry of a kinematically-conceived relational transfer, so that bodies can now be assigned individual speeds by means of a specific physical hypothesis (e.g., conservation of mv^2), does not inspire much confidence in this latest gambit to salvage relational space (nor does his insistence that all motion is rectilinear or that mv^2 is conserved; e.g., "Specimen Dynamicum", AG 135): "For

---

[3] It is uncertain if this minimalist interpretation of relationism, super-sophisticated relationism, is included within Earman's (R2) category (1989, 12), where (R2) represents anti-substantivalism, but that minimalist conception is assumed in Slowik (2006) and earlier works when referencing Earman's (R2). Instead of the (quite tedious) appellation, "super-sophisticated", other (more entertaining) candidates are "dapper", "swanky", "posh", or "ritzy".

when the immediate cause of the change is in the body, that body is truly in motion, and then the situation of other bodies, with respect to it will be changed consequently, though the cause of that change is not in them" (L.V.53). Leibnizian dynamics, as the analysis of motion under the action of forces, is thus as relationally problematic as Leibniz' universal place. Although these issues are not the focus of this chapter, the reality of force, or cause of change, resides in substances/monads, with the reality of motion reducible to force. In the mid 1690s, he writes that "if motion, or rather the motive force of bodies, is something real,…it would need to have a *subject*" (AG 308). What has not been previously noted, however, is that passage (a), which posits universal place, would seem to represent the spatial framework relative to which the true motion of a body can be measured, that is, once the true cause of the motion is determined: after claiming that "change of place, [or] motion is not something entirely real", Leibniz adds "[b]ut the force or proximate cause of these changes is something more real, and there is sufficient basis to attribute it to one body more than another" (AG 51). Hence, rather than attribute a concept of "absolute speed" to Leibniz (e.g., Roberts 2003), perhaps a "universal speed" concept could be used to resolve the problem of Leibnizian motion, where the universal speed of a body is determined relative to universal place. Nevertheless, the very idea that speed is absolute (or universal) contradicts the texts (e.g., "motion is not something absolute but consists in a relation", AG 92), a point that will be explored in depth in Chap. 4.

3.2   The Ontological Foundations of Leibnizian Space

In this section, the non-relational features latent in Leibniz' natural philosophy will gain more clarity once the metaphysical underpinnings of his theory of space are explored, an investigation that will also encompass large portions of the next chapter.

3.2.1   *The Immensum and Its Immense Importance*

Among the many historical gems brought to light in Richard Arthur's revelatory study (LoC), one of the most intriguing is that Leibniz' early work on space, circa 1670s, was predisposed towards a view that resembles Spinozism or Neoplatonism. Like Newton's *De grav*, Leibniz holds that God serves as the underlying ontology of space in the literal sense, such that bodies are "in" God in some manner. Material bodies form the geometric contours of space, as in standard Cartesianism, with the continuous flux of the matter in the plenum resulting in space's constant change — but the ontological foundation of space, termed the "immensum", does not change, unlike its body-dependent individuation into diverse shapes: "But there is something in space which remains through the changes, and this is eternal: it is nothing other than the immensity of God, namely an attribute that is one and indivisible, and at the same time immense" (LoC 55). In a slightly later work, he adds:

[(b)] Space, by the very fact that it is dissected into parts, is changeable, and variously dissected; indeed, it is continuously one thing after another. But the basis of space, the extended per se, is indivisible, and remains during changes; it does not change, since it pervades everything. Therefore place is not its part, but a modification of it arising from the addition of matter.... [I]t is the immensum which persists during continuous change of space....[T]he immensum is not an interval, nor is it a place, nor is it changeable; its modification occur not by any change in it, but by the superaddition of something else, namely of bulk, i.e., mass; from the addition of bulk and mass there result spaces, places, and intervals, whose aggregates give Universal Space. But this universal space is an entity by aggregation, and is continuously variable; in other words, it is a composite of space empty and full, like a net, and this net continuously receives another form, and thus changes; but what persists through this change is the immensum itself. But the immensum itself is God insofar as he is thought to be everywhere" (LoC 119–121).

Importantly, Leibniz' tendency to view God as the direct foundation of space persists at least into the mid-1680s, despite the rise of both his force-based notion of matter (e.g., endeavor or appetite) and the categorization of space as a "real relation":

[(c)] Time and place, or duration and space, are real relations, i.e. orders of existing. Their foundation in reality is divine magnitude, to wit, eternity and immensity. For if to space or magnitude is added appetite, or, what comes to the same thing, endeavor, and consequently action too, already something substantial is introduced, which is in nothing other than God or the primary unity. That is to say, real space in itself is something that is one, indivisible, immutable; and it contains not only existences but also possibilities, since in itself, with appetite removed, it is indifferent to different ways of being dissected. But if appetite is added to space, it makes existing substances, and thus matter, i.e., the aggregate of infinite unities. (c. 1686, LoC 335).

In the transition from (b) to (c), bodies no longer dissect the one, indivisible, and immutable "real space" (which is associated with "divine magnitude") into its changeable, aggregate structure; rather, endeavors now fill this role, with bodies and their real relations (space, time, etc.) regarded as further derived results.

One might think that these Spinozistic or Neoplatonic tendencies would have been long since abandoned by the time of Leibniz' mature monadic writings (post mid-1690s), yet there are a number of discussions in this later period that are strongly reminiscent of the immensum. However, given the importance of universal place, passage (a), in this investigation, the ensuing sections will be largely confined to the examination of the *New Essays*. After commenting that space's "truth and reality are grounded in God, like all eternal truths", Leibniz responds to the question whether "space is God or that it is only an order or relation" by having his spokesman, Theophilus, state: "the best way of putting it is that space is an order but that God is the source" (NE II.xiii.17). At greater length, he argues:

[(d)] If God were extended he would have parts. But duration confers parts only on his operations. Where space is in question, we must attribute immensity to God, and this also gives parts and order to his immediate operations. He is the source of possibilities and of existents alike, the one by his essence and the other by his will. So that space like time derives its reality only from him, and he can fill up the void whenever he pleases. It is in this way that he is omnipresent. (NE II.xv.2)

The upshot of this illuminating passage is, first, that space (and time) obtain their "reality only from [God]", second, that God's "essence" is responsible for the possibility of any existing thing in space, and third, that God's will can fill up any void. A bit further, he adds that "absolutes are nothing but the attributes of God", so that the "idea of the absolute, with reference to space, is just the idea of the immensity of God and thus of other things" (NE II.xvii.3; with "absolute" defined as "an attribute with no limits", NE II.xvii.18). Additionally, it is God's "immediate operations" that can be assigned spatial position and parts, and not God's immensity *per se*. In the correspondence with Clarke, he insists that "God is not present to things by situation but by essence; his presence is manifested by his immediate operation" (L.III.12). He also invokes a number of arguments against Clarke's attempts to equate space with either a substance or attribute of God, which Leibniz perceives as lessening God's independence: "[t]he immensity of God is independent of space as his eternity is independent of time", and "[t]he immensity and eternity of God are things more transcendent than the duration and extension of creatures"; nevertheless, "[t]hose divine attributes do not imply the supposition of things extrinsic to God, such as are actual places and times" (L.V.106). In declaring that God is an *intelligentia supramundana*, he adds that "[t]o say that God is above the world is not denying that he is in the world" (L.IV.15), and, in criticizing Clarke's notion that space is God's place, he concludes that "[o]therwise there would be a thing [space] coeternal with God and *independent* of him" (L.V.79, emphasis added). Therefore, while God is independent of space, space is not independent of God.

What lessons can be extracted from the evolution of Leibniz' conception of God's immensity, from (a) to (d)? While many important aspects will be examined further in Chaps. 4, 7 and Part III, when the focus turns to the concept of ubeity, the most prominent change is the avoidance of language that refers to the immensum as "the extended per se" in (b), or "divine magnitude" in (c). Similarly, there are no longer any references that spatialize God as a form of container of body/force properties, such as "adding" mass to the immensum in (b), or that endeavors are "in nothing other than God" in (c). Whatever the influence of Neoplatonist ideas on Leibniz' natural philosophy, it is likely that the continuing disputes with the Neoplatonist crowd in England prompted his more careful, and more Cartesian, disavowal of attributing spatial features directly to God (with the exception of God's operations, of course). On most other issues, however, immensum-like notions persist in Leibniz' mature natural philosophy, from the 1676 (b) through the *New Essays* and the Clarke correspondence. First, God's immensity remains the *ontological* basis of space. Second, in (d), God's essence grounds the possibilities that are so often implicated in his later analysis of space (e.g., L.V.104), just as God's immensity is the unchanging basis of all possible changes in space brought about by either bodies or endeavors. Consequently, the contemporary neglect of the importance of the immensum, i.e., as less central to his mature thought, simply does not do justice to its continuing relevance throughout the Leibnizian corpus—to summarize, a more Cartesian version of the immensum is a central feature of his late natural philosophy of space as well. The specific details of Leibniz' new interpretation of God's grounding role regarding space will be deferred to the next chapter, however, in addition to Chap. 7 and, especially, Chap. 9.

To return to the central topic of this chapter, the examination of God's relationship to space in Leibniz' work has revealed many features that have truly damaging ramifications for any relationist interpretation. Among these features, Leibniz holds that God's essence grounds the capacity of an empty space to receive bodies. Interestingly, there was another natural philosopher, also associated with Newton and More, who held almost identical views to Leibniz on space—namely, Isaac Barrow! Although the details are lacking, Barrow regards space, like Leibniz, as dependent on God (MWB 154): e.g., "there was Space before the World was created, and...there is now an Extramundane, infinite Space, (where God is present;)" (LG 203). This last quote may appear to suggest that space exists prior to bodies, contra Leibniz' view, but Barrow's explanation of space's existence is put entirely in terms of the possibility of existing bodies, since he concludes that space is the mere capacity to receive bodies (and thus he follows a long Scholastic tradition associated with imaginary space; see, Grant 1981, chap. 6): "Time therefore does not imply an actual existence, but only the Capacity or Possibility of the Continuance of Existence; just as space expresses the Capacity of a Magnitude contain'd in it" (LG 204). That is, Barrow's nominalist-leaning account denies that space has quantity or magnitude; rather, only bodies possess magnitude. Space, which is not an actual entity, is described as "nothing else than a simple pure potency, mere capacity,...of some magnitude" (MWB 158). Barrow's spatial theory is, accordingly, nearly identical to Leibniz' view in that space merely signifies a capacity for material quantities. Specifically, while there is much that is absolutist in Barrow's approach, his claim that space exists prior to bodies matches Leibniz' identical assessment, since the intended meaning for both is that there exists a God-based possibility for bodies to fill a vacuum. Nonetheless, unlike Leibniz, Barrow is usually grouped with the absolutists: see, e.g., De Risi (2007, 564), who sides with absolutism based on his reading of "space before the world"; and, Hall (1990, 210), who claims that Barrow's "space exactly anticipates Newton's conception of absolute space".[4] In all likelihood, it is the unlucky conjunction of Leibniz' relational theory of motion and the contemporary substantival/relational controversy that has continued to distract commentators from the very close similarities between Barrow's and Leibniz' natural philosophies of space (although their overall natural

---

[4] Yet, as Futch explains (2008, chapter 2), Barrow's treatment of time is much closer to the absolutists; in particular, he separates time from bodily change, much like Newton, whereas Leibniz remains somewhat wedded to the older Scholastic tradition. Nevertheless, on the issue of space, Barrow and Leibniz are nearly identical, save for Leibniz' possibly having attributed quantity to empty spaces that are bounded by matter or at least measurable (NE II.xiii.22). Likewise, Barrow claims that space, i.e., as a capacity, exists prior to bodies—but, as argued above, Leibniz holds the very same view, although he does not refer to this sheer possibility using the term "space" (rather, space only co-exists with bodies). Barrow does accept the absolutist scenario that the whole world could move through space (UML 172), but Leibniz' "universal place" concept is nearly identical in it absolutist implications—and, although Leibniz denies the uniform motion of the material world, it is for reasons linked to his nominalism and his force-based conception of matter (see §3.3.2 below), unlike Barrow. As regards the later analysis in §3.2.3, it is interesting to note that Barrow claims that numbers, like points, have position, since position requires a multiplicity (MWB 62).

philosophies differ on many other points, of course). In Chap. 9, a more systematic investigation of the conceptual presuppositions that underlie Barrow's brand of spatial ontology, as well as the spatial hypotheses of many other Early Modern natural philosophers (including Leibniz) will be presented.

3.2.2 Substance, Accident, and Relations in Leibniz' Metaphysics

Needless to say, one of the principal reasons for assuming that Leibniz' theory of space is relational is that he often uses the label "relation" (*relatio*) in characterizing space—nevertheless, this terminological usage is deceptive, for Leibniz' ontological classification of space stems from a different source than the default body-based relationism often ascribed to his natural philosophy. It is true that a number of passages seem to endorse relationism; e.g., "[e]xtension or space, and the surfaces, lines and points one can conceive in it are only relations of order or relations of coexistence, both for the actually existing thing and for the possible thing one can put in its place" (1695, AG 146). Yet, a closer examination of Leibniz' concepts undermines any brand of sophisticated relationism that these passages might seem to endorse.

First of all, as regards the substance/accident dichotomy, he denies that space is a substance for one of the same reasons that Newton offers in *De grav*, specifically, that space cannot act upon things (NE II.xiii.17; cf. N 21). Space is not a property (accident, affection) that inheres in *individual* material beings, at least in the traditional Scholastic sense of "inheres", since "the same space will be sometimes the affection of one body, sometimes of another body, sometimes of an immaterial substance,..... But this is a strange property or affection, which passes from one subject to another" (L.V.39). In contrast, he does support a key element of a substance/accident metaphysics, namely, that there can be no ontologically unsupported spatial properties: "if [a] space is empty, it will be an attribute without a subject, an extension without anything extended" (L.IV.9). Accordingly, while space/place is not an individual property that flits from one substance to another, a possibility that would violate the commonly accepted, Scholastically-influenced approach to substances/accidents, he does leave room for place to be a property of a different sort. Besides Leibniz' claim, examined above, that links God's essence to empty spaces (and his frequent assertions that space is not independent of God), a similar property-oriented demand for an ontological grounding for space is evident in the remainder of the section just quoted from the Clarke correspondence: "Thus, by making space a property, the author falls in with my opinion, which makes it an order of things and not anything absolute" (L.IV.9). In a well-known passage from the fifth letter, Leibniz' claim that focuses on the ideality of place or space (as opposed to the property-like implications of "relation of situation") provides more detail on the relevance of individual properties in Leibniz' theory:

And here it may not be amiss to consider the difference between place and the relation of situation which is in the body that fills up the place. For the place of A and B is the same, whereas the relation of A to fixed bodies is not precisely and individually the same as the relation which B (that comes into its place) will have to the same fixed bodies; but these relations agree only. For two different subjects, such as A and B, cannot have precisely the same individual affection, since it is impossible that the same individual accident should be in two subjects or pass from one subject to another. But the mind, not contented with an agreement, looks for an identity, for something that should be truly the same, and conceives it as being extrinsic to the subjects; and this is what we call *place* and *space*. But this can only be an ideal thing, containing a certain order, in which the mind conceives the application of relations. (L.V.47) [5]

Therefore, because the "relation of situation" is "in the body that fills in the place", it would seem to follow that "relation of situation" is an internal feature of bodies. But, how is that possible? How can situation be both an internal property (accident, affection) of bodies and a relation that involves other bodies? Leibniz' metaphysics of individual bodies/substances is certainly in play in this passage, although a long aside on the "complete concept" notion is beyond the bounds of this investigation (see, e.g., Mates 1986, 58–69, for more details). Put briefly, Leibniz holds that the spatial relations among all bodies are packed into the complete concept of each individual, thereby validating Leibniz' numerous claims that each body/substance/monad "mirrors the world" (e.g., AG 42).

In a tract entitled, "On the Principle of Indiscernibles" (c. 1696), he states that "there are no purely extrinsic denominations", and that we erroneously "conceive position as something extrinsic, which adds nothing to the thing posited, whereas in fact it adds the way in which that thing is affected by other things" (MP 133–134). In more detail, he argues:

[(e)] To be in a place seems, abstractly at any rate, to imply nothing but position. But in actuality, that which has a place must express place in itself; so that distance and the degree of distance involves also a degree of expressing in the thing itself a remote thing, either of affecting it or receiving an affection from it. So, in fact, situation really involves a degree of expressions. (MP 133)

While the implications of (e) for Leibniz' dynamics will be examined in later sections, the same tendency to view place, position, etc., as akin to an internal property is evident in the often cited example from the fifth letter to Clarke that involves the ratios of lines:

---

[5] As an aside, this portion of Leibniz' analysis in L.V.47 (i.e., "concerning the place of A and B") has much in common with the empiricist conception of space that one finds in, say, Locke's *Essay* (II.13.2-4), i.e., that "same place" and other spatial features are abstractions or generalizations from experience. Yet, as will be discussed below, Leibniz denies that the continuous, ideal nature of space can be derived from experience; rather, it is derived from our pure understanding. Nevertheless, he may have used an empiricist-sounding explanation at this point in the correspondence with Clarke to make his spatial hypotheses more palatable to a wider audience (or, this discussion may simply amount to an alternative manner of understanding space without having to accept its status as an independently existing entity). Overall, this part of the L.V.47 discussion has probably generated more confusion as regards the nature of Leibniz' view of space than any other: in particular, since an empiricist conception of space is largely equivalent to the modern relationist conception, it is not surprising that modern relationists normally (and erroneously) quote this part of L.V.47 to support their interpretation of his spatial ontology.

I shall adduce another example to show how the mind uses, on occasion of accidents which are in subjects, to fancy to itself something answerable to those accidents out of the subjects. The ratio or proportion between two lines L and M may be conceived three several ways: as a ratio of the greater L to the lesser M; as a ratio of the lesser M to the greater L; and lastly as something abstracted from both, that is, as the ratio between L and M without considering which is the antecedent or which is the consequent, which is the subject and which is the object....In the first way of considering them, L the greater, in the second, M the lesser, is the subject of that accident which philosophers call relation. But which of them will be subject in the third way of considering them? It cannot be said that both of them, L and M together, are the subject of such an accident; for if so, we should have an accident in two subjects, with one leg in one and the other in the other, which is contrary to the notion of accidents. Therefore we must say that this relation, in this third way of considering it, is indeed out of the subjects; but being neither a substance nor an accident, it must be a mere ideal thing, the consideration of which is nevertheless useful. (L.V.47)

Since this passage follow the previous citation from L.V.47 on the concepts of place and space, the conclusion to be drawn is that the relations of place and situation function in much the same manner as an internal property, a point that has been raised by previous commentators, such as Auyang (1995, 247–251). Not only are relations likened to internal accidents (properties) of subjects L and M on the first two ways of considering this relation, but these subject-based accidents are also deemed to be relations *between* the two objects, with space (as the third way) comprising a mere ideal notion abstracted from the two ways that the accident figures in each subject's set of predicates, either from L or M's perspective. More importantly, the abstracted third way of considering the relation is not put forward as negating the first and second interpretations, rather, it is just presented as a different manner of understanding the relation.[6]

A relationist might attempt to read an eliminativist or super-eliminativist construal of spatial relations into the many passages, like L.V.47, where Leibniz employs relationist and abstractionist terminology; that is, they may strive to abstract a conception of space from the extension of each individual body, which in an infinite plenum gives an infinite space once the bodily boundaries have been subtracted—all of this is in keeping with a relationist orthodoxy that denies any reality to space apart from the non-eliminative extension possessed by each body. But, as revealed above, the spatial relations between bodies are included within a body's accidents, alongside bodily extension. This alone indicates that Leibniz cannot be reconciled to contemporary spatial relationism, and that he is working with a different conception of how space is linked to bodies and, ultimately, individual substances/monads. In short, Leibniz does not use "relation" in the sense employed by spacetime theorists, and hence it fails to meet the demands of any modern relationist theory of space.

Other problems for a strict eliminativist interpretation of relations lie in Leibniz' comments on the intricate interrelationship between bodily extension and space. On a straightforward interpretation of eliminative relationism, the internal extension

---

[6] For a more cognitive-based account of relations in Leibniz, see, e.g., Mugnai (1992) and Futch (2008, chapter 7).

and relative configuration of bodies can remain invariant whereas the actual distance relations among bodies, i.e., the geometry, can vary significantly; likewise, one could employ the observations of various rigid body motions as a means of determining the geometrical structure of space as a whole: e.g., Sklar's (1985, 234–248) relational method for explicating the orientation (handedness) property of spacetime. Yet, in opposition to these strategies, there are a number of discussions in Leibniz' late corpus that identify only one possible spatial structure (which happens to be Euclidean geometry; see footnote 7). For instance, on the many ways that the world could be filled with matter, Leibniz comments that "there would be as much as there possibly can be, given the capacity of time and space (that is, the capacity of the order of possible existence); in a word, it is just like tiles laid down so as to contain as many as possible in a given area" (AG 151). Besides disclosing the existence of a predetermined geometric structure prior to the introduction of matter, this explanation also reveals that spatial structure is not determined by matter—instead, spatial structure determines the possible material configurations, a view that not only violates modern eliminative relationism, but seems to wholeheartedly embrace a type of spatial absolutism.[7]

More specifically, Leibniz insists that "although it is true that in conceiving body one conceives something in addition to space, it does not follow that there are two extensions, that of space and body" (NE II.iv.5). He continues:

> [T]here is no need to postulate two extensions, one abstract (for space) and the other concrete (for body). For the concrete one is at it is only by virtue of the abstract one: just as bodies pass from one position in space to another, i.e. change how they are ordered in relation to one another, so things pass also from one position to another within an ordering or enumeration—as when the first becomes the second, the second becomes the third, etc. In fact, time and place are only kinds of order; and an empty place within one of these orders (called 'vacuum' in the case of space), if it occurred, would indicate the mere possibility of the missing item and how it relates to the actual. (NE II.iv.5)

The same theme is asserted in a famous passage from the Leibniz-Clarke correspondence:

> I do not say, therefore, that space is an order or situation, but an order of situations, or (an order) according to which situations are disposed, and that abstract space is that order of situations when they are conceived as being possible. Space is therefore something merely ideal. (L.V.104)

---

[7] Euclidean geometry appears to be the determinate structure of Leibnizian space. After defining "distance" as "the size of the shortest possible line that can be drawn from one [point or extended object] to another", he comments that "[t]his distance can be taken either absolutely or relative to some figure which contains the two distant things", but adds that "a straight line is absolutely the distance between two points" (NE II.xii.3; as opposed to the arc of a great circle on a spherical surface). By defining a straight line as the absolute distance, in contrast to the relative distance— i.e., *relative* to the various figures or surfaces that can be delineated *within* Euclidean (three dimensional) geometry—this implies that the overall structure of Leibnizian space is Euclidean. Interestingly, it would seem to follow that a limited material world with, say, a spherical shape would have a non-Euclidean metric on that surface, given his notion of relative distance.

On the construal of relations assumed in modern spacetime interpretations of strict eliminative relationism, in contrast, there are two extensions of sorts that can vary independently of one another: on the one hand, space or extension as a quantitative measure, i.e., distance, and on the other hand, the extension within bodies as well as their relative configuration (or, if unextended point masses or events are utilized, then it is simply their relative configuration). Leibniz' view differs quite substantially from this modern relationist approach. First, he collapses the difference assumed in modern relationism between relative order and spatial distance in favor of the former notion, order of situations, which he declares has quantity/distance. Second, as disclosed previously, he then treats order of situations as an internal property of each body, an internal property which when idealized apart from bodies becomes space. Third, the statement that bodily extension "is at it is only by virtue of the abstract one [space]", signifies that even bodily extension cannot be separated from the order of situations (space), a conclusion that is also in line with his view that both bodily extension and space are internal to each body. All of this, of course, runs counter to the priority assigned to bodily extension and the relative configuration of bodies, in opposition to space (distance, geometry), in modern eliminative relationism. Incidentally, an analogous misreading of Leibniz' view was disclosed in Clarke's contention (C.IV.41) that the order of bodies can remain the same despite a difference in the quantity of space between the bodies, whereupon Leibniz responds: "As for the objection that space and time are quantities, or rather things endowed with quantity, and that situation or order are not so, I answer that order also has its quantity: there is in it that which goes before and that which follows; there is distance or interval" (L.V.54; see also §7.4.1 for further commentary on the dispute between Leibniz and Clarke on this issue). Once again, Leibniz holds that space cannot vary independently of bodily extension and relative order, contra an eliminativist, external relationism—apparently, these three concepts (bodily extension, relative bodily order or situation, and the quantity, space) come as a kind of package deal for Leibniz, and this further implies something like a property theory of space, whether of the P(loc), P(O-dep), or some other variety, regardless of how one parses its ontological implications.[8]

But, if Leibniz' account of space is close to a property theory, it is a rather strange property theory: on this account, a body's extension and its relative order with other bodies functions like an internal property of each body, with space comprising a mere idealization of these two. Apparently, this is Leibniz' method of overcoming

---

[8] In the work previously quoted from 1676, Leibniz comments that "Space is only a consequence of [the Immensum], as a property is of an essence." (LoC 55). Compare with: "Thus, by making space a property, the author falls in with my opinion, which makes it an order of things and not anything absolute" (L.IV.9). Of course, Leibniz does not sanction the ontology of a property theory of space in the way that, say, More's late *Enchiridium* seems to espouse (see, EM 1995, 56–57, where space becomes God's internal property). Yet, as argued above, P(loc) is viewed as a key requirement of the traditional property view, and thus it is imperative that the manner by which Leibniz' theory resembles a property theory is investigated on this point—indeed, as argued in this chapter, Leibniz' theory of space more accurately reflects a property theory orientation than modern relationism.

the problem of migrating spatial properties, such as place, as well as an expression of the complete concept notion and the prohibition on extrinsic denominations, etc. More precisely, if you build the entire order of situations into each body, then the Scholastic ban on an accident existing outside a substance is upheld, as when one erroneously presumes that the "same place" transfers from body A to body B (as in L.V.47). Yet, as all Leibniz scholars know, the price to be paid for such a scheme is astonishingly high: *apparently* renders the whole of space internal to each body/ substance, with phenomenalist or idealist consequences difficult to evade: e.g., "situation really involves a degree of expressions" in (e) (more on this below). This predicament can be taken as the spatial incarnation of the old adage that Leibniz replaced Spinoza's single substance (space) with an infinity of isolated, non-interacting substances (spaces) in pre-established harmony. Likewise, it provides a unique twist on the holenmerism prevalent in his day (see Chap. 2), although rather than posit a God who is whole in every part of space, as holenmerists such as Gassendi or Charleton support, Leibniz makes space whole in every body/substance (of which there are an infinite number), all founded upon God's immensity.

3.2.3   Leibniz' Physical-Geometric Holism

However, if the ban on a purely idealist or immaterialist interpretation of Leibniz' spatial theory is upheld, such as that presented in Furth (1967)—and thereby limit the investigation to those ontological concerns relevant to both a corporeal substance/accident metaphysics and the modern substantival/relationism dispute—then a species of material world "holism" is the only plausible conclusion that can be drawn from the inclusion of external bodily relations within each body. That is, the corporeal manifestation of Leibniz' claims that the order of situation is an internal property, or are "expressed" by each body, is just his deeper metaphysical hypothesis about the interconnectedness of all material bodies. This exact point is made in citation (e) and its accompanying passages, where the PII is utilized to deny the relevance of spatial position or location: "there are no purely extrinsic denominations, *because* of the interconnection of things", and, "if place does not itself make a change [i.e., no external denominations], it follows that there can be no change which is merely local" (MP 133, emphasis added). The interconnection of the material universe is a major theme in Leibniz' work, needless to say: "each corpuscle is acted on by all the bodies in the universe, and is variously affected by them" (MP 176, c.1712; also, AG 221). Even so, the link between the PII, space and time, and the world's bodily interconnection has been more often overshadowed by a Furth-like idealist reading of the internal (force-based) denominations that usually take center stage in discussions of Leibniz' theory.

But, in a very instructive early passage (from 1678), Leibniz explains that the term "expression", which has idealist overtones, signifies a structural identity between any two things (which, in the modern parlance, is dubbed a partial isomorphism): "That is said to express a thing in which there are relations which correspond to the relations of the thing expressed" (Lm 207). While this work does not offer examples of perceptions expressing material bodies, that term is used in a perceptual context in other works; e.g., "it is very true that the perceptions or expressions of all substances mutually correspond in such a way that each one,…coincides with others", and although all substances "express the same phenomena, it does not follow that these expressions are perfectly similar; it is sufficient that they are proportional" (AG 47). In the work from 1678 quoted above (Lm 207), however, a number of examples are given that pertain to the material world alone, and not mental content, such as models/machines, linear projections/solids: "What is common to all these expressions is that we can pass from a consideration of the relations in the expression to a knowledge of the corresponding properties of the thing expressed. Hence it is clearly not necessary for that which expresses to be similar to the thing expressed, if only a certain analogy is maintained between the relations" (Lm 207). If one takes "expression", therefore, as denoting a structural relationship between any two things, and *not* just between perceptions and the material world, then the holistic interconnectedness of Leibniz' material world is strongly implied: "that which has a place must express place in itself; so that distance and the degree of distance involves also a degree of expressing in the thing itself a remote thing, either of affecting it or receiving an affection from it" (MP 133). Not surprisingly, each body/substance, taken "in the thing itself", expresses distance relations to "remote things" by means of dynamic change, "affecting it or receiving an affection from it": consequently, the holistic interconnection of the material world is both dynamical and spatial.

The holistic interconnection of the physical world is matched, additionally, by the holism of Leibniz' classical conception of geometry, which conceives geometric elements or structures as equally interconnected. For instance, the situations that form the basic component of his novel geometric theory, *analysis situs,* are determined by the space's distance relations, an approach that runs counter to the modern division between the topological and metrical aspects in modern differential geometry (see, De Risi 2007, 176, for an overview of the metrical basis of Leibniz' *analysis situs*). On a similar theme, the most classical feature of his geometric outlook is his stance on geometric/spatial points, which "strictly speaking, are extremities of extension, and not in any way, the constitutive parts of things; geometry shows this sufficiently" (AG 228). That is, points are merely the boundaries of lines, and so they cannot exist apart from lines (just as situation is inseparable from the quantitative distance relations). All of these views, which are characteristic features of his philosophy, are consistent with an approach that takes the whole of space/geometry, which is ideal, as prior to its parts. In real entities, on the other hand, the part precedes the whole:

> For space is something continuous, but ideal, whereas mass is discrete, indeed an actual multiplicity, or a being by aggregation, but one from infinite unities [monads]. In actual things, simples are prior to aggregates; in ideals things, the whole is prior to the part. (September 6, 1709; LDB 141)

In a Rationalist critique that even Descartes would probably have admired, Leibniz uses a similar line of reasoning to argue, against Locke, that space and time cannot be grasped by an imagination-based construction from the experience of particular discrete entities: "Ultimately one can say that the idea of the *absolute* is, in the nature of things, prior to that of the *limits* which we contribute, but we come to notice the former only by starting from whatever is limited and strikes our senses" (NE II.xiv.27). As attributes without limits, the absolute "precedes all composition and is not formed by the addition of parts" (NE II.xvii.1), "is internal to us", and "these absolutes [i.e., space and time] are nothing but the attributes of God; and they may be said to be as much the source of ideas as God himself is the principle of beings" (NE II.xvii.5).

3.3   Final Assessment

A synthesis of the various separate themes in this chapter can now be offered, all centered on the evolution of Leibniz' spatial ontology and the corresponding viability of a relationist interpretation. The discussion of the role that nominalism plays in Leibniz' spatial ontology will also set the stage for a more comprehensive investigation of this topic in Chap. 7.

3.3.1   Space, Property and Nominalism

Throughout Leibniz' natural philosophy, God's immensity is directly linked to the unity, oneness and indivisibility of space—i.e., the holism or monism of space: for instance, space and time, which are the holistically-conceived absolutes put forward in the *New Essays* (as above), "are nothing but the attributes of God". Nonetheless, in his writings up through the 1680s, such as in (b) and (c), bodies or endeavors are directly *added* to the divine unity, thereby resulting in the discrete, aggregate structure that we actually experience. By the time of his later output, though, God's earlier quasi-Spinozistic or Neoplatonic role as the holistic ontological platform for space, to which endeavors and bodies are directly added, has been drastically altered, leaving only the holistic, ideal conception of space that is *posterior* to the experience of the material world of discrete, bodily phenomena. God's immensity remains as the foundation of space, of course, but in a more transcendent manner, so that only God's operations can be straightforwardly given a spatiotemporal location (a point that will be discussed at length in Chaps. 4, 7, and Part III).

Nevertheless, how can a world composed of discrete, individual bodies, no matter how dynamically interconnected, serve as sort of ontological platform for a spatial or geometric property that is holistic and non-discrete? In answering this question, it is important to bear in mind his provisos disclosed above, that bodily extension "is at it is only by virtue of the abstract one [space]" (NE II.iv.5), and that "there is no need to postulate two extensions, one abstract (for space) and the other concrete (for body)" (NE II.iv.5). Given these frequently asserted claims as well as the analysis of expression in (e), the best inference is that the dynamically interconnected world of discrete bodies *instantiate* or *exemplify* the *truths* of the ideal holistic structure of space and geometry, a nominalist perspective that is predicated on God's role as the transcendent source of both actual and possible existents. Put differently, the truths of geometry are obtained by the world's dynamic material interconnections, with God grounding the possibility of these truths, and where the monads play an analogous grounding role as regards the existence of matter (see §3.3.3). As used in this context, the "instantiation of geometric truths" simply means that the behavior and analysis of material bodies and their interactions upholds or follows the truths of (Euclidean) geometry, a conclusion that, at least on this issue, is in accordance with a geometric nominalism that rejects the existence of the platonist's uninstantiated mathematical objects.[9] In short, one of the components of Leibniz' rejection of absolute space that has been curiously overlooked by commentators is his tendency to equate spatial absolutism (substantivalism) with platonism, an outlook manifest in his assertion that "there is no need to postulate two extensions, one abstract (for space) and the other concrete (for body)" (NE II.iv.5). Spatial absolutists, conversely, do posit two extensions, one for space and one for body—but so do platonists: extension (space) as an abstract object that exists independently of all material instantiations, and extension (space) as a property of matter that constitutes the instantiation of that abstract object. Unlike the substantivalist/relationist dichotomy, consequently, the platonism/nominalism dichotomy does offer a way of securing a consistent interpretation of important features of Leibniz' natural philosophy of space; e.g., for why space is not an "absolute reality" (L.V.62), or as he more carefully puts it, "an absolute being" (L.III.5), *but* does expresses "real truths" (L.V.47). In essence, to regard space as an "absolute reality" or "absolute being" is "to postulate two extensions, one abstract (for space) and the other concrete (for body)", hence "space as absolute reality/being" can be interpreted as positing space as an entity that functions like a platonist's abstract object, i.e., space as a really existing (geometric) object that exists in addition to the existence of spatially-extended bodies.[10] On the other hand, to insist that "abstract space is that order of situations when they are conceived as being possible. Space is therefore something merely ideal" (L.V.104), while simultaneously claiming that space involves "real truths", is to employ a type of explanation that correlates with nominalism in the precise sense that it rejects space's status as an independently existing abstract object. In Chap. 7, nominalism and platonism in the context of spatial ontology will be examined in greater detail, including a lengthier analysis of Leibniz' nominalist conception of space.

---

[9] As noted in Belot (2011, 184), Leibniz holds that, somehow, "the ideal governs the real", a point that Leibniz' discussion of the mathematics of infinitesimals would seem to support: "Yet one can say in general that though continuity is something ideal and there is never anything in nature with perfectly uniform parts, the real, in turn, never ceases to be governed perfectly by the ideal and the abstract" (2 February, 1702; Lm 543). To clear up any potential confusion, the type of nominalism endorsed in this investigation as it pertains to Leibniz' theory of space refers to the instantiation of mathematical truths, and not the instantiation of mathematical entities.

[10] In the correspondence with Des Bosses, Leibniz states that "I hold every absolute to be substantial [a substance]" (Lm 608). Hence, "space as absolute being" is identical to "space as substance" as well.

Nominalism is likewise similar to a property theory of space, especially the P(O-dep) version introduced in Chap. 1, although the nature of that spatial property is quite unique in Leibniz' system. For instance, take his claim that "[t]ime and space are of the nature of eternal truths, which equally concern the possible and the actual" (NE II.xiv.26). Overall, these sentiments would appear to be more in harmony with the idea that space is a property of *both* God and the material world, albeit in different ways. As eternal truths, space and time concern the possible via God's essence, so that there can be no space and time apart from God's essence, and where an essence is a necessary property or attribute of a being, thereby explicating his assertion that space and time "are nothing but the attributes of God" (NE II.xvii.5). But, space and time also include the actual, via the material world's instantiation of these spatial truths; specifically, since space is an ideal whole, while bodies are discrete, the instantiation of space thus amounts to the instantiation of the geometric truths associated with Euclidean space, so that the behavior of bodies is governed by the truths of Euclidean geometry. More carefully, while a more straightforward brand of nominalism would likely dictate that only the whole interconnected dynamical world can properly instantiate the holism of geometric truths, Leibniz' novel approach circumvents this limitation since his conception of relations entails that each individual body/substance "knows about", *expresses*, the dynamic interconnections of the world, and thus each body/substance has access to the holism of geometric truths via the dynamic/spatial correlation surveyed in §3.2.3 (see, especially, the quotes from MP 133).[11] In evading the problem of migrating individual places discussed in §3.2.2, it would seem that Leibniz' response to this predicament is to render absolute space, i.e., universal place in (a), a truth that is instantiated by the world's holistic interconnections. Accordingly, while space is not a material property in the standard ontological, substance/accident sense of the term, it still nonetheless fits a crucial criterion of both a property theory and nominalism, namely, that it is instantiated by matter and is not a platonic abstract object, the latter being an entity that exists even in the absence of matter. Of course, relationists would claim that space is also instantiated by matter; but, as argued above, the many non-relational features of Leibniz' spatial hypotheses undercut the case for a relationist interpretation: e.g., P(loc), and the fact that bodily extension, relative bodily situation, and the quantity, space, cannot vary independently of one another, as they do for modern eliminative relationists.

In summary, the material world's function as the instantiating basis of an ideal, holistic space explains many of the distinctive features of Leibniz' theory: namely, "that there is no space where there is no matter and that space in itself is not an absolute reality" (L.V.62; i.e., matter instantiates space but space does not exist apart from that instantiation), and, "if there were no creatures, space and time would

---

[11] The use of such terms as, "instantiate", "exemplify", etc., is drawn from structuralist conceptions of mathematics (see, Shapiro 1997). Given God's role as the basis of these eternal (geometric) truths, the non-reductive, *in re* (nominalist) version of mathematical structuralism would seem to best fit Leibniz' conception, although more traditional (non-fictionalist) strands of nominalism would also be applicable (and hence this investigation does not rely on a structuralist reading). See, once again, Chap. 7.

be only in the ideas of God" (L.V.41; i.e., pre-instantiation, space is merely a possibility guaranteed by God's essence). The universal place in quotation (a), which presents such a serious obstacle to relationism by deviating from bodily-defined place, is hence akin to both his holistic notion of space/geometry and his dynamical, holistic understanding of material world change: universal place is "determined by reasoning" and "is related to everything, and in terms of it all changes of every body whatsoever are taken into account" (i.e., it is an ideal notion that allows us to grasp the dynamics of bodily interactions). In addition, this conception of how geometry relates to bodies contrasts sharply with Newton's stance, examined in Chap. 2, which reified the holistic structure of spatial geometry in a somewhat platonic fashion (i.e., independent of matter): "For the delineation of any material figure is not a new production of that figure with respect to space, but only a corporeal representation of it, so that what was formerly insensible in space now appears before the senses" (N 22). Leibniz' nominalist conception of geometry, conversely, only regards spatial geometry as the property of the whole corporeal world, with God's essence securing this basis, and which includes the possibility of bodies existing in a vacuum (assuming there are empty spaces), much like Barrow. Therefore, the universal place in (a), although not an absolute being, is an eternal truth grounded in the material world as a whole (by way of God) — this entire conception and approach, it should be noted once again, bears little resemblance to a modern relationist theory of space.

3.3.2   Reconsidering the Leibniz Shift Scenarios

By way of conclusion, it is worth pointing out how the preceding discussion also helps to shed light on the nature of Leibniz' espousal of P(loc), introduced in §3.1.2, as well as the shift arguments that appear in the Leibniz-Clarke correspondence. As previously discussed, Leibniz would deny that the material world could either occupy a different position in absolute space (static shift) or have a different uniform velocity in absolute space (kinematic shift), although this contradicts P(loc). Does this fact undermine the assertion that Leibniz' theory resembles a property theory of space?

While the shift scenarios demonstrate that Leibniz does not sanction a straight-forward property theory, there are reasons for his exclusion of the shift cases that connect with his other conceptual priorities, namely, his nominalism and his force-based conception of matter. As demonstrated in §3.2.2, relative bodily configuration and extension cannot vary independently of space, although the shift arguments assume this independence. Specifically, since Leibniz, in (e), regards space (distance) as instantiated by the dynamical interactions among bodies, the hypothetical scenarios presented by Clarke that involve different non-dynamical relationships between the entire world and space—position (static shift), velocity (kinematic shift)—are simply inapplicable given Leibniz' nominalist-influenced conception of space. Indeed, as demonstrated in the passages that accompany (e), the PII is actually employed to argue for the irrelevance of extrinsic denominations, and in support of a dynamic material holism as the instantiating basis of space. Therefore, the many arguments in the correspondence against absolute space (as a being entirely independent of body) that utilize the PII are totally in keeping with the analysis provided above, even though Leibniz frames his use of the PII in ways that differ from (e); e.g., if space "is nothing at all without bodies but the possibility of placing them, then those two states, the one such as it is now, the other supposed to be the quite contrary way [static shift], would not at all differ from one another" (L.III.5). That is, because space is merely an idea in God's mind prior to the material world's instantiation of space—i.e., "the possibility of placing them", secured via God's essence—then stipulating that the world's inhabitants could have possessed different spatial locations in a uniform way (such as each body in the universe being three feet further to the left) is to make the claim that space could have been instantiated by the material world in a different fashion. But, given (i) a plenum, (ii) that Leibnizian space concerns the possible and the actual, and (iii) seems inexorably tied to Euclidean structure, there is only one space to be instantiated (but, see footnote 7 for the non-plenum cases that might provide non-Euclidean results within a larger Euclidean structure). Accordingly, a uniform and unobservable static shift would amount to nothing more than a relabeling of the places already instantiated by the material world, i.e., it is a mere difference in scale or gauge, which is no difference at all, as Leibniz insists. The same holds true for a kinematic shift, since any uniform addition or subtraction of speed, or difference in direction, as applied to all bodies does not affect the dynamical interconnections among those bodies. A relationist would likely resist the shift arguments with an analogous argument, of course, but that simple fact does not render Leibniz' unique brand of property theory identical to modern relationism, since they differ on many other issues, as has been often noted above.

Finally, the static and kinematic shift arguments also explain why Leibniz' theory does not exactly correlate with a strict or thorough property theory, P(loc). On a strict property theory, a static or kinematic shift of the world would allow new positions of the world's shifted bodies to be obtained, just as they do as regards passage (a), where a body's dynamic change brings about a new spatial position in universal place. However, since Leibniz sees space as arising from the dynamic interconnections of the whole world, this fact limits the changes in spatial location to individual bodies, or sets of bodies, relative to other bodies, but not all bodies in unison. Hence, Leibniz' dynamic holism provides a conclusion that is quite similar to traditional relationism on this issue alone—but, as argued above, this aspect of his natural philosophy stems from a commitment to his blend of nominalism and dynamic holism, which is quite different from traditional body-centered relationism.

3.3.3 A Monadic Conclusion

Although it will be examined at greater length in the next chapter, a few words are in order regarding how the monadic realm in Leibniz' spatial theory correlates with the main conclusions of the present chapter. First, the world's dynamic interconnections arise from monads: "And since everything is connected because of the plenitude of the world, and since each body acts on every other body, more or less, in proportion to its distance, and is itself affected by the other through reaction, it follows that each monad is a living mirror or a mirror endowed with internal action, which represents the universe from its own point of view and is as ordered as the universe itself" (AG 207). Second, monads ultimately are grounded on God: "And a monad, like a soul, is, as it were, a certain world of its own, having no connections of dependency except with God" (5 February 1712, AG 199). Consequently, one of the main themes in this investigation, the dependence of space upon both God and the world's dynamic interconnections, finds a correlate at the level of monads. In the notes to this letter, Leibniz offers one of his favorite metaphors for the different way that things "appear", whether as judged from bodies/monads or God:

> If bodies are phenomena and judged in accordance with how they appear to us, they will not be real since they will appear differently to different people. And so the reality of bodies, of space, of motion, and of time seem to consist in the fact that they are phenomena of God, that is, the object of his knowledge by intuition. And the distinction between the appearance bodies have with respect to us and with respect to God is, in a certain way, like that between a drawing in perspective and a ground plan. For there are different drawings in perspective, depending upon the position of the viewer, while a ground plan or geometrical representation is unique. Indeed, God sees things exactly as they are in accordance with geometrical truth, although he also knows how everything appears to everything else, and so he eminently contains in himself all other appearances. (AG 199)

The assertion that God sees things "in accordance with geometric truth", as opposed to the perspectival view of bodies/monads, would seem indicative of the difference between the whole of Euclidean space, and a perspective from a single location within Euclidean space. This distinction is captured in Leibniz' analogy between a ground plan and a perspective drawing, since the ground plan, presumably, contains all of the different perspectives, and he also adds that God "eminently contains" the perspectival appearances as well (cf. N 22–30). Moreover, the ground plan is both "unique" and the "phenomena of God", hence it is not surprising that geometric *truth* is linked with this assessment (as argued above in §3.3.1).

Finally, monads share an essential feature with Leibniz' God that helps to explain their distinctive role within the metaphysical scheme surveyed in this chapter. Like God, monads are not in space ("[f]or monads, in and of themselves, have no position with respect to one another"; AG 201), although they have a sort of derivative location, via bodies: "[f]or even if they are not extended, monads have a certain kind of situation in extension, that is, they have a certain ordered relation of coexistence to other things, namely, through the machine in which they are present" (AG 178). Furthermore, monads are the link between God and the material realm: monads, like God, are not in space *per se*, but they are the means by which God "brings about" matter, and hence, space. As Leibniz puts it, "properly speaking, matter is not composed of constitutive unities [monads], but results from them" (AG 179). So, as will also be demonstrated in subsequent chapters, if one truly desires a modern analogue of Leibniz' theory within contemporary philosophy of space and time, then the details of his monadological thesis seem much more closely allied with the many recent attempts to explain how the macroscopic level of reality, i.e., material bodies and the large scale structure of space, arises from a manifestly different, and more fundamental, level of reality, e.g., quantum gravity. The continuing preoccupation with relationism, consequently, has only had the unfortunate effect of distracting philosophers from this more fruitful line of investigation.

Chapter 4
Motion, Matter, Monads, and Their "Forced" Relationship

After the examination of the metaphysics of Leibniz' theory of space as it pertains to material bodies and relationism—a theory that, as we have seen, contravenes core elements of the relational conception of space—it is time to turn to a more extensive examination of two further metaphysical issues related to space: motion and monads.

Like his hypotheses of space, Leibniz puts forward several hypotheses on the phenomena of motion that appear to both sanction and oppose modern relationism, a point first mentioned in §3.1. As will be revealed in this chapter, the traditional classical physics inspired reconstructions of Leibnizian motion have not been successful in resolving these contradictions, a verdict that is based not only on the evidence of the texts, but also due to the insufficiency of their results. Instead, an interpretation will be developed that, besides gaining textual support, draws upon a more metaphysical line of argument that bears a closer resemblance to non-classical theories of modern physics, such as quantum mechanics or the contemporary development of quantum gravity hypotheses. In order to meet this challenge, it will be argued that Leibniz endorses a view that allows a host of divergent hypotheses to remain consistent with the experience of motion, a view that is both hierarchical and perspectivalist in orientation but grounded in an invariant feature, force, that lies at a deeper level of ontology incompatible with classical physics. The interpretation advanced in this chapter, consequently, neither saddles Leibniz with a crude form of idealism or immaterialism, i.e., that only incorporeal beings (God, souls, etc.) and their mental content exist, nor does it equate with an objectivity-denying brand of subjectivism or relativism about material phenomena. In short, a realism concerning the external world is consistent with a form of "invariantism", as we will call it, centered on force.

If force is the invariant underlying Leibniz' conception of reality, then an examination of the other central component of this chapter, monads, must come into play, since force is grounded in the monads. Unfortunately, there are numerous, and severe, textual and philosophical challenges tied to the monad-matter relationship, in particular, Leibniz' insistence that monads are non-spatial but somehow have a

© Springer International Publishing Switzerland 2016
E. Slowik, *The Deep Metaphysics of Space*, European Studies in Philosophy
of Science, DOI 10.1007/978-3-319-44868-8_4

derived situation in bodies. Likewise, how does force bring about matter? In order to provide an account that addresses the issues associated with matter and motion, as examined in the first half of the chapter, it is thus imperative to better grasp the relationship between monads, force, and matter. In pursuit of this goal, it will be argued that Leibniz may have utilized a variant of the "extension of power" doctrine, an hypothesis that Descartes and many Scholastics had employed as a means of explicating how an unextended immaterial being relates to extended matter. In various late metaphysical discussions, Leibniz refers to this doctrine and other related concepts within the context of discussing extension and the primitive and derivative force of monads. Given this added information, some of the key questions concerning the spatiality of monads, as well as the motivation underlying various aspects of Leibniz' theory, gain considerable insight and clarity. Furthermore, insights into the foundational level of Leibniz' metaphysical system will prove invaluable in later parts of the investigation, not only for understanding matter and motion in the seventeenth century, but for the purpose of constructing successful analogies with non-classical theories in modern physics. Overall, the discussion in this chapter and Chap. 3 will provide an alternative ontology for Leibnizian physics that likewise challenges the adequacy of the substantival/relational dichotomy, not only for Leibniz but for modern conceptions as well.

After introducing the details of Leibniz' theory of motion in §4.1.1 and §4.1.2 will offer an interpretation of one of his most contentious doctrines concerning motion, the equivalence of hypotheses, while also demonstrating the inadequacy of recent readings that rely on classical physics, namely, the absolute speed hypothesis. An alternative interpretation of Leibniz' theory, which relies on an invariantist view of physical theories that differs significantly from the classical physics based approaches, will then be outlined in §4.1.3 and §4.1.4. Turning to the second half of the chapter, §4.2.1 will survey the problem of monadic situation as well as various solutions that have been put forward by several prominent commentators, while §4.2.2 and §4.2.3 will present evidence for the extension of powers doctrine alongside an argument that relies on the important function of primitive and derivative force to explain how matter and motion can arise from monads. In §4.3, a final summary and integration of the material of chapter will be presented.

4.1 The Interpretive Challenge of Leibnizian Motion

In response to the obvious difficulties associated with Leibnizian motion, in particular, the problem of reconciling his alleged commitment to relational motion with a conservation law that is quite inhospitable to relational motion (as will be explored in greater depth below), commentators have espoused various strategies, such as the use of reference frames in a spacetime structure congenial to relational motion (e.g., Earman 1989, Slowik 2006), or the endorsement of absolute motion or speed (e.g., Cook 1979, Roberts 2003). Nearly all of these modern reconstructions, including the examples just cited, attempt to reconstruct Leibniz' physics within the context of classical mechanics or classical gravitation theories, specifically, by comparing Leibniz' conception of body, velocity, force, etc., against their modern usage, and then striving to find an equivalent notion within classical theories of physics that can represent Leibniz' usage (Westfall 1971, Bernstein 1984, and Arthur 1994, comprise further instances of this approach). Nevertheless, the attempts to interpret Leibniz' physics using the conceptual resources employed by contemporary theories of classical physics have achieved little, if any, success, since the troublesome inconsistencies in Leibniz' conception still remain. Some commentators have questioned the adequacy of a spacetime reconstruction Leibnizian motion, it should be noted (namely, Huggett 2006a), yet no substitute interpretation has been offered so far.

4.1.1   Overview of Leibnizian Motion and the Equivalence of Hypotheses

As first introduced in Chap. 3, a nice encapsulation of the dilemma(s) associated with Leibnizian motion can be found in the *Discourse on Metaphysics* (1686):

> [I]f we consider only what motion contains precisely and formally, that is, change of place, motion is not something entirely real, and when several bodies change position among themselves, it is not possible to determine, merely from a consideration of these changes, to which body we should attribute motion or rest,.... But the force or proximate cause of these changes is something more real, and there is sufficient basis to attribute it to one body more than another. Also, it is only in this way that we can know to which body the motion belongs. (AG 51; also, L.V.53)

The first half of this quotation would seem to fit nicely with the strict relationist account of motion: i.e., "it is not possible to determine, merely from a consideration of these changes [in position], to which body we should attribute motion or rest". In the second half of the above quotation, however, Leibniz seems to retract this potential relationist commitment by assigning individual states of motion to these bodies via their force: "there is sufficient basis to attribute it [force] to one body more than another", and "it is only in this way that we can know to which body the motion belongs". In the later *Specimen Dynamicum* (1695), he refers to these force-based motions as "true" motions: "[F]orce is something absolutely real in substances, even in created substances, while space, time, and motion are, to a certain extent, beings of reason, and are true or real, not per se, but only to the extent that they involve either the divine attributes (immensity, eternity, and the ability to carry out works), or the force in created substances" (AG 130–131). Put in the modern terminology, motion conceived kinematically (without considerations of force), is apparent or geometric motion, whereas motion conceived dynamically (under the action of forces) is true or real motion. Hence, by declaring that there is a difference between true and apparent motion, Leibniz' view parts company with relational motion, since a strict account of relational motion cannot admit any such distinction.

A further impediment for a relationist construal of Leibnizian motion centers on his conservation principle, mv^2 (roughly, the size of the body times the square of its velocity; see, e.g., AG 128). Since the conservation law demands a series of privileged reference frames or perspectives for measuring this conserved quantity, namely, inertial (non-accelerating) reference frames, not all attributions of motion and rest to the bodies that undergo a relative change in position will uphold mv^2. For instance, an accelerating reference frame will likely attribute motions to these bodies that fail to conserve mv^2, thus it would appear that the assignments of motion or rest are not entirely arbitrary, contra the strict relationist view of motion. In the parlance of spacetime theories, the dilemma imposed by Leibniz' physics is that his dynamical symmetries, i.e., the perspectives that preserve his conservation law, do not match his spacetime symmetries, i.e., the perspectives allowed for determining motion per se (see, also, §1.1). A neo-Newtonian (or Galilean) spacetime, which possess an inertial structure that can delineate accelerating and non-accelerating motions, is the minimal structure required to uphold mv^2, for it is only in inertial frames that the conservation of mv^2 can be guaranteed. Yet, Leibniz' insistence that all assignments of motion are possible when force is absent would seem to side with the so-called Leibnizian spacetime, a setting that lacks inertial structure and thus cannot distinguish accelerating and non-accelerating motions (see, once again, Earman 1989, chap. 2, for a presentation of these spacetime structures). The spacetime theorist, accordingly, must attempt to steer a course between the Scylla of too much structure (neo-Newtonian spacetime), which violates strict relational motion by picking out preferred trajectories through space, and the Charybdis of too little structure (Leibnizian spacetime), whose inability to distinguish inertial paths undermines the application of the conservation law (see, Slowik 2006, for an examination of several reconstructions along these lines, although all of these reconstructions are ultimately deemed unsuccessful for the general reasons just outlined).[1]

Lastly, the dichotomy between true and apparent motion would likely seem less puzzling if it were not for an additional doctrine known as the "equivalence of hypothesis" (hereafter, EH). The significance of EH for Leibniz' overall natural philosophy has long been a source of dispute, for it seems to implicate the arbitrariness of the assignments of motion given a geometrical change of position, as described above, in addition to the principle of Galilean relativity, the hypothesis

---

[1] One of the biggest problems for a relationist interpretation of Leibnizian motion is rotational motion, although in depth examination would take us too far afield. In brief, Leibniz' approach to rotational motion, like his handling of inertial motion and his conservation principle, is incompatible with Leibnizian spacetime structure, but the adoption of Newtonian or neo-Newtonian spacetime structures seem to contradict the tenets of his relational account of rotation. See, Slowik (2006, 2009), for analysis: these articles can be seen as first steps in the evolution of the ideas presented in this chapter. Slowik (2009) also critiques Jauernig's (2008) use of spacetime structures to treat Leibnizian motion; but her reply, (2009), does not address the main criticisms, especially concerning rotation (although, in fairness, perhaps several terminological and conceptual miscommunications are in play here). For earlier insightful examinations of Leibnizian rotational motion, see Stein (1977b), and Earman (1989, chapter 4).

that the laws of physics are the same in all inertial frames (i.e., whether the frame is at rest or moving rectilinearly with a uniform speed):

> [W]e must hold that however many bodies might be in motion, one cannot infer from the phenomena which of them really has absolute and determinate motion or rest....From this follows something that Descartes did not notice, that *the equivalence of hypotheses is not changed even by the collision of bodies with one another,* and thus, the laws of motion must be fixed in such a way that the relative nature of motion is preserved, so that one cannot tell, on the basis of the phenomena resulting from a collision, where there had been rest or determinate motion in an absolute sense before the collision. (AG 131)

Leaving aside the possible conflation of relational motion and Galilean relativity, perhaps the most controversial aspect of EH is that Leibniz apparently sanctions something akin to a pragmatic or instrumentalist account of the truths associated with the individual states of bodily motion, a pragmatism that stems from a free choice of the approach or "point of view" adopted to describe the phenomena. More carefully, in the 1689 tract, "On Copernicanism and the Relativity of Motion", Leibniz argues that the determination of the "greater intelligibility and simplicity" of an hypothesis, Ptolemaic and Copernican, is *relative* to a point of view, respectively, spherical astronomy and planetary theory—but, since the Ptolemaic and the Copernican hypothesis can assign different states of motion to the same body, such as the earth, it follows that the truths of an individual body's state of motion are thus relative to a pragmatic or instrumentalist choice to adopt either spherical astronomy or planetary theory as one's point of view of material phenomena. After recounting the nature of motion and the equivalence of hypotheses in the same manner as the quote offered above, he then proceeds to address the question of bodily motions:

> But since, nevertheless, people do assign motion and rest to bodies,...we must look into the sense in which they do this, so that we don't judge that they have spoken falsely. And on this matter we must reply that one should choose the more intelligible hypotheses, and that the truth of a hypothesis is nothing but its intelligibility. Now, from a different point of view, not with respect to people and their opinions, but with respect to the very things we need to deal with, one hypothesis might be more intelligible than another and more appropriate for a given purpose. And so, from different points of view, the one might be true and the other false. Thus, for a hypothesis to be true is just for it to be properly used....[T]he truth of a hypothesis should be taken to be nothing but its greater intelligibility, indeed, that it cannot be taken to be anything else, so that henceforth there would be no more distinction between those who prefer the Copernican system as the hypothesis more in agreement with the intellect, and those who defend it as the truth. For the nature of the matter is that the two claims are identical; nor should one look for a greater or a different truth here. (AG 91–92)

Furthermore, Leibniz seems to rank these points of view according to their overall intelligibility: "the Ptolemaic account is the truest one in spherical geometry, on the other hand the Copernican account is the truest theory, that is, the most intelligible theory and the only one capable of an explanation sufficient for a person of sound reason" (AG 92).

Overall, it is difficult to gauge the scope and intended function of Leibniz' "truth=most intelligent hypothesis" gambit. As persuasively argued by Lodge (2003, 300), since there are gradations of truth, it is more likely that Leibniz equates intelligible hypotheses with empirical adequacy, and not literal truth. For example, if the Copernican hypothesis is the truest account, then some bodies will be judged to be at rest from the perspective of this hypothesis (such as a body or particle situated at the center of gravity of the solar system), even though Leibniz often claims that no body is entirely at rest (Lm 706; see also footnote 4). Therefore, to avoid contradiction, the empirical adequacy of an hypothesis seems the most plausible interpretation of his use of the term "truth" in this context. A passage that supports this interpretation can be found in a 1694 letter to Huygens: "I hold, of course, that all hypotheses are equivalent, and when I assign certain motions to bodies, I do not and cannot have any reason other than the simplicity of the hypothesis, since I believe that one can hold the simplest hypothesis (everything considered) as the true one" (AG 308).

4.1.2   Leibnizian Motion: An Alternative Account Contra Absolute Speed

Among the strategies for interpreting Leibniz' complex conception of motion, a recent approach is to accept the distinction between true versus apparent motion but without requiring an absolute space to ground true motion. Dubbed "absolute speed" in Roberts (2003), the idea is that, given a change in position among several bodies (apparent motion), some of the bodies possess a real, force-induced motion (true motion) that is responsible for that observed change in position (hence true motion=absolute speed/motion). However, since absolute speed (hereafter, AS) does not necessitate absolute space, the AS interpretation does not violate Leibniz' antipathy to absolute space; see, Roberts (2003) for an elaborate treatment, but it was earlier put forward by Cook (1979), and probably several others. Unfortunately, AS runs counter to the textual evidence, which regularly dismisses the concept of absolute motion: for example, in the piece on Copernicanism quoted above, Leibniz states that "since space without matter is something imaginary, motion, in all mathematical rigor, is nothing but a change in the positions of bodies with respect to one another, and so, motion is not something absolute but consists in a relation" (AG 91); and, in more detail, he adds:

> Since we have already proved through geometrical demonstrations the equivalence of all hypotheses with respect to the motions of any bodies whatsoever, however numerous, moved only by collisions with other bodies, it follows that not even an angel could determine with mathematical rigor which of the many bodies of that sort is at rest, and which is the center of motion for the others. And whether the others are moving freely or colliding with one another, it is a wondrous law of nature that no eye, wherever in matter it might be placed, has a sure criterion for telling from the phenomena where there is motion, how much motion there is, and of what sort it is, or even whether God moves everything around it, or whether he moves that very eye itself. (AG 91)

Given his frequent renunciations of absolute motion (e.g., A.VI.iv.1968, GM VI 251), and his above claim that EH is akin to an invariable law of nature that applies to collisions as well as planetary motions, the tenability of the AS hypothesis is thus called into question.

In response to these doubts, Roberts attempts to separate Leibniz' endorsement of EH from his various references to true motion:

> What, then, are we to make of the opening passage of "On Copernicanism"…? There, as we have seen, Leibniz argues that all frames of reference are equivalent for the purpose of describing the motions of bodies, and that we are justified in using the reference frame that makes the phenomena out to be most intelligible. But very early in this paper, Leibniz makes it explicit that in this context, by "motion" he means nothing more than "change of place" [AG 91]. As we have seen, Leibniz makes it clear, in other writings spanning the time of publication of "On Copernicanism", that this is not the only way of regarding motion. In the essay on Copernicanism, Leibniz is working with a relatively thin conception of motion by his own standards, and what he says there shouldn't be taken to trump what he says elsewhere about motion in the fullest sense. (Roberts 2003, 559–560)

In brief, Roberts claims that the scope of EH is restricted to only a "thin" kinematical/geometric conception of motion, whereas true motion, and hence absolute speed, constitutes a different account ("motion in the fullest sense"). Yet, Leibniz explicitly mentions bodily collisions in the passage concerning EH that Roberts' cites from "On Copernicanism" ("moved only by collisions with other bodies"), and it is these types of bodily interactions that serve as the basis for the introduction of the concept of true motion in the *Discourse* and *Specimen Dynamicum.*

Furthermore, as persuasively argued by Arthur (2012), the assignment of true motion based on the notion of intelligibility, and thus EH, is a common theme in much of Leibniz' mature work. An analysis of the motion of a body in a fluid, for example, prompts the following conclusion: "[a]nd granted that motion is a relative thing, nonetheless that hypothesis which attributes motion to the solid, and from this deduces the waves in the liquid, is infinitely simpler then the others, and for this reason the solid is adjudged to be the cause of the motion. Causes are not derived from a real influence, but from the providing of a reason" (LoC 311; c. 1686). Since causes are obtained by finding a reason, i.e., the most intelligible hypothesis, and not through the transmission of a force, this passage confirms that EH applies to the types of phenomena covered in the *Discourse*, the work most often cited for introducing the true motion distinction. The same point is raised in the Huygens letter quoted above: "when I assign certain motions to bodies, I do not and cannot have any reason other than the simplicity of the hypothesis" (AG 308). Leibniz' rationale for employing EH in these cases ultimately stems from his conception of substance, a conception which rejects any interaction or causal influx among substances since each substance contains an expression of the entire world as an internal feature (i.e., the complete concept notion, see §3.2.2): "no real influence of [substances] on one another is intelligible. For whatever happens to each of them would flow forth from its own nature and notion even if all of the others were imagined to be absent, since each one expresses the world" (LoC 311).

Additionally, Leibniz' interpretation of bodily interactions, wherein the grounding substances as well as the bodies do not really causally interact, plays a key role in his rejection of absolute motion.[2] The non-interaction of bodies, he insists, provides the only explanation for the "phenomena of motion", for "every single body must be supposed to have a motion in common with any other, as if they were in the same ship, as well as its own motion, reciprocal to its bulk; how this could be so could not be imagined if motions were absolute and each body did not express all others" (LoC 335; c. 1686). Importantly, this passage equates absolute motion with a motion that does not express all of the other bodily motions in the world, i.e., that is independent. Therefore, absolute motion violates his pre-established harmony doctrine (i.e., that non-interacting substances are, as it were, synchronized by God to give the impression of interaction). Leibniz makes a similar point in an earlier essay (from 1677), commenting that "if motion is an affection, its subject will not be any one individual body, but the whole world. Hence all its effects must also necessarily be relative. The absolute motion we imagine to ourselves, however, is nothing but an affection of our soul while we consider ourselves or other things as immobile, since we are able to understand everything more easily when these things are considered as immobile" (LoC 229).[3]

Finally, it would be useful to examine a longer passage from the *New System of Nature* (1695), for it brings together Leibniz' conception of substance, true motions, and EH, all within the context of the bodily interactions that comprise his force-based physics:

> For we can say that the substance, whose disposition accounts for change intelligibly, in the sense that we may judge that the other substances have been accommodated to this one in this regard from the beginning, according to the order of God's decree, is the substance we must consequently conceive as *acting* upon the others. Furthermore, the action of one substance on another is neither the emission nor the transplanting of an entity, as commonly conceived, and can reasonably be taken only in the manner just stated. It is true that we readily conceive emissions and receptions of parts in matter, by which we reasonably explain all of the phenomena of physics mechanically. But since material mass is not a substance, it is clear that action with respect to substance itself can only be as I have just described....And as for absolute motion, nothing can fix it with mathematical rigor, since everything terminates in relations. This makes for the perfect equivalence of hypotheses, as in astronomy, so that no matter how many bodies we take, we may arbitrarily assign rest or a particular degree of speed to any body we choose, without being refuted by the phenomena of rectilinear, circular, or composite motion. However, it is reasonable to attribute some true motions to bodies, in accordance with the assumption that accounts for the phenomena in the most intelligible way, this denomination being in conformity with the notion of action we have just established. (AG 145)

---

[2] In striving to bridge the divide between his non-interacting simple substances (or monads, in his later work) and the phenomenal level of material bodies, Leibniz often appeals to bodily elasticity as a means of modeling the simple substance's internal expression of the world. He reasons that "no impetus is transferred from one body to another, but each body moves by an innate force, which is determined only on the occasion of, i.e. with respect to, another. For...the cause of the impulse one body gets from another is the body's elasticity itself, by means of which it recoils from the other" (LoC 311).

[3] In this early discussion, Leibniz may have accepted the notion of a world body or world motion, an idea that would ultimately be replaced by an infinity of simple substances in pre-established harmony. Either way, absolute motion is judged to violate his underlying metaphysics.

After rebuffing the suggestion that bodies communicate motion, and denying absolute motion, he asserts that the most intelligible account of the true motions of bodies stems from the "notion of action". Specifically, the concept of action mentioned in the first part of the quotation concerns substances in pre-established harmony, but he later adds that the most intelligible hypotheses of explaining the phenomena of bodily motion is "in conformity" with his theory of the action of (non-interacting) substances. In this context, the phrase, "the assumption that accounts for the phenomena in the most intelligible way", likely refers to the practical ability of the mechanical approach to determine the true motions of bodies (from the apparent) via his conserved quantity, mv^2, and the prior history of the system's motions and relationships. Returning to the earlier (LoC 311) example, the choice to regard the solid as in motion, and not the liquid, constitutes the most intelligible way of explaining the phenomena (i.e., simplest) for Leibniz, with the force of the moving body judged to be "in conformity" with his underlying metaphysics of non-interacting substances. To recap, Leibniz' use of EH in this passage is based on the idea that the true motions that we assign to bodies resembles or mirrors the action of substances, even though substances do not truly interact, and bodies and their motions are not truly real.

Accordingly, given the abundant evidence of the texts, Leibniz' account of true motion is, in fact, an instance of his EH doctrine, and not an alternative or contrary conception of motion. Since the expression, "most intelligible way", stands as yet another instance of Leibniz' pragmatist approach to scientific explanation, it follows that the assignment of true motion is simply the best method of describing bodily phenomena *from the perspective of the mechanical hypothesis*—but, of course, there are many other hypotheses of motion. As he asserts in the Copernicanism essay, "one hypothesis might be more intelligible than another and more appropriate for a given purpose. And so, from different points of view, the one might be true and the other false" (AG 91). As a result, *if* we desire to designate individual bodily motions from the perspective of the mechanical hypothesis and the conservation of mv^2, then we can appeal to the activity of the substances that ground those material bodies as a deciding factor—however, that assignment of individual bodily motions does not entail that motion is something real or absolute, i.e., as real as substance and primitive force, or that the mechanical account of motion based on mv^2 is the one true account of bodily motion, and that all other accounts of bodily motion, such as the Tychonic, are false. The fact that the mechanical hypothesis of bodily motions based on mv^2 falls under EH is crucial in this respect, since it reveals that there is, so to speak, a hierarchy of hypotheses of motion constructed for different purposes. Presumably, the mechanical hypothesis is the most intelligible hypothesis of all, since it more closely reflects, or is in conformity with, the action of simple substances. But, this "hierarchical" of "perspectivalist" conception of Leibniz' metaphysics of motion, as we will occasionally refer to it, runs counter to the AS hypothesis. In particular, the AS hypothesis holds that true motions are absolute and that EH applies only to the kinematic/geometric class of motions, hence the motions determined from the perspective of the mechanical hypothesis are metaphysically privileged, i.e., really true, whereas the motions determined from the Ptolemaic point of view are really false. Not only does Leibniz never make this claim, i.e. that Copernican motions are really true while Ptolemaic motions are really false, but he asserts instead that "from different points of view [spherical astronomy versus theory of the planets], the one might be true and the other false" (AG 91). The textual evidence, therefore, points toward a metaphysics of motion different than AS, a metaphysics whereby all of the different hypotheses of motion fall under EH in a sort of hierarchical series that includes both the mechanical hypothesis and the purely kinematic: it is only by this interpretive route that Leibniz' use of such terms as "truer" and "truest" make sense, e.g., "the Ptolemaic account is the truest one in spherical geometry, on the other hand the Copernican account is the truest theory, that is, the most intelligible theory" (AG 92).

Defenders of the AS hypothesis, furthermore, construe the complexities associated with Leibnizian motion as stemming from the denial of absolute space rather than absolute motion: "I've argued that Leibniz's doctrine of force commits him to the doctrine of absolute speed, that absolute speed doesn't entail absolute space" (Roberts 2003, 569). Yet, besides the denial of absolute space, Leibniz repeatedly rejects absolute motion, and one of the rationales offered, as mentioned above, is that absolute motion does not "not express" all of the other bodies and their motions. The AS hypothesis not only fails to account for this facet of Leibniz' system, but a body's absolute motion would, in fact, appear to be independent of all the other bodies and their motions, contra Leibniz' demand. For these reasons, an interpretation that is more consistent with the textual evidence is required, such as the perspectival, hierarchical conception explored above.

4.1.3   Realism, Invariantism, and Leibnizian Motion

Under an idealist or immaterialist interpretation of Leibniz' system, the instrumentalism or pragmatism that is seemingly embodied in EH is perfectly feasible, since only incorporeal beings (God, monads) and their perceptions exist (see, Furth 1967, for a contemporary idealist reading of this ilk). In contrast, a realist interpretation of Leibniz' theory, i.e., a theory that posits the existence of a mind-independent material world of some sort, confronts a much more difficult task given the instrumentalism inherent to EH. A realist interpretation of Leibniz' theory, additionally, can accept the fact that the true nature of corporeal existence is not identical to our mathematical and geometrical conceptions of that reality, or, to put it differently, that there is a degree of idealism in our mathematical grasp of matter. As Leibniz often remarks, the continuous, holistic structures employed by mathematics are ideal, whereas matter is ultimately composed of discrete elements that aggregate to form bodies (see, e.g., G II 379).[4] Nonetheless, even if bodies do not resemble our

---

[4] The mathematical basis of all of the hypotheses of motion, whether Ptolemaic, mechanical, etc., is clearly one of the motivating factors behind EH. For example, in explaining how force can single out true motion (in the Leibniz-Clarke correspondence), Leibniz claims that "there is not any one body that is perfectly and entirely at rest, but we will frame an abstract notion of rest by considering

mathematical notions precisely, a realist interpretation must accept the existence of a material world apart from our minds/souls—to deny this inference is to abandon realism for some version of idealism (and Leibniz does, in fact, reject Berkeley-style idealism; see, e.g., Hartz 2007, 81–82, and, Rutherford 2008, for the anti-idealist case).

Yet, is it possible to reconcile a pragmatic approach to bodily motions, as exemplified in the hierarchical conception of EH defended in §4.1.2, with a realist interpretation of physics, even granting this concession to the idealistic component latent in our mathematical conceptions? Specifically, if one adopts the thesis that the assignment of individual bodily motions are relative to the overall theory used to explain the phenomena, such as the Tychonic, Ptolemaic, or Copernican theories—and that none are actually true in the literal sense (and the others false)—has one simply abandoned any plausible realist interpretation for a full-blown instance of idealism (since, e.g., the Tychonic and Copernican theories posit different states of motion to the earth)?

The dilemma imposed by Leibniz' handling of EH can be overcome by the realist if one is willing to accept a more nuanced and complex conception of his ontology than the standard classical physics-based strategies put forward thus far. The chief problem with the approach to Leibniz' physics based on theories of classical physics, such as classical mechanics or spacetime theories, is that they rely, not surprisingly, on the conception of space, time, and matter associated with classical mechanics or spacetime theories, but with the additional proviso that these structures do not support absolute space. While denying absolute space entails that these reconstructions are consistent with Leibniz' rejection of the so-called static and kinematic shift arguments in the Leibniz-Clarke correspondence—i.e., the possibility that the entire world could have, respectively, a different position in space or a different velocity (AG 325; see Chap. 3)—an ideal space alone does little to resolve the EH puzzle. If one substitutes an ideal space for a real space, but retains the remainder of the classical physics ontology centered upon a realist construal of bodies and their motions, then one remains mired in the non-realist consequences of EH surveyed above, namely, that individual states of motion become relative to the chosen theory.

A better strategy for dealing with the EH quandary is to render all of the *geometrical* aspects of the ontology of bodies, including both motion and bodily extension, mind-dependent to some degree, and thereby accommodate the subjective or instrumentalist implications of EH via a corresponding subjective or instrumentalist form of ontology as regards those geometrical features. This interpretation of Leibniz' material world, where space and time, bodies and their motions, are all phenomenal or ideal to a varying extent, has the added advantage that it is, in fact, Leibniz' oft-repeated view. In the *Discourse*, he states that it is "possible to demonstrate that the notions of size, shape, and motion are not as distinct as imagined and that they contain something imaginary and relative to our perception, as do (though to a greater extent) color, heat, and other similar qualities, qualities about which one

---

the thing mathematically" (Lm 706). Hence, while no body is actually at rest, mathematical hypotheses can still admit that notion.

can doubt whether they are truly found in the nature of things outside ourselves" (AG 44; see also, Lm 365). In effect, Leibniz rejects a sharp primary/secondary property distinction, placing the so-called primary properties, such as shape, size, and motion, alongside the secondary properties, like color and heat. This viewpoint, which Garber (2009, 297) dubs "primary-quality phenomenalism", is related to, but more general than, Leibniz' belief (mentioned above) that our mathematical notions do not accurately correspond to material existents. By reckoning that bodily extension and motion "contain something imaginary and relative to our perception", one can thus consistently uphold or defend the subjective element in EH since the individual motions of bodies can now be included within the category of the perceptually imaginary and relative. Therefore, given this perceptual relativity as regards all of the geometrical aspects of bodies, the determination of the individual states of motion can be left to a pragmatic choice of theory (Ptolemaic, Copernican) that best fits a point of view (spherical astronomy, theory of the planets). This interpretation avoids idealism or immaterialism by accepting the existence of the material realm, although this material realm can only be characterized in objective terms by way of force: it is the geometrization of force, by means of bodies, motions, etc., that is ideal and subjective, but not force itself—and so there is more to the world than just immaterial beings, God and monads, and their perceptions, contra idealism.

Taking this interpretational route, however, means abandoning any strategy based on a realist construal of classical mechanics or spacetime structures, since these realist strategies cannot allow their structures, nor the bodies within these constructions, to be relative or subjective in a way similar to the relativity of the perception of motion. The same holds for the absolute speed interpretation, since that reading relies on neo-Newtonian spacetime structure (as in, e.g., Roberts 2003).[5] If bodies and their accompanying spacetime structures are themselves relative or pragmatic, then these classical mechanical and spacetime strategies obviously cannot provide the necessary realist grounding to resolve the EH puzzle.

What is required to overcome this dilemma is an invariant or absolute ontological item at a level of reality deeper than classical mechanics and its classical spacetime structures, an ontology that, while different from classical physics, can consistently explain the relative perspectives at the ontological level of bodies and motions normally associated with classical physics—and, once again, this is exactly the strategy that Leibniz offers: "We must realize, above all, that force is something

---

[5] As regards the special and general theories of relativity, allowing the shape of bodies (but not volume, rest mass, etc.) to be perspectival in the same fashion as the judgments of velocity (but not acceleration) is a consequence of its four-dimensional Minkowskian spacetime structure and its all-important invariant, the spacetime interval (Lorentzian metric). The attempts by various Neo-Kantians and logical positivists to regard Leibniz as presaging Mach and Einstein can be traced in part to Leibniz assertions on EH. Briefly, these twentieth century thinkers viewed Leibniz' EH as supporting the type of relational theory of motion championed by Mach, as well as Einstein's own early belief (later rejected) that general relativity upholds Machian relationism (see, Bernstein 1984, De Risi 2012, for some of this history). De Risi's investigation of EH may draw conclusions similar to those put forward in this essay, moreover, although he approaches these issues from a phenomenalist angle (whereas our study has strived to remain within the ontological): in particular, De Risi seems to deny that absolute motion applies to the noumenal (real) world (2012, 163), and this is one of the major conclusions advanced in this chapter. The temporal component in Leibniz' theory is a separate topic, but see, Futch (2008), for a detailed study.

absolutely real in substances, even in created substances, while space, time, and motion are, to a certain extent, beings of reason, and are true or real, not *per se*, but only to the extent that they involve either the divine attributes (immensity, eternity, and the ability to carry out works), or the force in created substances [monads]" (AG 130). This invariant force is, naturally, the primitive force of monads, but it is manifest at the bodily level as derivative force, the force measured by mv^2 from the mechanical perspective.

Although a more in depth analysis will be provided in §4.2, a brief discussion is required concerning the nature of Leibnizian force and its relationship with mv^2. As first revealed at the end of Chap. 3, the force that brings about motion is not located at the level of his simple substances, or monads, since there is no space at this level: "there is no spatial or absolute nearness or distance between monads" (Lm 604). Instead, a monad's primitive force is manifest at the level of well-founded phenomena (i.e., the bodily level) as derivative force: a monad is "endowed with primitive power" such that the "derivative forces [of bodies] are only modifications and resultants of the primitive forces" (AG 176). And, it is the derivative forces that are associated with motion: "I, however, do not consider motion to be a derivative force but think rather that motion, being change, follows from such force" (Lm 533). If derivative forces are, as Leibniz explains, a determinate value of the primitive forces, and motion "follows from" derivative force, then it would seem natural to conclude that the forces that are implicated in motion, i.e., mv^2, also exist at the phenomenal level of bodies. Furthermore, whereas monads are non-spatial, derivative force is coupled to the phenomena of extended bodies, which are aggregates of monads, or secondary matter: "the derivative force of being acted upon later shows itself to different degrees in secondary matter" (AG 120). So, both bodily extension and the derivative force of motion, mv^2, are brought about by primitive force. Since Leibniz regularly characterizes derivative force as a limitation or modification of primitive force, an interpretation that deems both bodily extension and motion as perspectival in nature thus gains support. As he explains to De Volder, "unless there is some active principle in us, there cannot be derivative forces and actions in us, since everything accidental or changeable ought to be a modification of something essential or perpetual, nor can it contain anything more positive than that which it modifies, since every modification is only a limitation, shape a limitation of that which is varied, and derivative force a limitation of that which brings about the variation" (AG 179–180). Lastly, from within the mechanical perspective based on the conservation law of mv^2, the only constraint on the perspectival assessments of bodily extension and motion is that they combine to uphold mv^2, with the invariance of that quantity reflecting the absoluteness of primitive force at the deeper monadic level.

Unlike Descartes, who apparently assumes that the application of his laws of motion are not undermined by the relativity involved in the assessment of motion (see §1.3.2), Leibniz builds the relativity of motion directly into his overall ontology, with mv^2 representing, in an indirect and partial manner, an invariant feature that underlies all of the conflicting estimations of the bodily motions that constitute his mechanically-conceived system. Hence, despite their similarities on a purely mathematical basis, Descartes' quantity of motion and Leibniz' mv^2 operate in vastly different ways. For Descartes, the reciprocal nature of motion (i.e., the transfer between a body and its containing bodies) and his conservation law for the quantity of motion (roughly, size times speed) are two separate issues that are only accidentally, and problematically, conjoined in his physics. Leibniz, in contrast, treats the relativity of motion and its attendant conservation law as two interrelated effects of a deeper cause, force, whose reality is only indirectly and incompletely expressed through the quantity mv^2 employed by the mechanical perspective. A quick rationale for this strategy stems from his belief that force is real, whereas the geometrical properties of bodies are ideal—but, since mv^2 concerns those geometrical features, it automatically follows that it cannot reveal the full or real nature of force. This reading is consistent with Leibniz' explanation that motion is linked to derivative force, but that derivative force is an instantiation of the ontologically deeper, and "absolute", primitive force.[6]

At the level of material bodies, it is true that Leibniz hit upon a key discovery, represented by EH, that conjoins his conserved physical quantity with a group of spatiotemporal symmetries (i.e., the symmetries encoded within neo-Newtonian spacetime and its inertial structure, a framework that constitutes the modern spacetime setting for modeling the Galilean relativity implicit in EH). Nevertheless, any interpretation of Leibnizian force that limits its ontological scope to the material plane of bodies and motions fails to grasp its true significance. A case in point are the modern physics-based reconstructions of Leibniz' theory that treat his conception of force and mv^2 as the components of a larger equation: force$=mv^2$. On this understanding of mv^2, Leibniz' distinction between an invariant (non-relative) force and relative motion leads to serious difficulties, since force *is* the product of the square of velocity and the size of the body; so, if motion is relative, then force is relative. Russell notes that this objection is "unavoidable on any relational theory of space" (1992, 101), and Earman expresses the same misgivings (1989, 132). The answer to this conundrum, also put forward by Lodge (2003, 286–287), is to jettison the conception of Leibnizian force based on modern physics, i.e., force$=mv^2$, when interpreting Leibniz' *ontological* use of that term, and the textual evidence supports this maneuver, as argued above.

4.1.4 An Outline of a Leibnizian Invariantist Ontology

In the literature on the metaphysics of objectivity and invariance, Leibniz' theory could stand as an early instance of what Debs and Redhead (2007, 61) call "invariantism" (a view inspired by the work of Weyl 1982 [1952], and Nozick 2001, among many others). On the whole, an invariantist reckons that objectivity is best understood as a feature of the world that remains fixed under changes in perspective, and

---

[6] It is outside the bounds of our investigation to examine the full range of interpretations of Leibnizian force and motion; rather, given limitations of space, the goal is to show how Leibnizian force in his mature period (mid-1680s onward) can fit the invariantist interpretation of Leibniz' physics that will be sketched in §4.1.4.

incorporates both (i) inter-subjectivity, a purely epistemological feature that many observers share identically, and (ii) ontological independence, whereby it is stipulated that the observed invariant feature actually stems from a deeper ontological item that can exist apart from all observers (i.e., it is not a completely mind-dependent phenomena). By distinguishing the epistemological and ontological aspects of invariance, this interpretation also provides a convenient means of separating a fully idealist reading of Leibnizian force, which corresponds with (i), from a realist construal which must incorporate both (i) and (ii). Specifically, the demand for inter-subjectivity is consistent with an idealist or immaterialist ontology that only accepts the existence of immaterial entities, whereas the stipulation that there is an ontological feature that exists in the external world, and which is responsible for the perspectival variance in our ascriptions of the individual states of motion, supports a corresponding materialist or physicalist inference (since force lies on the materialist side of the material/immaterial divide). Among the modern physics-inspired interpretations of Leibniz, Earman sides with route (i), contending that Leibniz' later ontology includes "only monads and their perceptions", wherein "the physical world is reduced to such perceptions" (Earman 1977, 223).[7] Whether or not Earman favors a purely immaterialist ontology, most other commentators of Leibniz' physics and EH must include component (ii) alongside (i), since most of these commentators do not accept an idealist interpretation a la Furth. As noted previously, most interpretations of Leibnizian motion strive to locate a feature within his physics that can resolve the relational motion dilemma, such as the concept of absolute speed discussed in §4.1.2, but this type of strategy only makes sense if one accepts the existence of the physical world along realist lines.

If one does accept the invariantist interpretation of Leibnizian force, and strives to synthesize both (i) and (ii), then what are the ontological implications for his physics, as well as for his larger metaphysics? The strategy adopted by most commentators, especially those who favor AS, has been to employ the spatial translation symmetries intrinsic to neo-Newtonian spacetime as the means of consistently measuring mv^2 among inertial frames (see §4.1.1), where the invariant quantity mv^2 is interpreted as corresponding to Leibniz' conception of force. Unfortunately, spatial symmetries alone do not capture the type of ontology advanced in our analysis, namely, where all of the properties of bodies, save force, are perspectival or relative to an instrumentalist or pragmatic choice of theory. As argued previously, the classical mechanical or spacetime frameworks only allow differences in the determination of position or velocity (and shape, in the setting of relativity theory), but the Leibnizian theory that we are proposing allows all of the non-force aspects of bodies, such as volume, to be perspectival in the same manner. Furthermore, since mv^2 is the invariant quantity from the perspective of the mechanical hypothesis, but other hypotheses of motion are guaranteed via EH, it follows that the spacetime and classical mechanical approaches to Leibnizian invariantism are limited to just their perspective of the ontology of Leibnizian force, namely, mv^2. Accordingly, a differ-

---

[7] Earman also claims that Leibniz "gives up a realist interpretation of space", and that "he did not pursue the middle path of constructing the physical world by assuming monads to have relative spatial relations" (1977, 226).

ent conception of the invariant associated with Leibnizian force is needed that is more general than the standard conception based on mv^2.

In modern physics, a closer match to the ontology of Leibniz' overall system is the various quantum gravity proposals that posit physical processes that do not exist in space (spacetime) but which generate both matter and space at a higher level of reality. As will be examined in greater depth in Chap. 10, the "foundational" level physical processes put forward in many quantum gravity hypotheses are responsible for the emergence of both matter and space at the "secondary" material/bodily level, with the foundational and secondary levels possessing drastically different sets of properties—and this entire conception bears an obvious affinity with Leibniz' ontology, where the foundational non-spatial monads and their primitive forces bring about matter and space at a different scale or level of reality. Although the quantities that might remain invariant in these quantum gravity theories, such as mass-energy, do not equate with Leibniz' mv^2, the general strategy is the same: namely, to explicate secondary level phenomena via invariant foundational level quantities and processes. While the examination of the best analogues to Leibniz' view among these modern physical theories will be the subject of Chap. 10, the main goal of our current discussion is to demonstrate that the apparatus of spatial translations and symmetries utilized in classical mechanics and classical gravitational theories is not adequate to the task of explicating the ontology implicit in Leibniz' understanding of EH. Leibniz' theory is much more radical than classical physics, and thus a different stock of physical theories is required as a basis of comparison. In fact, Leibniz often describes the material level of well-founded phenomena as akin to a rainbow (e.g., G II 436), and rainbows are not a feature of the world that can serve as a fundamental ontological item—rather, rainbows result from a different layer of physical ontology. Similarly, the holographic principle that informs many quantum gravity hypotheses contends, roughly, that our familiar three-dimensional material world (secondary level) is akin to a hologram image of quantum information "projected" from the more fundamental realm of quantum processes and entities (foundational level). As Susskind explains, "the combination of quantum mechanics and gravity requires the three dimensional world to be an image of data that can be stored on a two dimensional projection much like a holographic image" (Susskind 1995, 6377).

Leibniz' rainbow analogy, along with the relationship between the foundational and secondary levels of his ontology, also draws into the discussion issues that pertain to supervenience and emergence. Given a foundation in force at the level of monads (simple substances), as well as our realist construal of his overall system, does the matter, and hence, motion, that arises at the secondary level of well-founded phenomena count as a form of supervenience or emergence? The answer to this question must ultimately reside in the process by which matter "results" from monads, a term often used by Leibniz to describe this relationship: e.g., "properly speaking, matter is not composed of constitutive unities [monads], but results from them" (AG 179). As briefly mentioned in §3.3.3, the main idea behind Leibniz' "results" description would appear to be a type of grounding or dependence relationship, so that if one thing is posited (monad), a different thing follows (matter).

As Rutherford comments, "[r]esulting is best interpreted as a relation of ontological determination" (1995, 221–222). All told, the relationship of ontological dependence between monads and matter might be best captured via the concept of supervenience, which can be defined in simple terms as: a property X supervenes on a property Y if and only if X-properties covary with Y-properties; or, alternatively, there can be no difference in the X-property without a difference in the Y-property (see, McLaughlin 2008 for supervenience in general). Consequently, since the geometrical aspects of body are straightforwardly dependent on monads in the manner required to satisfy the definition of supervenience, monads constitute the subvenient base (Y), and matter serves as the supervening property (X). There have been previous supervenience-based interpretations of Leibniz' monadic system (e.g., Cover 1989, Hartz 2007), but these past attempts do not address, or to take into account, the EH quandary and its instrumentalist consequences. The invariantist conception of Leibnizian force advanced above, on the other hand, might offer a method of bringing the EH component of Leibniz' system into a larger supervenience interpretation.

As for the prospects of a Leibnizian metaphysics of emergence, however, the question is more uncertain and open to conflicting interpretation, especially given the elusive nature of emergence itself. If the intuitive notion of an emergent property is employed—as a novel or unique supervenient property that arises from the subvenient base but that is not strictly reducible to that base (see, once again, McLaughlin 2008)—then Leibniz' monadic metaphysics does not support emergence. Leibniz often states that derivative force does not "contain anything more positive than that which it modifies [i.e., primitive force]" (AG 180), and that bodies are a mere aggregate of real unities; hence, it is hard to justify any interpretation that would regard the phenomena of monads (i.e., derivative force, bodies, space and time) as somehow novel or fundamental features of his system that surface at a higher ontological level than the monads.[8]

---

[8] Nevertheless, in the late correspondence with Des Bosses, Leibniz' metaphysical musings that relate to the "substantial chain" raise the intriguing possibility that he contemplated a genuine emergent property. Although the extent to which Leibniz seriously entertains the substantial chain is open to debate—some (Garber 2009, 380–382) offer a mild defense, but others are much more skeptical (Rutherford 1995, 276–281; Hartz 2007, 107–108)—this new metaphysical item would seem to function as a non-ideal link among monads. A substantial chain is defined as "something substantial which is the subject of [the monads'] common predicates and modifications, that is, the subject of the predicates and modifications *joining* them together" (AG 203; emphasis added). Like the relationship between monads and matter, "monads are not really ingredients of this thing [substantial chain] which is added [to the monads], but requisites for it" (AG 198). Overall, the substantial chain appears responsible for bringing about real extension in bodies, i.e., not merely phenomenal, for he argues that if these unifying substances were "supernaturally to cease, the extension which belongs to [a] body will also cease", and "[w]hat will remain is only a phenomenal extension, grounded in the monads" (AG 198). Leibniz also explains that "[r]eal continuity can arise only from a substantial chain"; thus, trying to obtain bodily extension without a substantial chain is like claiming that "a continuum [can] arise from points, which, it is agreed, is absurd" (AG 203). The substantial chain, accordingly, is tantamount to real continuity or extension in corporeal substance, and, since there is apparently nothing like real continuity or extension in the monads

To summarize, the invariantist interpretation that we have explored in this chapter would seem to be consistent with a form of supervenience, but it probably falls short of emergence. This conclusion, furthermore, challenges any ontological appraisal of Leibniz' theory of matter and motion that depends solely on the conceptual resources of classical mechanics and classical spacetime constructions (including the AS strategy). In short, a supervenience relationship between a fundamental level and a secondary material level plays no role in the type of impact mechanics that commentators often ascribe to Leibniz' EH, i.e., where the bodies that conserve mv^2 are often envisioned as like billiard balls on a flat plane. Classical mechanics is all secondary, material level phenomena for physical interactions of that kind—and this demonstrates, once again, that the standard Newtonian-inspired physics of bodily interactions is not the proper framework by which to interpret Leibniz' ontology of matter, motion, and force.

Lastly, how does Leibniz' instrumentalist or pragmatic account of truth, first examined in §4.1.2, factor into this story? The example of a rainbow can be useful in this regard (as in, G II 436). The cause of our perception of a rainbow, rain drops and light, are real entities that exist in world, of course, but it is also true that the properties that we ascribe to rainbows are utterly different (e.g., the rainbow's color, continuity, position in space, etc.). Because the observed phenomena do not accurately represent the true nature of the grounding entities and processes, one might conclude that there exists a certain amount of latitude in explaining the behavior of that observed phenomena. That is, Leibniz may have reasoned that, since none of the geometrical hypotheses that treat material level bodily phenomena actually correspond to the true ontology at the foundational level of monads, it follows that an array of different explanatory hypotheses are consistent with the phenomena at the material level, and thus the only criterion for deciding among these competing hypotheses must be their simplicity from a particular standpoint (e.g., Ptolemaic in spherical geometry, etc.). Put differently, if the secondary level of extended bodies is not the true ontology, but is closer to the phenomena of rainbows, then truth cannot be assigned to these hypotheses in any straightforward fashion, and so a different means of evaluation must be procured, such as instrumentalism.

4.2 The Mystery of Monadic Situation

With the force-based, invariantist interpretation of Leibniz' system now developed, it is imperative to secure a more detailed account of the relationship between the monadic realm and the observable world of bodily phenomena, both in terms of textual support and with regard to specific metaphysical problems. Regarding the latter, one of the most perplexing, long-standing puzzles in Leibniz' late metaphysics concerns the spatial status of the monads: as previously discussed,

---

that are the requisites of these chains, the substantial chain could thus be judged to qualify as a full-blown emergent property.

monads bring about extended matter, and hence space, but are not themselves in space or spatially related to one another. This quandary has prompted some commentators to deny the spatiality or extension of Leibnizian bodies altogether, while simultaneously rejecting a purely idealist or Berkeley-style reading that treats bodies as entirely mental items: we will call this the "non-spatiality" reading of Leibnizian matter.[9] Other commentators, in contrast, have nonetheless strived to uphold extension as a real bodily feature in addressing the perplexing difficulties associated with the spatiality of monads. Yet, how can a commitment to the view that Leibnizian bodies exist in the external world and are really extended in length, breadth, and width (although perhaps not identical to continuous geometrical extension)—a view we will dub the "real extension" hypothesis—reconcile the non-spatiality of monads with the spatially extended bodies that arise from monads? In what follows, a defense of the real extension hypothesis will be offered, although the non-spatiality interpretation would seem to better complement the invariantist conception explored in §4.1.

4.2.1   Monads and Spatiality

In his late metaphysics, Leibniz holds that monads are without parts, non-extended, and form composites or aggregates, which are merely collections of monads, i.e., bodies ("bodies are only aggregates", AG 319), and possess merely an ideal unity (G II 256). Yet, Leibniz also insists that bodies are the results of monads: "properly speaking, matter is not composed of constitutive unities [monads], but results from them" (AG 179). As we have seen, despite the fact that extended matter results from monads, Leibniz repeatedly denies that monads are spatial: e.g., "there is no absolute or spatial nearness or distance between monads" (LDB 255; June 16, 1712), and, "monads in themselves do not even have situation with respect to each other—at least one that is real, which extends beyond the order of phenomena" (LDB 241–243; May 26, 1712). Yet, Leibniz also insists that monads retain a sort of derived position within matter: "although monads are not extended, they nevertheless have a certain ordered relation of coexistence with others, namely, through the machine which they control" (Lm 531).[10]

---

[9] As used with reference to monads (or God), "spatiality" concerns the relationship between a monad's (God's) being/substance and space; thus, the non-spatiality of monads means that their being/substance is not situated in space, although their actions/operation can be in space (and the same for God). As used with respect to bodies, spatiality refers to their extension in length, breadth, and width; hence, to declare that bodies are non-spatial means that bodies are not really extended in the external world, but only appear extended.

[10] While the topic of our investigation concerns the question of the spatial situation of Leibniz' monads in his later metaphysics, roughly from the late 1690s onward, it is worth noting that Leibniz' earlier work seems to support the same non-spatial status for souls/minds as one finds in the later output. In a tract from 1668 to 1670, he writes (in Cartesian vein) that "[w]hatever is not a body is not in space; for to be in space is the definition of a body" (Lm 113). The same outlook is likewise in evidence in the middle years, 1680s and 1690s: in a work from 1695, Leibniz claims

Leibniz' puzzling claim, that monads have a certain type of situation in extension, has prompted various interpretations. One possibility is to lean heavily on the mind-based aspects of Leibniz' theory, as argued in Futch (2008):

> The solution...is to see Leibniz as assigning a monad the position of its body considered representationally, not realistically as an aggregate....But it is the body as represented, as an intentional object, that confers on its representing monad a spatial position. (Futch 2008, 159–160)

Variants on this phenomenalist solution to the problem of monadic situation can be found in, among others, Rutherford (1995, 192) and Adams (1983, 242), with the emphasis placed on linking the "derived position" of monads in bodies, as we will call it, with the monad's own intentional, mind-dependent states. This interpretation is thus in accordance with the ideal status that Leibniz attributes to the continuous and holistic notions, space and time: "For space is something continuous, but ideal, whereas mass is discrete, indeed an actual multiplicity, or a being by aggregation, but one from infinite unities [monads]. In actual things, simples are prior to aggregates; in ideals things, the whole is prior to the part" (LDB 141; September 6, 1709). Yet, while there is much merit to holding that monads have only a derived position in space, these types of responses offer few details, and seem better suited to a fully idealist/immaterialist reading of Leibnizian matter. In addition, a realist about Leibnizian extension is left with little guidance in explaining how this hypothesis fits into their interpretation, i.e., the real extension hypothesis.[11]

In contrast to these cognitive-centered interpretations of the monadic situation puzzle, Daniel Garber has tried to rescue something like the view that monads possess a primary, as opposed to derivative, spatiality by recourse to several passages from the Des Bosses correspondence: first (July 21, 1707), "a simple substance, even though it does not have extension in itself, nonetheless has position, which is the foundation of extension" (LDB 99); and, second (April 30, 1709), "extension indeed arises from situation, but it adds continuity to situation. Points have situation, but they neither have nor compose continuity, and they cannot subsist by themselves" (LDB 125). Garber argues that these passages signal a transition in Leibniz' thinking from a view that bases extension on impenetrability and resistance, in the earlier metaphysics, to a new conception that utilizes monadic position

---

that "[m]inds thus have special laws that place them beyond the revolutions of matter" (Lm 455), i.e., after explaining that his pre-established harmony thesis forbids souls from "disturbing the laws" of matter, he nevertheless concludes that "[t]his makes it clear how the souls has its seat in the body by an immediate presence" (Lm 458).

[11] In Cover and Hartz (1994), the monadic situation puzzle is presented in the guise of a circularity argument: "having monads with spatial position is an essential part of the story about what it takes to have an aggregate, but having an aggregate with spatial location is an essential part of the story about what it takes to have spatially located monads" (308). This criticism would seem to be apt as regards Adams' account of aggregation (1983, 1994), where a body is reckoned to be an aggregate of "the substances whose positions are within some continuous three-dimensional portion of space", and hence "[t]his spatial togetherness is a necessary condition for any corporeal aggregation" (1994, 248–249). Yet, as Cover and Hartz note (1994, 308), if monads have no spatial position with respect to one another, then how can they partake in the "spatial togetherness" required for an aggregate?

as the source of material extension, and which is coupled with the phenomenal/ideal perception of mathematical extension examined previously:

> The position or situation of an infinity of monads now replaces the impenetrability and resistance of the earlier corporeal substance view,..... In this way we can hold that extension arises from situation. But the infinity of monads situated with respect to one another is discrete, and not continuous, of course. In imposing a full-blown Euclidean geometrical structure onto the world of situated monads, we are adding continuity. (Garber 2009, 361–362)

Consequently, it would seem to follow that monads exist in a sort of discrete ur-space, with a discrete distance among monads rather than a continuous Euclidean distance, although the contribution of the mind is responsible for our perceptions of a continuous Euclidean space.

There are numerous difficulties with this view, however. First, as Cover and Hartz argue (1994, 300), the 1707 quotation (LDB 99) that Garber employs as the basis of his interpretation only discusses extension, and not space, and so the position of the simple substance mentioned in the quote is likely a reference to its position in a body's extension, the latter comprising, of course, a well-founded phenomenon. Therefore, Leibniz' analysis in the 1707 passage is perfectly in keeping with his earlier explanations to De Volder that deem monads as having a derived spatiality in bodily extension (such as Lm 531, quoted above). Second, if Garber's analysis is correct, and monads have position in a discrete ur-space, then monads would seem to reside in the points of that discrete space (since spatial points have situation in that discrete space, and are the basis upon which Euclidean extension is phenomenally imposed). But, in the 1709 letter to Des Bosses cited by Garber, Leibniz rejects the view that souls, which are often associated with monads (e.g., G IV 512–513), are in points: "I do not think it appropriate to regard souls as though in points" (LDB 125). In response, Garber could claim that Leibniz is here rejecting the placement of souls in the points of Euclidean space, but is instead advocating that they are situated in the points of a discrete space—yet, as a counter-reply, it is not clear that a non-dimensional spatial point in a discrete space really differs at all from a non-dimensional point in a continuous space; nor is there any textual evidence to back up Leibniz' use of any such distinction. More importantly, not only does Leibniz specifically rebuff the idea that monads are in space in the subsequent correspondence with Des Bosses (see the various 1712 entries cited above), but he explains at great length that any assignment of spatiality to monads—nearness, distance, or that they are in points—is purely fictional and misguided:

> [T]here is no absolute or spatial nearness or distance between monads. To say that they are crowded together in a point or disseminated in space is to employ certain fictions of our mind when we willingly seek to imagine things that can only be understood. (LDB 255; to Des Bosses, 16 June, 1712; also, G III 623)

In short, the generality of Leibniz' argument would appear to cover all cases of the assignment of spatiality to monads, whether discrete or continuous, and thus a position or distance among monads in a discrete space would likely run afoul of this critique as well.

But the question remains, if monads are non-spatial, as the evidence of the texts indicates, then how do aggregates (i.e., bodies), which result from monads, acquire spatiality? One possibility is to simply deny that aggregates are spatial, rather, aggregate spatiality is a further contribution of the mind. This interpretational strategy does not lead to Berkeleyan-style idealism, argues Hartz, since an aggregate "is real, active, and has force" (Hartz 2007, 133). The reality posited to the monads in this reconstruction of Leibniz' system hence relates in some manner to force, which is physical (contra idealism). A somewhat different realist strategy that also denies the real extension of bodies can be found in Rutherford (1990), an approach that places the emphasis on the constitutive relationship between monads and aggregates. While aggregates are "necessarily mind-dependent", Rutherford adds that "[i]t does not follow from this, however, that aggregates are nothing real; on the contrary, Leibniz maintains that in terms of their reality aggregates are to be identified with the plurality of things from which they result [monads]" (1990, 20). Put simply, the fact that monads are constitutive of matter and bodies, a point that Leibniz consistently invokes, is hard to square with idealism: if bodies are merely mental content, then why demand that "an aggregate is nothing other than all those things taken at the same time from which it results" (G II 256)?

Yet, for those who embrace the real extension hypothesis, Hartz and Rutherford's interpretations are unacceptable. For these realists, an interpretation of Leibniz' system must uphold the real extension of bodies, even if the continuous Euclidean extension by which we perceive and understand bodies is an ideal contribution of the mind. Given this presupposition, Garber's thesis that (secondary) matter is really discrete and non-continuous, despite our perceptions that impose a continuous structure, is a more plausible method of preserving real bodily extension. The trick, consequently, would then be to preserve something like Garber's notion of a discrete extended material world alongside Cover and Hartz' persuasive denial of monadic spatiality—a very tall order indeed. In what follows, we will examine important clues that can assist in developing an account of Leibnizian bodies consistent with the real extension hypothesis, an aspect of Leibniz' metaphysics that, moreover, has seldom received the attention it deserves.

4.2.2 Monads and the Extension of Power

This section develops an interpretation of Leibniz' later metaphysics that accepts that bodies are really extended, although the form of that extension is not identical with geometrical/mathematical extension. It should be noted, however, that the interpretation offered in this section is only one possible strategy for upholding the real extension of Leibnizian bodies, and hence its success or failure does not in itself affect the relevance of the key doctrine that we shall introduce, the extension of power, for addressing the monadic situation problem. In contrast, the extension of power doctrine probably offers little advantage for those realist interpretations that deny the spatiality of both monads and bodies, i.e., the non-spatiality hypothesis, although it might be useful in understanding the historical and conceptual backdrop to Leibniz' comments on the non-spatiality of monads.

Important clues as to what exactly may be driving Leibniz' puzzling conception of the non-spatiality of monads can be found in many late period works, including a discussion in the *New Essays* on the ways that a being can be related to place or space:

> The Scholastics have three sorts of *ubeity,* or ways of being somewhere. The first is called *circumscriptive.* It is attributed to bodies in space which are in it point for point, so that measuring them depends on being able to specify points in the located thing corresponding to points in space. The second is the *definitive.* In this case, one can "define"—i.e. determine—that the located thing lies within a given space without being able to specify exact points or places which it occupies exclusively. That is how some people have thought that the soul is in the body, because they have not thought it possible to specify an exact point such that the soul or something pertaining to it is there and at no other point. Many competent people still take that view....What should be said about angels is, I believe, about the same as what is said about souls. The great Thomas Aquinas believed that an angel can be in a place only through its operations [upon what is there], which on my theory are not immediate and are just a matter of the pre-established harmony. The third kind of ubeity is *repletive.* God is said to have it, because he fills the entire universe in a more perfect way than minds fill bodies, for he operates immediately on all created things, continually producing them, whereas finite minds cannot immediately influence or operate upon them. (NE II.xxiii.21)

Many aspects of this discussion will be taken up in Chaps. 9 and 10, but there are a few points that can be examined directly. Whereas circumscriptive ubeity maps bodies to space over an extended region in a point by point manner, and definitive ubeity only links a spiritual being to a specific place or point within that region, Leibniz opts for repletive ubeity, wherein God "operates immediately" by continually producing things that exist in space. Much in this discussion, as the context makes clear, concerns finite souls and angels and how they relate to material bodies, whereupon Leibniz worries that the definitive account entails that souls can act immediately upon the things in space, with "immediately" pertaining to the soul's acting directly upon things, presumably due to the fact that the soul is co-located with the body in space (as will be discussed below). In contrast, Leibniz prefers a view where "finite minds cannot immediately influence or operate upon" bodies, and he offers his theory of pre-established harmony as an instance of this better strategy.

In the particular correspondence with Des Bosses that we have often explored (April 30, 1709), many of these issues resurface in the context of material extension and souls/monads:

> Nevertheless, I do not think it appropriate to regard souls as though in points. Perhaps someone might say that souls are not in place but through operation, speaking here according to the old system of influx; or rather, according to the new system of preestablished harmony, that they are in place through correspondence, and that in this way they are in the whole organic body that they animate. On the other hand, I do not deny a certain real metaphysical union between the soul and an organic body...according to which it can be said that the soul is truly in the body....You realize, though, that until now I have been speaking here not of the union of an entelechy or active principle with primary matter or passive power, but the union of the soul or of the monad itself (which results from both principles) with mass, or with other monads. (LDB 123–127)

Once again, Leibniz offers his notion of pre-established harmony as preferable to the view that souls are "in place but through operation", which he equates with the "old system of influx". What is important, in these last few quoted passages, is that Leibniz does not openly reject the operation of monads, i.e., that a being can be in space only through its operation, a doctrine also known as "extension of power" (hereafter, EP)[12]; rather, as is more clearly stated in the prior citation from the *New Essays*, what Leibniz rejects is the *immediate* operation of soul on body, which he associates in the Des Bosses letter with the system of physical influx. As is well-known (and previously discussed), a basic principle of Leibniz' philosophy is the denial that substances can causally interact: e.g., "[s]trictly speaking, one can say that no created substance exerts a metaphysical action or influx on any other thing" (AG 33).

To summarize, it is for reasons relating to his denial of inter-substance or inter-monadic causation that Leibniz sides with pre-established harmony. Yet, leaving aside the inter-monadic causation issue, Leibniz' reference to the "the old system of influx" would seem to draw a close analogy between, on the one hand, the monad-matter relationship, and, on the other, the immaterial being-matter relationship in those older instances of EP. Leibniz seems willing to concede the general point that a finite entity, soul, angel, or monad, can be conceived as in place through its operations, but only on the condition that there is no influx or real causal interaction, since the influx has been replaced by the mediation of God's providence in establishing the harmony between the soul and its operations. In contrast, since God "operates immediately on all created things" by "continually producing them" (NE II. xxiii.21), EP straightforwardly applies to God. There are several other notable instances in the later Leibnizian corpus where God's immediate operation is addressed, some which were introduced in Chap. 3:

> God is not present to things by situation but by essence; his presence is manifested by his immediate operation. The presence of the soul is of quite another nature. To say that it is diffused all over the body is to make it extended and divisible. To say it is, the whole of it,

---

[12] In use during the Medieval period, the origin of the term "extension of power" is unclear, but other descriptions include "presence of power", "virtual extension", and "virtual presence". An issue pertaining to EP concerns whether the essence that operates immediately must be spatially present where it acts, as Aquinas and many others had held, or whether the essence need not be really present, as Scotus had argued (see, Grant 1981, 146–147, for a brief survey). In the 1692 correspondence with Pellisson, Leibniz defends the former thesis, stating that "everything that operates immediately in several places also is in several places by a true presence of its essence, and that the immediate operation cannot be judged to be distant from the individual that operates, since it is a manner of being of it" (A.I.vii.294; Adams 1994, 357). In these letters, Leibniz even defines EP as the essence acting non-immediately, i.e., at a distance: "A presence by power [*presence virtuelle*], as opposed to a real presence, must be without that immediate application of the essence or primitive force, and happens only by actions at a distance or by intermediate operations" (A.I.vii.249; Adams 1994, 356). Yet, he later reverses his position, and claims that an immediate operations could be at a distance: "if God should bring it about that something immediately operates at a distance, by that fact he would bring about its multipresence" (LDB 171; May 2, 1710). As will be explained shortly, and in Chap. 9, Leibniz' understanding of "presence by essence" rejects the spatial presence or situation of a being's substance. See, also, footnote 13.

in every part of the body is to make it divisible of itself. To fix it to a point, to diffuse it all over many points, are only abusive expressions, *idola tribus*. (L.III.12)

Where space is in question, we must attribute immensity to God, and this also gives parts and order to his immediate operations. He is the source of possibilities and of existents alike, the one by his essence and the other by his will. (NE II.xv.2)

In short, God is "not present to things by situation but by essence", yet "his presence is manifested by his immediate operations", i.e., his immediate operations are given "parts and order" in space even though God is not actually situated in space (and where God's "immensity", as used by Leibniz, would seem to pertains to the onto-logical dependence of matter and space on God; see, L.V.106, and §3.2). As will be addressed in Chap. 9 as well, Leibniz accepts an unorthodox reading of the Scholastic relationship between a being's essence and a being's situation in space. On Aquinas' authority (ST I.Q8 Art.3), if God's essence is situated in space, then God's substance is, to use Leibniz' phrase, "present to things by situation". Yet, as is apparent in the L.III.12 quotation above, Leibniz reasons instead that a presence by essence situates God's operations alone, but not God's substance or being; and, in the late correspondence with More, Descartes may opt for the same interpretation (see, footnote 15, and §9.3.6). Leibniz, and possibly Descartes, have thus conflated the view developed by Scotus and others, where God's essence need not be situated in space to act in space, with the Thomistic account and its terminology that does require God's essence to be situated (see footnote 12). Therefore, on Leibniz' admit-tedly unique (if not idiosyncratic) understanding of these concepts, to claim that God is present to things by essence is to claim that God's immediate operations are situated in space, but that God's substance or being is not situated in space, with God's essence serving the more general metaphysical role of grounding the possi-bility of any existing thing. This reading seems to be upheld later in the correspon-dence with Clarke, for Leibniz rejects the view that "God discerns what passes in the world by being present to the things", rather, he reasons that God discerns things "by the dependence on him of the continuation of their existence, which may be said to involve a continual production of them" (L.V.85). Accordingly, since God's immediate operation correlates with the continual production of the material world, the world's spatial order thereby situates that continual act of production. Returning to the Leibniz-Clarke passage examined above (L.III.12), Leibniz then goes on to deny that either a soul is diffused "all over a body", which doubtless equates with circumscriptive ubeity in NE II.xxiii.21, or that "the whole of it, [is] in every part of the body", i.e., holenmerism (see Chap. 2), which is consistent with, although not identical to, his account of definitive ubeity in the same *New Essays* passage (to be specific, the holenmerist, or "whole in every part", doctrine would include definitive ubeity as used by Leibniz; see, Grant 1981, 343, n.67, which also links definitive ubeity with holenmerism in the later Scholastic period).[13]

---

[13] Another idiosyncratic aspect of Leibniz' interpretation of these doctrines concerns the scope of EP. As developed by Aquinas (ST I. Q.8. Art.2), EP is linked to a non-extended (or non-dimen-sional) being, and thus both definitive and repletive ubeity, as defined by Leibniz in NE II.xxiii.21,

For understanding the vexed subject of monadic situation, the ramifications of EP are quite significant, although few commentators have ventured into this territory. Since Leibniz uses the terms "soul" and "monad" interchangeably in his late period, *if* "souls are not in place but through operation, speaking here according to the old system of influx" (LDB 125), then monads are, like God, not situated in space due to the fact that they fall under the EP doctrine as well, albeit a non-influx construal of EP. More carefully, while Leibniz does not strictly sanction EP for monads, his comments would seem to admit that the relationship between his non-spatial, non-situated monads and extended bodies is like the relationship between the non-spatial, non-situated immaterial beings and extended bodies in the older influx EP theory, but excluding the influx component of the older theory, of course. This interpretation, which would uphold the non-situated component of immaterial beings in the original influx formulation of the EP hypothesis, thereby explains why monads only have a derived position in space, i.e., through the body which they control.

In brief, one of the main arguments of this chapter is that Leibniz' puzzling reference to the derived position of monads in extended bodies is best understood as a non-influx, pre-established harmony version of EP. Contra Garber, monads are not in space *per se*, i.e., situated in space, although their operations are situated in space, just as God is not in space but God's operations are situated in space. Specifically, because Leibniz states in LDB 127 that "the union of the soul or of the monad itself…with mass, or with other monads" is the context under which he entertains the idea that monads are in place through their operation (under the influx construal), and since mass is associated with extended secondary matter or aggregates (e.g., AG 177), the monads are only in place by means of mass/secondary matter. This inference correlates perfectly with his claims concerning the derived situation of monads: "although monads are not extended, they nevertheless have a certain ordered relation of coexistence with others, namely, through the machine which they control" (Lm 531). Furthermore, the fact that monads themselves are not situated is consistent with the real extension hypothesis presented in §4.2.1, where Leibnizian bodies are really extended but perhaps lack the continuous structure of geometrical extension. Hence, while rebuffing Garber's view that monads are actually situated in space, the EP doctrine can provide support for Garber's more general notion that there are non-continuous, discrete extended bodies. To sum up, given the non-influx version of EP suggested above, the only aspect of a theory that posits real extension to bodies that must be sacrificed is the real or actual situation of monads in matter, and hence in space: like God, monads are not situated in space although

---

would meet the definition (since definitive ubeity situates a being in the non-extended points of space). Leibniz, on the other hand, equates EP with his interpretation of repletive ubeity alone, since the L.III.12 quotation specifically rules out definitive ubeity (More's holenmerism) when he concludes that "God is not present to things by situation". For these reasons, and given their general similarity, EP will be taken as synonymous with repletive ubeity (and More's nullibism) throughout the remainder of the book, i.e., that the being/substance of God/monads is not in space but the actions of God/monads are in space. Likewise, our usage of the repletive ubeity concept need not include the recreation of the world.

their operations are situated. In the next section, we will take up monadic operation in more detail, but further textual evidence will be examined first.

Besides the *New Essays* and Des Bosses correspondence, there is additional support for the above interpretation of EP in the transubstantiation debate with Pellisson in 1692, an issue that Adams addresses at length (1994, 350–358). Despite its early date, and the specific worries associated with the multi-location of the Eucharist, the Pellisson correspondence is worth quoting for the extra details it supplies:

> [I]t is by the application to several places of this [higher] principle [of action and resistance], which is nothing but the primitive force of which I have spoken, or (to speak in more ordinary terms) the particular nature of the thing, that the multipresence of a body is to be saved. It is true, however, that the substance *in concreto* is something other than the Force, for it is the subject taken with that force. Thus the subject itself is present, and its presence is real, because it emanates immediately from its essence, as God determines its application to the places....I would even say that it is not only in the Eucharist, but everywhere else, that bodies are present only by this application of the primitive force to the place; but this occurs naturally only in accordance with a certain extension, or size and shape, and in regard to a certain place, from which other bodies are excluded. (A.I.vii.249; Adams 1994, 355)

The story that Leibniz tells is that God applies primitive force to a place (or places) in order to bring about the presence of a body or substance in that place (or multiple places); i.e., the subject *in concreto* that, besides being a part of this force, has a "real" presence that emanates from its essence, with essence identified with primitive force. While this topic was first introduced in §4.1.3, and will be discussed again in the next section, how the subject comes about from primitive force, and obtains a real presence, must implicate derivative force, but the Pellisson letters leave this process unexplained. At this stage in his thinking, consequently, not only is the essence (i.e., primitive force) of a body present in the body's place, but, in fact, the same essence can be in several different places simultaneously, thereby demonstrating that the normal restrictions on location and spatiality do not apply to primitive force (with primitive force roughly equivalent to a substantial form; see, e.g., AG 162–163). Whether Leibniz continues to insist that the essence needs to be present in this manner to operate immediately seems doubtful given the evidence of later texts, it should be noted (see footnote 12 once again). Nevertheless, there are obvious similarities here with Leibniz' later claim, in L.III.12, that God's essence is present to things but only his immediate operations are situated in space—and this lends support to the conclusion that his later monadic metaphysics is roughly analogous to his conception of God's EP, as argued above.

Confirmation of this reading of the evidence can likewise draw upon Adam's insightful commentary on the Pellisson correspondence, although it is interesting to note that he overlooks the relevance of these issues to the problem of monadic situation. After observing that, in these texts from the 1690s, Leibniz had "already rejected the system of influence in favor of that of pre-established harmony, but in which he nonetheless ascribed to primitive forces (doubtless including souls) a local presence by immediate operation", Adams concludes that, by the later Des Bosses correspondence, "being in a place by (immediate) operation, as affirmed in the 1690s, is reduced to being in a place by correspondence" (Adams 1994, 357). Not only is this inference justified, but, as argued above, it holds the key to understanding the monadic situation puzzle. However, rather than apply these findings to his own discussion of monadic situation (1994, 248–255), Adams offers a problematic "spatial togetherness" criterion instead (see footnote 11).

There is also evidence to support the view that monadic operation, or a surrogate notion, is a factor in other well-known Leibnizian tracts. In these works, various enigmatic discussions that pertain to the actions of monads, or monadic change, assume the role that the monadic operation idea had played under the influx theory. For instance, in the June 20, 1703 letter to De Volder, which contains his oft cited endorsement of the derived position of monads (within extended matter), he states:

> I had said that extension is the order of possible coexistents and that time is the order of possible inconsistents. If this is so, you say you wonder how time enters into all things, spiritual as well as corporeal, while extension enters only into corporeal things. I reply that the relations are the same in the one case as in the other, for every change, spiritual as well as material, has its own place, so to speak, in the order of time, as well as its own location in the order of coexistents, or in space. For although monads are not extended, they nevertheless have a certain ordered relation of coexistence with others, namely, through the machine which they control. (Lm 531)

That "every *change,* spiritual as well as material" has a situation in space is quite significant, for what can spiritual change mean, in the *context* of a discussion of monads, if not monadic change or monadic activity? An objection that might be raised is that spiritual change refers to God's activity in this excerpt. Nevertheless, there is another piece of evidence, from 1714, that more directly cites monadic change:

> There are simple substances everywhere, actually separated from one another by their own actions, which continually change their relations; and each distinct simple substance or monad, which makes up the center of a composite substance (an animal, for example) and is the principle of its unity, is surrounded by a *mass* composed of an infinity of other monads, which constitutes the *body belonging to* this central monad, through whose properties the monad represents the things outside it, similarly to the way a center does. (AG 207).

To insist that monads are "actually separated from one another by their own actions" provides further support for the spatiality of monadic activity or change, especially when it is recalled that by this date, 1714, Leibniz has repeatedly claimed that monads themselves are not in space. Put differently, how can the non-situated monads be *actually separated* by their own actions?: the answer, of course, is that he is still wedded, to some degree, to the extension of powers doctrine, EP, although the powers assumed in his theory now refer to monadic activity. One might reply that the term "separation" employed in this last quote may signify a mere difference in internal properties, with no spatial connotations intended. Yet, in addition to the context, which implies a straightforward spatial interpretation, such a reading is difficult to justify given the many other spatial terms utilized throughout the discussion—e.g., "everywhere", "center", "surrounded", "outside"—all of which strongly suggests that "separation" is meant in its normal spatial sense.

4.2.3   Monadic Activity and Derivative Force

At this point in our analysis, it is worthwhile to more closely examine the analogy between God's operation and monadic operation. As we have seen, God's continuous production of the world situates that act in space, even though God's substance or being is not in space. For the advocates of a realistic account of Leibnizian extended bodies, such as Garber's discrete body hypothesis, it would seem that this analogy should serve as the foundation of an EP-centered account of the monadic-matter relationship: monads are not in space but their operations are situated in space via extended matter. But, turning to the operation of monads, since Leibniz accepts the pre-established harmony view, whereby everything that happens is internal to a monad, it might appear that there is little similarity between God's operation, which creates matter, and monadic operation, which only involves the internal properties of monads. As noted above, Leibniz brings up the older influx-based notion of the operation of monads when discussing "the union of the soul or of the monad itself...with mass, or with other monads" (LDB 127), and the Leibnizian concept that fits this aspect of Leibniz' theory is aggregation. But, while bodies depend on the aggregation process, aggregation is obviously not an internal aspect of monads. Nevertheless, besides aggregation, there is one aspect of the story of how extended bodies result from monads that would appear comparable to an internal monadic operation or activity, namely, the role of force, both primitive and derivative. In the remainder of this chapter, we will strive to elucidate how primitive and derivative force might function as a monadic operation within the context of the rise of extended matter, and thereby provide a means for understanding how a monad's operation might be comparable to God's (immediate) operation. Overall, this portion of our investigation is quite speculative, largely due to the difficult nature of the relationship between primitive and derivative force, an aspect of Leibniz' system that is itself quite tentative and seems to have been constantly evolving. Consequently, what follows is merely a suggestion as to which elements in Leibniz' system might correlate with a monad's activity or operation, although this discussion can also serve the subsidiary goal of shedding more light on the important metaphysical function of the primitive/derivative force distinction.

In several later works, including the correspondence with De Volder, there are tantalizing hints that incorporate the function of derivative force in Leibniz' aggregation hypothesis:

> [T]he nature which is supposed to be diffused, repeated, continued [i.e., to form extension of bodies], is that which constitutes the physical body; it cannot be found in anything but the principle of acting and being acted upon, since the phenomena provide us with nothing else....But when force is taken for the principle of action and passion, and is therefore something modified through derivative forces, that is, something modified through that which is momentary in action, you can understand well enough from what has been said that this principle is bound up with the very notion of extension,....[U]nless there is some active principle in us, there cannot be derivative forces and actions in us, since everything accidental or changeable ought to be a modification of something essential or perpetual, nor can it contain anything more positive than that which it modifies, since every modification is only

a limitation, shape a limitation of that which is varied, and derivative force a limitation of that which brings about the variation. (AG 179–180; June 30, 1704)

Primitive force, as the principle of action and passion, is "modified through derivative force" in such a way that derivative force is "that which is momentary in action"; and, since primitive force is "essential or perpetual", derivative force is a mere limitation, adding nothing positive, as shape is a "limitation of that which is varied" (see also §4.1.2 and §4.1.3). For understanding how monadic operation or activity might relate to the formation of extended matter, the references to primitive force as the principle of action, and derivative force as what is momentary in action, is crucial, for it brings together monadic action or operation and primitive/derivative force, the latter implicated in the account of extended matter ("this principle [primitive/derivative force] is bound up with the very notion of extension").

Accordingly, Leibniz' somewhat perplexing idea of the derived position of monads in extended matter, first explored in section §4.2.1, gains a great deal more clarity when the role of derivative force is incorporated into the picture: "For although monads are not extended, they nevertheless have a certain ordered relation of coexistence with others, namely, through the machine which they control" (Lm 531). While monads are not spatial, derivative force is coupled to the phenomena of extended bodies, which are aggregates of monads, or secondary matter: "the derivative force of being acted upon later shows itself to different degrees in secondary matter" (AG 120). So, monads are not situated in space, but their effects or results are spatial via the extended bodies that come about from derivative force. This last inference would seem to explain Leibniz' statement that "I relegate derivative forces to the phenomena" (AG 181), a claim which has puzzled commentators (e.g., Garber 2009, 363), but which makes perfect sense given that extended bodies, i.e., well-founded phenomena, are ultimately manifestations or instantiations of primitive force (via derivative force). Derivative force, in turn, is then associated with the diffusion process: "[T]he nature which is supposed to be diffused, repeated, continued, is that which constitutes the physical body; it cannot be found in anything but the principle of acting and being acted upon [i.e., primitive force], since the phenomena [i.e., a determinate value of primitive force=derivative force] provide us with nothing else" (AG 179).[14] Consequently, the means by which the non-spatial monads obtain spatiality involves derivative force, since extended secondary matter is ultimately linked to derivative force. To return to the God-monad analogy, a parallel case can thus be made since both God's operation and a monad's operation or activity are associated with the rise of extended matter, although in different ways.

---

[14] On diffusion, Leibniz states: "[E]xtension is only an abstract thing, and…it requires something extended. It needs a subject; it is something relative to that subject, like duration. It even presupposes something prior to it in this subject, some quality, some attribute, some nature in this subject, which is extended, is expanded with the subject, and is continued. Extension is the diffusion of this quality or nature. For example, in milk there is an extension or diffusion of whiteness, in a diamond, an extension or diffusion of hardness, and in body in general, an extension or diffusion of antitypy or materiality. In this way you see there is something prior to extension in bodies" (AG 261).

Unlike God's operation, a monad's activity does not involve the creation of an entirely new entity, but this is consistent with a non-influx version of monadic operation conceived along the lines of an internal feature of monads—i.e., derivative force as a determinate value of the primitive force internal to each monad.[15]

4.3 Final Synthesis

By way of conclusion, it would be useful to bring together the various arguments and issues that have been explored in this chapter. As argued in §4.1.1, Leibniz' analysis of motion defies relationism in various ways, especially in the context of his conservation law and collision hypotheses, and thus an alternative interpretation is required that avoids the pitfalls of the absolute speed hypothesis and other constructions predicated on notions appropriated from classical physics. In §4.1.3 and §4.1.4, an invariantist conception of Leibnizian matter and motion was developed to meet this challenge, an approach that envisages the realm of well-founded phenomena as perspectival in nature, although arranged in a hierarchical series on the basis of intelligibility, and grounded in an invariant, non-idealist aspect of reality, namely, force. Much like the dual aspect theory that is often attributed to Spinoza's metaphysics, the invariantist interpretation allows different assignments of the geometrical features and motions of bodies, such as the Ptolemaic or Copernican hypotheses, without singling out any one of them as the correct view and the others as false. In short, while there are many different hypotheses that are equivalent in this manner, despite their differences in intelligibility, they are all manifestations of monadic force, which is the unchanging and non-perspectival foundation of Leibniz' physical world.

In order to fill in the details of the invariantist interpretation explored in the first half of the chapter, the remaining sections of the chapter explored the relationship between the non-spatial monads, force, and matter against the backdrop of the puz-

---

[15] Although it will be examined in more detail in Chap. 9, it is also worthwhile briefly comparing Leibniz and Descartes on EP, for Descartes had sanctioned that doctrine with respect to God, angels, and minds (souls). In his late correspondence with More, Descartes asserts that "[f]or my part, in God and angels and in our mind I understand there to be no extension of substance, but only extension of power. An angel can exercise power now on a greater and now on a lesser part of corporeal substance; but if there were no bodies, I could not conceive of any space with which an angel or God would be co-extensive" (CSMK 372–373). There is a certain similarity here in that the power of Descartes' spiritual beings and the primitive force of Leibniz' monads still remain even if there are, respectively, no actual Cartesian bodies or no limitation imposed on the primitive force, i.e., derivative force as a particular value of primitive force. On the other hand, while Cartesian matter can exist apart from a spiritual beings' EP, Leibniz' force-based conception of matter, especially secondary matter, denies this possibility. Leibniz was familiar with various works of More (e.g., *The Immortality of the Soul*, see, A.VI.iv.1678–1680), as well as the Descartes-More correspondence (see Lm 342). Nevertheless, given Leibniz' knowledge of Scholastic metaphysics, he was almost certainly well acquainted with the EP doctrine and its alternatives apart from the Descartes-More correspondence.

zle of monadic situation. With respect to this issue, the extension of power doctrine, EP, provides one of the most plausible explanations for the conjunction of Leibniz' claims about the non-spatiality of monads with his additional stipulation that monads have a derived position in bodies: e.g., the conjunction of (i), "monads in themselves do not even have situation with respect to each other—at least one that is real, which extends beyond the order of phenomena" (LDB 241–243); and (ii), "for every change, spiritual as well as material, has its own place, so to speak, in the order of time, as well as its own location in the order of coexistents, or in space. For although monads are not extended, they nevertheless have a certain ordered relation of coexistence with others, namely, through the machine which they control" (Lm 531). For those realists who merely accept the existence of an external world apart from the mind (i.e., who reject the Berkeleyan fully idealist interpretation but deny really extended bodies), the extension of power doctrine explicates the historical and metaphysical background to these issues, but probably little else, since a world without extension need not worry about the relationship between non-spatial monads and extended bodies. In fact, the invariantist conception of matter and motion outlined in this chapter, where force is real and all of the geometrical aspects of bodies are ideal and perspectival, is already straightforwardly compatible with those interpretations that deny the reality of extension. On the other hand, for those commentators who accept the reality of extension (and even granting the ideality of geometrical extension), the Leibnizian version of EP developed in this chapter presents a coherent and historically informed explanation for both the non-spatiality of monads and their derived situation in matter. While monads are not situated in space, their activity is situated in secondary matter, and hence monads possess a derivative spatiality via their activity—and this conclusion is in perfect accordance with the basic idea behind the extension of power doctrine. Of course, there is only indirect evidence in support of an EP conception of monadic activity, although, as argued above, that evidence is both mutually consistent and, in some discussions, compelling: e.g., "[p]erhaps someone might say that souls are not in place but through operation, speaking here according to the old system of influx" (LDB 123). That Leibniz does not openly reject this interpretation is important, and, since his main objections concern an influx of powers among substances, this suggests that a non-influx variant of EP, as a sort of pre-established harmony version of EP, is compatible with his monadic metaphysics.

Part II
Third-Way Spatial Ontologies: Past and Present

Chapter 5
From Property to Structure: Exploring Contemporary Third-Way Conceptions of the Ontology of Space

In the search for alternative conceptions of space that transcend the traditional substantivalist and relationist dichotomy, a host of different strategies stand out in the recent literature: the positivist-inspired definitional approach favored by Stein and DiSalle; the various formulations of a property theory defended by Sklar and Teller, to name only a few; and the structural realist conception of space put forward by Dorato, Rickles, and several others. While some of these third-way approaches have been previously discussed, this chapter will provide an examination of their content and overall viability in greater detail, in particular, the property theory, §5.1, and the structural realist conceptions of space (spacetime), §5.2. In keeping with the stated goals of our inquiry, the emphasis will be placed on exploring aspects of these alternative hypotheses that both correlate with the key elements of Newton and Leibniz' views as well as improve upon the standard substantivalist and relationist categories that they seek to replace. However, the relationship between these contemporary alternative spatial ontologies and seventeenth century spatial ontologies will not be taken up until later in Part II (Chap. 7, in particular) and Part III.

5.1 The Property Theory of Space

"A mongrel view" is how John Earman described the property view of spacetime, although he was quick to add "but like many cross breeds, this one displays a hardiness" (1989, 14). The modern property view of spacetime, as its name implies, judges space and time to be a property of a physical entity, whether that entity is defined as a macroscopic material object or a physical field. Unlike substantivalism, which treats spacetime as a substance, or relationism, which deems spacetime to be the mere relations among material substances, the property view attempts to steer a unique path by selectively endorsing various components from both of these more traditional ontological camps. The advantage of the property view is that it strives to avoid the deficiencies of both substantivalism and relationism while simultaneously

© Springer International Publishing Switzerland 2016
E. Slowik, *The Deep Metaphysics of Space*, European Studies in Philosophy of Science, DOI 10.1007/978-3-319-44868-8_5

endorsing their best features. On the one hand, the property theorist benefits from accepting the relationist's quite plausible contention (against substantivalism) that space and time cannot sustain a separate, independent existence in the absence of physical entities and processes. Yet, on the other hand, it sides with the substantivalist's equally persuasive belief (contra relationism) that space and time are more than mere material relations. For example, according to the spatial location version of the property theory, P(loc), it makes perfect sense to posit a monadic location property to a single object in an otherwise vacant universe, a conclusion that the relationist would deny since location is a relation among bodies. Likewise, the P(O-dep) and P(TL-dep) conceptions of a property theory reject strict eliminative relationism, and so it makes sense to discuss the state of motion of that lone object in an empty universe; see §1.2 for the definition of these property theory variants, but, as noted in that section, references to P(O-dep) will include the similar P(TL-dep) version as well.

To outline this section: after discussing an early variant of the property theory in §5.1.1, important recent entries will be discussed in §5.1.2, followed by an important dichotomy among property theories in §5.1.3. Finally, §5.1.4 will summarize the strengths and deficiencies of the property theory, and set the stage for the entry of structural realism into our investigation in §5.2.

5.1.1   An Historical Precedent: Philoponus

As is the case with most modern Western conceptions of space, the basis of the property theory can be traced back to Aristotle and a long tradition of Aristotelian/Scholastic conjecture and commentary. Indeed, of the many metaphysical puzzles that Aristotle's *Physics* bequeathed to succeeding generations of natural philosophers, few can claim to have imposed more difficulties than his somewhat obscure arguments for the concept of "place" and his subsequent rejection of a void. Aristotle rejects the notion that the place of a body is the three-dimensional extension within, and bounded by, the body's two-dimensional surface. Instead, he opts for a definition that utilizes only the surface of the containing bodies in the plenum (i.e., matter-filled universe) that enclose the contained body, a strategy that thereby delimits the place of the contained body without directly referring to its surface; see, §1.3 and §1.4 on more of this pre-seventeenth century history. The opacity of Aristotle's arguments against three-dimensional extended place have raised a great deal of speculation among commentators, but a brief summary of their probable intention can be provided.[1] First, since bodies are commonly defined as having length, breadth, and width (or three-dimensionality; W 204b.20, 209a.5), Aristotle seems to worry that if place (or void) were to be ascribed three-dimensionality, then it would

---

[1] For a more thorough discussion of Aristotle's complex concepts and arguments concerning place and void, see Morison (2002) and Lang (1998). For a penetrating analysis of the later development and reception of these ideas, see Sorabji (1988) and, of course, Grant (1981).

qualify as body; "but it is impossible for place to be body, because in that case there would be two coinciding bodies" (W 204a.4-7). A tacit presupposition of this line of reasoning lies at the very core of Aristotelian ontology, namely, the substance/property (substance/accident) dichotomy. In the *Categories*, Aristotle asserts that properties are "in a subject [i.e., substance]", and proceeds to list a variety of properties so contained, among them quality, quantity, and even place, which is defined as "one of the continuous quantities" (CWA 1a.20-25, 5a.5-10). Furthermore, since dimensionality comes under the category of "quantity" (*Metaphysics*, CWA 1029a.10–19), attributing dimensionality to a place may have suggested to Aristotle that place now possesses a bodily property, which leads to the absurd consequence that both a body and its place meet the classification of body. Second, in his long series of arguments against the void, Aristotle raises the objection that if the void possessed dimensional extension, then the dimensionality of a body, say, a wooden cube, and the dimensionality of the void space that the cube enters, would both exactly coincide. This leads to the troubling question: "What will be the difference between the body of the cube [which is defined by its dimensional extension] and the void and place which are equal to it" (W 216b.6-11)? Apparently, the interpenetration of a body's dimensional extension and a void or place of equal dimensional extension thereby threatens the very identity of the body, or, at the least, proves that the dimensionality of the void space is a superfluous addition to an analysis of the cube's motion.

Not only did Aristotle's analysis of place and void remain a pervasive influence on the future development of the concept of space, but, as Grant notes, "his explicit identification of three-dimensional void with three-dimensional material body was destined to play a significant role in the controversy over the possible existence of a separate space in the period from the Middle Ages to the seventeenth century" (Grant 1981, 6). Nevertheless, many later philosophers were not content with Aristotle's conclusions. The work of the sixth century Neoplatonist philosopher, John Philoponus, is merely one of a multitude of critical tracts from the ancient and medieval periods that followed in the wake of Aristotle's spatial hypotheses. In commenting on the two Aristotelian arguments against void space (examined above), Philoponus argues, respectively, that "[Aristotle] is wrong to identify the [three-dimensional] void with body" (CAP 687, 30),[2] and that the interpenetration of bodily extension and void extension is neither superfluous nor impossible: "it [does] not follow that body is in body, since the void is not a body at all…, neither will the body qua extension be in another [bodily] extension: rather, qua bodily-extension it will be in place-extension" (CAP 688, 28–35). The interpenetration of bodily-extension and place-extension is thus sanctioned, an admission that amounts to a straightforward dismissal of the thematic core of Aristotle's argument against the interpenetration of these two extensions.

On the whole, Philoponus' approach to place (space) and motion can be classified as broadly non-relational in its emphasis, since he clearly rebuffs the strict

---

[2]Philoponus seems to use the terms "void" and "place", and sometimes "vacuum", somewhat interchangeably: we will use the term "space" to cover these usages.

eliminative conception of space as well as the idea that motion is a purely relative phenomenon among bodies. First, Philoponus presents several well-known criticisms (as Henry More would later raise for Descartes, see §1.3) against the hypothesis that motion is merely the reciprocal (relative) change of Aristotle's two-dimensional containing surface; e.g., given a body at rest, the resting body could still be judged to be in motion if the surrounding air (in contact with the body's surface) is constantly changing, a possibility that demonstrates the inconsistencies inherent in the Aristotelian concept of bodily motion (CAP 564, 15–25). As an alternative, Philoponus contends that place is "a certain extension in three dimensions, different from the bodies that come to be in it, bodiless in its own definition" (567, 30–35). This notion of place, he reasons, can supply the volumetric (three-dimensional) measure of body, which is a well-known aspect of bodily existence not explained by the surface area (two-dimensional) notion of place favored by Aristotle (568, 1–30). Philoponus eventually concludes that "there could be no motion in place at all if there were not such an extension [i.e., place-extension]" (689, 28–32). Finally, the thesis that place is sort of a nominalist abstraction from three-dimensional bodily extension, as his predecessor, Themistius, had held, is likewise refuted: "[that] place-extension is like bodily-extension without bodily qualities, is altogether far from the truth" (577, 10–15). The extension of body and place are consistently differentiated throughout his analysis, a position that effectively dispels any relationist reduction of place to, or abstraction from, bodily extension.

Nevertheless, Philoponus does not seem to endorse the view that space is a separate substance, a form of entity that is independent of material bodies. Like many in the ancient period, moreover, he affirms that space can never be absent of body. In the course of developing an analogy between the dependence of properties (form) on material substance, he states:

> Just as, in the case of matter, when form [properties] are destroyed, another form at once takes over [e.g., whiteness is replaced by another color property], so too in this case the exchange of bodies never leaves the space empty: simultaneously one body departs, and another rushes in instead. And thus it is never possible to find even this kind of quantity without substance....Thus, then, we can also save what seems to have been agreed upon through our habit of continually saying it: I mean, that quantity cannot subsist without substance. For the void can never exist in separation from body. (CAP 579, 10–18)

Employing Aristotle's substance/property dichotomy, Philoponus classifies the quantity of place-extension, which is three-dimensionality, as a property of substance ("quantity cannot subsist without substance"). Therefore, since this quantity must subsist in some material substance, and since bodies are continuously in motion (as well as going into, and out of, existence), he seems to conclude that place-extension must be a quantity, i.e., a property of all material substance. This is the grounds for the analogy between bodily forms and the quantity of place-extension: just as a body will continuously exhibit forms (such as a color property), despite the fact that these properties change, so will the place-extension of the fluxing plenum always "subsist" in the same total amount (volume?) of material bodies.[3] In short, Philoponus would appear to endorse a view that puts space in the class

---

[3] See Sorabji (1988, 24), who also reckons that Philoponus' *Corollaries* deem space-extension to be a quantity and attribute of corporeal substance. See also Sedley (1987, 140–154), who reaches a similar conclusion.

of bodily property (as a quantity dependent on a grounding substance), but not as a property of a single material body—rather, it is a property of all material existence.

On the whole, Philoponus' *Corollaries* puts forward what can be deemed one of the first theories of space that upholds P(loc) and P(O-dep) via corporeal being: that is, space as a property of matter, or, to be more specific, the plenum as a whole, so that every body has a monadic spatiotemporal location property, P(loc), determined from the perspective of the whole plenum. It stands in sharp contrast to the type of God-based theories of space common in the late classic (or early Medieval) period of Philoponus' day, such as the views of Simplicius (see, e.g., Samburksy 1962, 1–9). Nevertheless, the matter-based property theory proved to be a somewhat unstable construct, at least in Philoponus' hands, for he eventually moved towards a quite different formulation of space (in his later work, *Contra Proclus*). Whereas the earlier *Corollaries* carefully separates the extension of bodies and the extension of space, his later work does seem to blur this distinction in favor of a tentative super-substantivalism (i.e., where space is the only predicable substance, like Newton's *De grav*), or even Cartesianism (where space is identified with extended body). In its later incarnation, three-dimensional extension is deemed to be the "substance of body", for he reasons that if extension were a property, a body could exist without it (which, he concludes, is absurd).[4] Philoponus' uncertainty, which he shares with many contemporary spacetime thinkers, lies in the classification of the quantity of spatial extension as regards the familiar Aristotelian substance/property dichotomy. And, although a host of sixteenth century natural philosophers would be influenced by his ideas to undertake very similar theories of space—e.g., Pico della Mirandola, Telesio, Patrizi, Bruno, Campanella (see Grant 1981, 19; and Chap. 9)—most of these theorists would side with a more Neoplatonist God-based ontology of space, as would Newton.

5.1.2   Contemporary Versions of the Property Theory

Despite a growing interest in third-way alternatives to the standard substantivalist/relationist dichotomy, indifference to the property theory continues unabated in the vast majority of current explorations of spacetime ontology, where a scant line or footnote is often allotted to the property theory (e.g., Belot 2000, 576, to name a recent example within a sophisticated analysis). However, influential attempts to resurrect the property theory, or a close cousin on the relationist side of the standard dichotomy, can be discerned in the work of Sklar, Teller, and others.

For Sklar, the motivation underlying the introduction of his "absolute accelera-tion" concept can be traced to a general dissatisfaction with the forms of explana-tion that both substantivalism and relationism offer with respect to non-inertial (i.e., accelerating) motion. In substantivalist-friendly neo-Newtonian spacetimes, for instance, acceleration is an absolute quantity that it is "real and empirically measur-

---

[4] J. Philoponus, *Contra Proclus,* 424, 35–425, 14, quoted in Sorabji (1988, 29).

able" (Sklar 1974, 205), whereas absolute position and absolute velocity are not (i.e., they are not invariant under the admissible coordinate transformations of the spacetime). Sklar is sympathetic to the relationist cause, moreover, thus he wishes to procure a form of relationism that is not susceptible to the types of criticism that can be leveled at the Machian variety. For either the strict or modal spatial relationists who accept relational motion, such as Mach, motions/accelerations are relative to something, in his case, "the mass of the earth and the other celestial bodies" (SM 287), thereby implying that a single body in an otherwise empty universe could not move/accelerate. This is an outcome Sklar would like to avoid (1974, 216), since the rotation or non-rotation of a lone body seems a meaningful state-of-affairs. For both camps, substantivalist and relationist, Sklar points out that accelerating bodies are accelerating *relative* to something else, whether conceived as substantival space or a material body. As an alternative, he posits absolute acceleration, not as a two-place predicate (e.g., "x is absolutely accelerating relative to y"), but as a one-place predicate; namely, "x is absolutely accelerating", an act of predication similar to "x is red", or "x is square", (1974, 230). In effect, Sklar reckons non-inertial phenomena to be a type of internal property—a property, moreover, that a body can possess independently of any other material object, contra the relational motion hypothesis: "Absolute acceleration is property that a system has or does not have, *independently of the existence or state of anything else in the world*" (1974, 230; original italics).[5]

An unequivocal instance of a property theory of space, as opposed to an internal property of acceleration, can be found in Teller (1987), where a monadic property of "spatial location" or "spacetime point" is directly assigned to material bodies: "space-time is a physical quantity or determinable, and space-time points are specific values of that quantity. All space-time points 'exist' in whatever (perhaps not very good) sense in which all values of quantities such as mass, including unexemplified values, exist" (1987, 427). Teller is likewise upfront in claiming that his "space as quantity" theory represents a third option to traditional substantivalism and relationism (425), and, since "substantivalism must postulate space-time points and the *occurs at* relation, both of which must be taken as primitive and which are otherwise mysterious" (431–432), his property view is claimed to be "more palatable and less mysterious" (432). Finally, it is interesting to note that Teller's later thoughts on the spacetime ontology debate show a marked preference for a sophisticated modal relationism that treats inertial structure as the "structure of the systems of actual and possible relative trajectories [of bodies, particles, etc.]" (Teller 1991, 381).

Another property theory of space can be found in the work of Dieks (2001a, b). Like Teller, he openly endorses a view that treats "space and time as physical quantities that have exactly the same status as mass, charge and similar direct quantities" (2001a, 6). Unlike Teller, however, Dieks' more rigorous construction displays a strong penchant for relational quantities, and thus his theory may fit more naturally

---

[5] As will be explained below, while Sklar's maneuver has attracted attention, the vagueness of his proposal has led to various criticisms, such as Earman (1989, 127–128)—although some commentators, like Huggett, have strived to fill in the details in a way favorable to relationism (see also footnote 10 for a substantivalist rejoinder).

into a modal relationist classification. For example, in order to bring the spacetime symmetry group of Newtonian mechanics into line with its dynamical symmetry group (and thereby rid the absolute position and absolute velocity quantities intrinsic to full Newtonian spacetime), Dieks singles out the "ratios of relative distances" (among the position values of an arbitrarily chosen coordinate system) as the invariant spatial property of the theory (where the Galilean transformations secure the invariant).[6] He concludes:

> Obviously, the position values themselves are not such invariants. But ratios of relative distances are (ratios rather than relative distances themselves because of the arbitrariness on the choice of a length unit). It follows that only quotients of relative velocities have an absolute status and, similarly, that the ratios of accelerations are invariant. (Dieks 2001a, 12).

Needless to say, given the complexities and uncertainties involved in their individual formulations, finding a common property theory classification for Sklar, Teller, and Dieks is rendered quite problematic. While various similarities can be detected, in addition to acknowledged similarities (e.g., Teller states that "in some respects [his view] is like Sklar's"; 1987, 446), there are important differences among these theories, differences that would seem to threaten the coherence of a common third-way classification: while Sklar and Teller deny relational motion, Dieks upholds relative quantities of motion; all three reject substantivalism; but only Teller (1987) accepts monadic spatiotemporal properties, P(loc). Sklar's "absolute acceleration" is a monadic property, but not necessarily a spatiotemporal one, and he accepts relationism, which is opposed to P(loc). Likewise, Dieks' ratios of relative distance are not monadic but relational.[7] Both Sklar and Dieks (and possibly Teller) would also seem to allow static and kinematic shifts via the Galilean transformations, and this possibility is inconsistent with a truly monadic spatiotemporal location property, P(loc). One might insist, therefore, that none of these theories constitute a true property theory since P(loc) may not be not upheld in all cases—but a better option, arguably, is to include all of these theories under the ontological dependence version of a property theory of space, P(O-dep), introduced in §1.2. All of these property theories reject substantivalism, i.e., that space is an entity that can exist in the absence of physical processes and entities, but space is not merely the relations among bodies either, hence they all fit the broader category represented by P(O-dep).

If many avowed property theories resemble relationism, then it is not surprising that many declared relational theories resemble the property theory. This point was noted at the outset of our investigation, in §1.2, and will form the basis of a relation-

---

[6] Dieks' work is also important in that he incorporates the lessons of non-relativistic quantum mechanics, and demonstrates how a monadic property view fits naturally into this formalism (with the hope that it can be extended to relativistic quantum field theory). In Chap. 8, aspects of Dieks' work on a property interpretation of quantum theory will be investigated in more detail.

[7] Despite its alleged status as an internal, non-relational property, Teller does acknowledge that "space-time location is in some ways relational, inasmuch as any space-time location gets fixed or 'measured' only relative to some perfectly arbitrary standard" (1987, 444). Whether this admission affects the monadic spatiotemporal property designation, P(loc), remains unclear.

ist counter-argument in §10.5.2, but it is worth demonstrating this structural resemblance before proceeding to an examination of various complications associated with the property theory in the next section. For instance, the many modal (non-eliminative) sophisticated relationist theorists that describe spacetime as a "disposition" or "possibilia" stand in an obviously close relationship to the property theory. As noted previously, the domain of the basic definitions of the strict eliminative brands of relationism are limited to existing material objects, e.g., "same place as" and "distance" can only be defined on material objects that actually instantiate those relations. Yet, leaving aside the possibility of a universal material field, this form of relationism seems unable to explicate the spatial relations that obtain in regions absent material bodies. To cover these cases, sophisticated relationists have invoked a dispositional construal of the basic definitions of the theory, thereby introducing possibilia/dispositions as a basic component of this form of relationism. For instance, the untranslated statement, "the distance between u and v is twice the distance between c and d", would not be given the relationist translation, "there is a (point-) object m between u and v, and um, vm, and cd are all equidistant" (since m may not exist); rather, it would be translated as (dispositional version), "if one were to mark off a copy of cd on uv starting at u, the resulting mark would be equidistant to u and v" (following Manders 1982, 585–586). Commenting on these modal constructions, Manders concludes:

> It is customary to explain physical possibility or dispositions in terms of physical laws and (non-dispositional) physical states. Our modal-dispositional relationist does not have this option. For him, the notion of a physically possible configuration is a primitive notion, in terms of which he formulates the laws of physics. (1982, 587)

While many critics have interpreted this type of maneuver as either a tacit endorsement of absolutism or as an attempt to obscure the difference between relationism and substantivalism (e.g., Earman 1989, 134–136, and Field 1985, 179–180),[8] the more important development, at this point in our analysis, is the incorporation of modality as a "primitive notion" of material bodies. As Manders himself acknowledges, "it would seem a key *methodological* tenet of scientific procedure to replace disposition statements by explanations in terms of actual physical states and properties" (1982, 587, original italics). Following this advise, consequently, entails a spacetime theory that regards spatial and temporal dispositions as *properties* of actual existing material bodies. In the absence of all material substances, furthermore, the modal relationist, like the property theorist, lacks a basis upon which to ground the dispositions, once more revealing the close similarities between the

---

[8] Mundy's relationist theory (1983) is also open to this charge, although he denies that his theory relies upon such modalities. Yet, as Belot comments, "he takes as given the structure of the inner product, which would appear to encode a large number of modal facts" (Belot 2000, 578, fn. 44). In brief, Mundy takes as the basis of his formal system the inner product function, $k(p, q, r)$, and then proceeds to show that any formalization of this same function as measured among existing point particles, i.e., the inner product function as an empirical law among real objects; $k(p, q, r) = a$, can be embedded up to isomorphism in the more general $k(p, q, r)$ (1983, 211–213). Overall, one could give a property theory interpretation of Mundy's project, with $k(p, q, r)$ as "the inner product property" (as a section of his essay is actually subtitled; 216).

respective theories. More specifically, dispositional talk lends itself naturally to an interpretation that involves the ontological dependence conception of a property theory, P(O-dep), where, in the case of sophisticated modal relationism, space is deemed a dispositional property of actual existing material bodies — and, to recall, these property theory variants falls under the third-way classification.

A general dissatisfaction with the standard substantivalist and relationist dichotomy has also motivated a number of philosophers to seek an intermediate position that, if not openly a property theory, closely resembles one. In articles by Stein (1977b) and Horwich (1978), for example, one could argue that a property view is implicitly sanctioned (although somewhat obliquely) in Stein's claim, first examined in Chap. 2, that "the structure of [a spacetime] manifold with [a tensor] field (Riemann-Minkowski metric)…is, in this sense, an 'emanative effect' of the existence of anything" (Stein 1977b, 397). DiSalle's version of the definitional interpretation, also explored in Chap. 2, fits the property mold as well. In trying to steer a middle course between the usual spacetime dualism, DiSalle is led to comment:

> Properly understood, space is *defined* by its regulation of the motions of bodies, and so we cannot meaningfully speak of space in the absence of the bodies which measure it. Yet we can at the same time assert its essential independence of bodies precisely insofar as the behavior of and interactions among the latter are conditioned by the structure of space. (DiSalle 1994, 274, original italics)

By claiming that space cannot be meaningfully discussed in the absence of body, although still somehow independent of body, the P(O-dep) version of the property theory might be the most plausible ontological categorization that fits DiSalle's philosophy, but DiSalle's reticence in discussing these ontological issues leaves much room for doubt.[9]

---

[9] In particular, DiSalle claims that the manner by which spacetime "forbids certain states of affairs and permits other… is more central to the metaphysical status of space and time than, say, the question of whether space is possible without bodies" (1994, 274). This verdict is debatable, however: if the main concern is with issues of objectivity and invariance, then he may be correct; but if one is interested in ontology, then the possible independent existence of space is of paramount importance — and the possibility that space might exist in the absence of bodies/fields would undermine the P(O-dep) classification. Overall, DiSalle avoids this vexed problem, and thus it is difficult to pin a natural or preferred ontology to his definitional approach, whether of the P(O-dep) variety or some other form. Furthermore, as mentioned in Chap. 1, whether Stein would accept the "definitional approach" label is unclear, since his conception of space is hard to determine, but it seems close enough to warrant that classification. Finally, while Huggett and Hoefer (2015) list Brown (2005) and Brown and Pooley (2006) alongside DiSalle as supporters of a dynamical approach, we regard these works (Brown, Brown and Pooley) as positing an emergent theory of spacetime (see Part III), and thus radically different from DiSalle's more epistemologically-oriented work.

5.1.3 Critiquing the Property Theory

One of the fundamental metaphysical obstacles that confronts the property theory, whether of the P(loc) or P(O-dep) sort, is the spatiotemporal domain of the property itself. Is the property internal to the three-dimensional boundaries of a body, much like the ordinary material properties of solidity, mass, charge, etc., whose domain lies within, or inside, the body? Or, is the spacetime property irreducibly external to a body's three-dimensional boundaries, as in the case of such properties as "taller than" or "is the brother of"? In what follows, we will dub these views, respectively, the "internal" and "external" property theories.

Between the two, an externalist interpretation of the property theory appears to be more natural, despite the internalist aims of certain versions, such as Sklar's. The rigorous relationist formulation of Sklar's theory in Huggett (1999) nicely demonstrates this point; i.e., that the actual construction of any property theory invariably favors a reading that involves the spatiotemporal relations and structures *among* bodies and/or reference frames, thus sanctioning an external construal of the spatial property. Huggett begins by utilizing "adapted relative reference frames" that are assigned to four points within a body $\{p, p_i\}$ (i = 1, 2, 3), with p as the origin of the coordinate frame, and each p_i extending a unit distance along one of the three spatial axes. The "sklaration" of any point q is then defined (roughly) as its acceleration *relative* to this material-based frame: "the sklaration of each point in the frame adapted to $\{p, p_i\}$...equals the affine acceleration in the rigid Euclidean frame" (Huggett 1999, 27).[10] As Huggett's construction indicates, whether the body's external relationships are with other bodies, inertial structure, possible trajectories, etc., these relationships seem to be external to the body. If the Stein-DiSalle definitional approach, or sophisticated modal relationist hypotheses, are deemed to be property theories, then an external property designation would apply as well.

---

[10] Skow (2007) and Pooley (2013), following Maudlin (1993), raise serious objections to these relationist efforts to regain neo-Newtonian spacetime structure. Skow claims that "[t]here can be no (relationalist) dynamical laws of motion based on Sklar's proposal that capture the content of Newton's theory" (2007, 777), since there are not enough resources in Sklar's system to account for the complexity of acceleration. He posits two systems, one with two bodies in a fixed circular motion around their center-of-mass, and another with two bodies following parabolic paths around their center-of-mass that, at an instant, possess the same values of mass, relative distance, velocities and absolute accelerations as the first pair of bodies. These two scenarios constitute an initial value problem for Sklar's theory since it lacks the ability to track the future evolutions of these systems and their diverging behavior (unlike the substantivalist). Skow admits that, for Sklar, "the property of undergoing an absolute acceleration is a fundamental intrinsic property" (Skow 2007, 782), but that Sklar gave few details on how this approach actually works (783–784). So, given this concession, why limit Sklar's use of his intrinsic acceleration property to just the second derivative of position (rate of change of velocity)? If acceleration is an intrinsic property, then higher derivatives of position (third order and higher) should be, too, especially since some of these higher derivatives may be coupled to observable non-inertial effects, just like acceleration: e.g., (I kid you not) "jerk" (third) and "jounce" (fourth), the former which is measured by a "jerkmeter". While this stratagem might allow Sklar to account for the different evolutions of the two systems in various cases, such as Skow's scenario just mentioned, these observations are not intended to diminish the severity of the obstacles to securing an account of Newtonian mechanics based on eliminative relationism and relational motion (see §1.1.2).

Nevertheless, Huggett's externalist approach seems to violate Sklar's demand that acceleration not be conceived as a relation between the body and something else, in this case, a relationally-palatable reference frame. Indeed, the goal of Sklar's maneuver is to break this habit of interpreting acceleration (for it tends to favor substantivalism; see footnote 5 as well). Another challenge for an external interpretation of the spatial property is that it fails to match the type of property theory implicit in both Newton and Leibniz' handling of spatial ontology. As discussed in Part I, there is a congruence of the domain of the ontological foundation of space, God (monads), and the domain of space. That is, space is not external to God in the sense that space exceeds the boundaries of God, whether God is situated in space (Newton) or only God's operations are situated in space (Leibniz; see Chaps. 3 and 4). If, as noted at the outset, one of our goals is to locate a third-way hypothesis that resembles the type of theory endorsed by these seventeenth century natural philosophers, then the external interpretation of the modern property theory fails (since the domain of the spatial property exceeds the boundary of the underlying ontology, i.e., the bodies). One could invoke a plenum (matter-filled) world in order to resolve this problem, but then it would be difficult to distinguish the internal version of the property theory from super-eliminative relationism (see §1.1.1); likewise, Huggett's construction is presumably neutral as to the material setting, plenum or no plenum.

Another rejoinder to these criticisms is to move to the internal version of the spatial property theory, and declare, as Huggett's formulation of Sklar seems to accept, that the mathematical description of this internal property merely utilizes such external features (path, positions, etc.) in an instrumentalist fashion—i.e., the paths and positions (and other structures) do not really exist outside the boundaries of material bodies or fields. So, although there really are no properties outside the three-dimensional spatial boundaries of, say, a lone rotating body, any mathematical account of that motion must incorporate structures that *seem* to transcend the confines of the body, such as the affine connection, ∇ (which determines the inertial paths of a body). The difficulty with this conception of the spacetime property, besides rendering it almost identical with modal relationism, is that it seemingly runs afoul of Earman's injunction against any "instrumentalist rip-off" of substantivalism (Earman 1989, 127). By declaring that the reference to external, absolute structures, like ∇, does not entail a commitment to the reality of ∇, this interpretation risks being seen as an imitation of the substantivalist position, but without the underlying ontological commitment that warrants the use of the absolutist structure. Earman quips that "it remains magic that the representative [of Sklar's absolute acceleration] is neo-Newtonian acceleration $(d^2x^i/dt^2)+\Gamma^i_{jk}(dx_j/dt)(dx_k/dt)$ [i.e., the covariant derivative, or ∇ in coordinate form]" (127–128).[11] He concludes: "the Newtonian apparatus can be used to make the predictions and afterwards discarded as a convenient fiction, but this ploy is hardly distinguishable from instrumentalism, which, taken to its logical conclusion, trivializes the absolute-relationist debate" (128).

---

[11] Modified quotation: sans an extra $1/dt$ term in the left hand side of the equation in Earman's text.

In response, the internal property theorist can attempt to turn the tables on the substantivalist by appealing to the virtues of a parsimonious ontology; i.e., the property theory has the advantage of *not* requiring the existence of a second, independent entity, substantival spacetime, in explicating structures like ∇. Of course, metaphysical simplicity is in the eye of the beholder, so it is difficult to adjudicate between the competing allegations of "instrumentalist rip-off" versus "bloated ontology". We will return to this issue in Chap. 7, when Earman's accusations of instrumentalism will be explored in greater depth.

On the other hand, a property theorist might find Earman's critique persuasive and therefore embrace the actual existence (in the ontological sense) of the positions and paths of ∇ that lie external to the body at time t. In other words, the property theorist could stick with the externalist version of theory, and affirm that, given the existence of any material substance, the whole of spacetime comes into existence, including all of the spatial position and trajectories that constitute its structure. This external version of the spacetime property, like all accounts, continues to uphold the dependency claim of the property view, with the existence of space or spacetime dependent on matter. Yet, given an externalist interpretation, the spacetime property is assumed to have the added trait of "extending beyond" the three-dimensional confines of any actually existing material body at a given instant, so it is meaningful to discuss the global properties of the spacetime (metric, topology, etc.) even in a world that includes only a single material object.[12] Throughout the remainder of our investigation, it is important to keep in mind the drastically different ontological implications of the competing internal and external formulations of the property theory: whereas the internal view holds that spacetime structures outside the confines of bodies or fields are just instrumentally useful idealizations (but correspond to real internal properties of bodies), the external view does indeed regard such structures, like ∇, as existing external to bodies in the full ontological sense (in addition to existing internal to bodies). Therefore, on the external property theory, the scope of the spacetime property need not coincide with the domain of material bodies, a view that sharply contrasts with the internal reading.

If, accordingly, the property theorist endorses the external version, then one of the oldest metaphysical assumptions would be effectively banished; namely, the metaphysical hypothesis that properties can only be internal to, or inside, a substance: we will label this view the "internal property restriction" (see also §2.2 and §3.2.2 on "inherence", which is the seventeenth century version of the internal

---

[12] A potential strategy for salvaging an internal reading of the spatial property in non-field physics is to adopt a four-dimensional ontology of material bodies. If material bodies are construed in this four-dimensional manner, i.e., as the world-line (world-tube) of a three-dimensional spatial body that endures over time, then the future spatial positions projected by ∇ does fall "within" the confines of the four-dimensional body (but only if the body does indeed coincide or occupy the trajectory laid out by ∇ at later instants of time). However, the fact ∇ can be used to indicate the inertial paths of any hypothetical, but not actual, test particle must then be interpreted as either a mere mathematical excess without ontological grounding (which raises instrumentalist worries again), or as a relation among the actual particle trajectories (which renders the view nearly identical to relationism).

property restriction endorsed by both Newton and Leibniz). The property theorist might happily accept this consequence, it should be noted. Since the property of spacetime has a unique ontological status—i.e., it plays a metaphysically foundational role for all material existence—it could be argued that the internal property restriction need not apply. The property theorist can claim that, despite its lack of intuitive appeal, the external manifestation of the spacetime property constitutes a fundamental, primitive concept of the theory, and hence cannot be analyzed further in terms of our normal spatial intuitions.

If the "externality" of the spacetime property is put forward as a primitive concept, the property theorist has an additional set of rejoinders to any criticism that is aimed at the seemingly incongruous nature of the spatial property. Since the critic would likely invoke the standard metaphysical conception of substances and properties (i.e., the internal property restriction) against the external type of property theory, the property theorist can claim that the internal/external (or inside/outside) distinction is *itself* a spatial distinction, and thus *presupposes without argument* a conceptual view of space that the external property view simply denies. Second, the proponents of the external property view can cite examples of other bodily properties that, like the spatial property, can violate the internal property restriction. While the relational properties between bodies are quite numerous (e.g., "is the father of"), there is at least one non-relational property that figures prominently in physics: namely, a body's center-of-mass point, which can have a spatial position outside (external to) various oddly shaped bodies. To object that center-of-mass points are not sufficiently spatial, since they are supposedly non-extended, would necessitate a criterion for the proper form of spatial existence for bodily properties—and this would be a very difficult task for the critic to meet. The spatial characteristics of most bodily properties are hard to categorize: Is the redness of the cube adjacent to the two-dimensional surface of the cube, identical to that surface, or is color merely a phenomenal, mind-dependent by-product of light and surface texture (so that color is not really in physical space, at all)? Presumably, our common-sense spatial intuitions dictate that color possesses a spatial location in the minimal sense that it can be fixed relative to other bodies and properties; but, the same claim can also be made with respect to a body's center-of-mass point. Overall, the relationship between space and *all* bodily properties is a deep metaphysical puzzle, so the burden of proof may unfortunately lie on the advocates of the internal property restriction to secure first a comprehensive and consistent explanation of this relationship before they can outlaw the external property view.

In putting forward the external type of spacetime property as a metaphysical primitive, another defensive strategy that the property theorist can implement is to expose the equally unsettling primitives that underlie both substantivalism and relationism. If all spacetime theories must employ a primitive concept, then the property theory is on an equal footing with its better known rivals as regards ontological grounding. Only a few of these metaphysical quandaries can be raised here. For the substantivalist, the primitive is substantival space itself, which is an unobservable substance that can exist apart from all bodies, yet bodies are somehow located and move "in" this substance, and their accelerated motions relative to this substance mysteriously affects those bodies (i.e., the existence of non-inertial force effects). There is also a regress problem for substantival space (which will be explored in greater depth in Chap. 6): if everything needs a place, or must occupy space, then space itself needs a place/space. For the relationist, the primitive notions crop up with respect to the spatial extension of body. In short, if space is a relation among bodies, then how can the relationist account for the space within a body? Answer: accept as a primitive notion either non-extended parts that somehow comprise extended bodies (but how?),[13] or, more likely, simply declare that these parts are themselves non-relationally extended (i.e., the extension within a body is akin to a mini-substantival space), an admission that weakens the relationist's case against substantivalism. All told, the primitive notions underlying both substantivalism and relationism are equally susceptible to metaphysical dilemmas, hence the unintuitive nature of the property theory's primitives does not inevitably favor these ontological alternatives in comparison.

Finally, it should be noted that the external property theory does not apply in the case of field physics, where physical properties, such as electric charge or gravity, are, as it were, "smeared out" over space, i.e., where each point of space bears a value of the property. On a field model, all properties are internal to the field, although it could be claimed that the external/internal distinction breaks down in the case of emergent properties of a field. If space is deemed a property of a field, more-over, then the property theory and the most strict form of relationism, super-eliminative (see §1.1), would be identical in one important respect: since both ground space on a physical thing, a field, there is a congruence of space and physical entity. On the other hand, since a super-eliminative relationist must reject absolute states of motion, a uniform rotation or motion of the entire field is automatically ruled out, unless that rotation is manifest relative to another field or entity.

---

[13] Harré (1986) offers a relationist theory of this type based on an exclusion principle, so that "two things [or two properties] cannot be at the same location in the spatial manifold" (1986, 125). He ultimately concludes that "the word 'space' names a set of non-causal relations between material bodies based on the principle of exclusion" (133). The spatial extension of a body is explained by the exclusivity of two properties, say, *Fa* and *not-Fa*, that a body possesses at the same time (with "exclusivity" allegedly derived from the principle of non-contradiction, thus precluding the need for a primitive concept of spatial extension): e.g., one can explain the spatial extension of a fire poker on the basis that one end has the property of being "hot" whereas the other end is "cold". However, in order to be empirically meaningful and detectable, these properties would have to be extended, so Harré's approach would seem to court circularity (or at least presuppose extension). Harré's theory also unwarrantedly presumes the internal/external distinction, since he limits the application of the exclusion properties (that constitute bodily extension) to only those properties located *inside* a body. Put simply, given Harré's definition, it should follow that the mutually exclu-sive properties of, e.g., the grass's property of "green" and the sun's property of "not-green", constitute the extension of a *single* body (since exclusion properties *define* a single body).

5.1.4 Concluding Assessment

Speaking of fields, the biggest obstacle for the property theory, at least for the version whose spatial property is grounded on material bodies, is the conjunction of the field concept in physics and our best current theory of the large-scale structure of space, general relativity (GR). The property theory may have represented a plausible option on an older model of Newtonian mechanics and gravitation theory, where bodies causally interact within a void, or near void, but GR has drastically changed the setting of modern physics. In GR, the metric or gravitational field pervades all of spacetime, so there is no "metric void", so to speak, either real or possible, where the manifold would lack a value (i.e., a region or point where the metric field is absent).[14] The metric/gravitational field, g, provides the geometry of spacetime, but g also incorporates the inertial-gravitational field, thus explaining the duality of its nomenclature (but "metric field" is the default choice). Within the context of GR, the problem for the property theory concerns a different field, the stress-energy field, T, which represents matter, radiation, and non-gravitational force fields, and which can vanish, unlike the metric field. These "vacuum solutions" to GR's field equations, consequently, undermine the credibility of a property theory grounded on *matter*, since it is meaningful to discuss the geometry of space, via the metric field in a matter-less vacuum. In contrast, given a Newtonian theory utilizing material bodies alone (or a vanishing material field without a GR-type metric field), a property theorist can coherently infer that a universe without matter is a universe without space, a conclusion that the relationist would endorse as well. Hence, GR's possible vacuum states represent a problem for both relationism and a property theory restricted to matter. In the second half of this chapter, the importance of the metric field will occupy much of the discussion, but, as regards the property theory in particular, the troubling repercussions of GR's vacuum solutions suggest that a viable spatial property requires a different ontological foundation (and the same for relationism). Given a GR setting, a plausible option is to ground space on the more basic concept of energy, as opposed to matter; e.g., the inertial-gravitational component of g (more on this below).

Teller cites Putnam as having suggested that "at bottom there is no real difference between calling something a 'property' and calling it a 'concrete thing'" (Teller 1987, 445), an insight that raises a troubling question that has been lurking in the wings since our initial foray into alternative ontologies in Chap. 1: Is there any non-trivial, non-conventional ontological difference between the property theory and sophisticated versions of substantivalism and relationism? The similarities between the rival internalist and externalist interpretations of the property theory and either sophisticated relationism or sophisticated substantivalism are a case in point: for non-field theories, the internal property theory's analysis of spacetime structures as

---

[14] In quantum field theory, the non-vanishing energy of the quantum field is thus similar to the lack of a metric void as regards the implications for spatial ontologies, but we will postpone that discussion until Chap. 10.

instrumentally-useful mathematical abstractions was seen as closely paralleling the similar attitude favored by sophisticated modal relationists; whereas the external property theory's commitment to an actual spacetime structure that is external to matter finds a direct analogue in substantivalism. As for the difficulties associated with both types of property theory, once again each hypothesis shares a nearly identical burden with one of its better-known rivals: the internal property view is similar to sophisticated modal relationism in that it must defend against the "instrumentalist rip-off" accusation, while the external property theory faces the difficult task, along with the substantivalists, of explicating the existence of spacetime structures that somehow transcend the material occupants of the spacetime (i.e., why not take the simpler view that mathematical structures are mere abstractions from physical processes?).

5.2 The Structuralist Aftermath of the Ontology Debate in General Relativity

As we have seen, there would appear to be few benefits, and many drawbacks, to conceiving space (or spacetime) as a property of matter. Yet, while the more straightforward matter-based conception of the property theory leaves much to be desired, a plethora of recent interpretations of the ontology of general relativity (GR)—in particular, sophisticated versions of both substantivalism and relationism, and spacetime structural realism—would seem to have much in common with the property theory interpretation of GR's metric field. The reason for this similarity is due to the fact that all of these interpretations (substantivalist, relationist, or structural realist) regard the mathematical structure of GR, in particular, the metric, or metric plus point manifold, as: (a) the key item in their respective ontological interpretation, but (b) treat that structure much like a property of the underlying ontology. Nevertheless, since structural realism adopts a specific mathematical structure of GR as the main ontological element, its interpretation of GR might now be seen as augmenting the property theory reading. That is, structural realism explicates the nature of the mathematical "properties" of GR, hence it can be argued that it more closely resembles the property theory than either substantivalism or relationism. Structural realism, as its name implies, places mathematical structure at the core of its realist conception of scientific theories, with varying emphasis on the epistemological and ontological aspects of theories. In the remainder of this chapter, consequently, a structural realist conception of spacetime theories will eventually take center stage, a conception that will ultimately prove to be one the most compelling third-way alternatives to traditional substantivalism and relationism, as well as embody many of the ideas implicit in the seventeenth century's natural philosophy of space.

5.2.1   The Hole Argument and Recent Substantivalist and Relationist Ontologies

Many recent investigations of spacetime structure can be traced to Earman and Norton's (1987) adaptation of Einstein's original hole argument against the contemporary substantivalist position dubbed "manifold substantivalism"; i.e., the view that that the substance of space/spacetime is the topological, differentiable manifold of points, M, which underlies all other spacetime structures (see §1.1). Put briefly, the hole argument concludes that substantivalism, in its modern GR setting, leads to an unacceptable form of indeterminism. By shifting the metric and matter (stress-energy) fields, g, and, T, respectively, on the manifold of points, M, with the latter representing the substantivalist basis of spacetime, one can obtain a new model $(M,\breve{g},\breve{T})$ from the old model (M,g,T) that also satisfies the field equations of GR. The new model is acquired via a "hole diffeomorphism", h: $h(g)=\breve{g}$, $h(T)=\breve{T}$. If the mapping is the identity transformation outside of the hole, but a non-identity mapping inside, then the substantivalist will not be able to determine the trajectory of a particle within the hole despite the observationally identical nature of the two worlds, (M,g,T) and $(M,\breve{g},\breve{T})$, i.e., the spacetime with, and without, the transformation. In simplest terms, the hole argument is akin to the static shift arguments first discussed in §3.1.2, since the shift of g and T inside the hole brings about an identical arrangement of the spacetime's matter/fields and their metric relationships but now situated in different parts/points of space relative to the manifold M.

   One of the more influential substantivalist solutions to the hole argument rejects a straightforward realist interpretation of the individuality of the points that comprise the manifold M. "A preferable alternative [to manifold substantivalism] is to strip primitive identity from space-time points: call this view *metric field substantivalism*. The focus of this view is on the metric tensor as the real representor of space-time in GTR" (Hoefer 1996, 24). Since the identity of the points of M are fixed by the metric g alone, any transformation of g, i.e., $\breve{g}$, does not result in the points of M possessing different g-values; rather, $\breve{g}$ simply gives back the very same spacetime points (since, to put it another way, a shift in g also shifts the identity criteria of the points of M along with it). In evaluating the various structural-role solutions to the hole argument (besides Hoefer (1996), there are similar hypotheses in Butterfield (1989), Brighouse (1994), Mundy (1992), and Healey (1995), to name just a few), the accusation can be made that metric field substantivalism bears a suspiciously close resemblance to relationism. In Belot and Earman (2001, 228), and as first discussed in Chap. 1, these structural-role constructions are somewhat pejoratively labeled "sophisticated substantivalism", with the charge being that the "substantivalists are helping themselves to a position most naturally associated with relationism" (Belot 2000, 588–589)—the rationale behind this estimate is that the identification of a host of observationally indistinguishable models with a single state-of-affairs is the very heart of relationism (e.g., Leibniz's rejection of the static and kinematic shift arguments; see, once again, §3.1.2).

A few additional worries concerning metric field substantivalism should be recorded at this point (see also Chap. 6). First, as Hoefer himself admits, the role of M in GR cannot be completely dismissed—namely, "it represents the continuity of space-time and the global topology" (Hoefer 1998, 24)—thus an alternative form of substantivalism, namely, "manifold plus metric substantivalism" (which Hoefer does not prefer), would seem to represent the mathematical structure of GR more faithfully. Here, the necessity of all of these mathematical structures for the entire space-time is the key point. Hoefer provides two reasons for singling out the g-field (1998, 24): (i), that the "empirically useful" work of GR is "primarily" carried by the metric (such as inertial structure); and (ii), that a metric field without a global topology is possible "for at least small patches of space-time", although a manifold M without a metric cannot capture even a portion of spacetime. Yet, the problem with (i) is that M provides a great deal of useful empirical work as well, such as the dimension and global topology of the spacetime. If these aspects of the spacetime were missing or different, it would, needless to say, make a vast difference in our experience of that world. As for (ii), a small patch of spacetime is an incomplete and inadequate representation of the overall spacetime (global topology being a major concern, once again), hence it does little to support the notion that g alone is somehow more privileged than M alone. Another problem with metric field substantivalism is that by singling out a particular mathematical structure as "the real representor of space-time", despite the fact that the structure in question is only one of an interrelated set of such mathematical structures, sounds suspiciously like a form of ontological conventionalism—that is, one structure would seem to have been arbitrarily singled out as ontologically privileged even though the other structures are necessarily required for the overall spacetime. In fact, this criticism of Hoefer's strategy is reminiscent of Quine's critique of Carnap's conventionalist-leaning doctrine of truth as regards geometry. Quine argued that the conventional element in Carnap's treatment of geometric truth arose, not in determining the truths of geometry (since "the truths were there"), but in selecting from that interrelated set of pre-existing truths those that serve as the fundamental Euclidean axioms, and those that serve as the derived results (Quine 1966, 108–109). In a similar manner, Hoefer seems to be arbitrarily privileging a single structure, the g-field, from within the set of interrelated structures of GR's spacetime, (M, g, T), to serve as the ontological foundation of (M, g, T)—even though one could have as easily adopted, and offered plausible arguments for, the structure M to serve as the privileged ontological basis. Earman (1989, chapter 9), for example, develops arguments that favor the identification of M with substantivalism: e.g., "fields are not properties of an undressed set of space-time points but rather properties of the manifold M, which implies that fields are properties jointly of the points and their topological and differential properties" (1989, 201).

Returning to the charge that sophisticated substantivalism steals a page from the relationist playbook, what is ironic about this allegation is that until recently they have usually gone the other way, with the substantivalists accusing many of the latest relational hypotheses of an illicit use of absolutist/substantivalist spatiotemporal structure. The allegedly non-relational structures can be classified broadly as those that pertain to space (topology, metric, etc.) and those that pertain to motion (in particular, acceleration and its accompanying force-effects). As regards space, a relationist who is confined to only existing material bodies may not be capable of constructing the full range of spatial structures that are freely available on the absolutist/substantivalist picture. To take a simple example, if congruence relations are founded upon actual material relationships, can the relationist consistently invoke the congruence of different regions of empty space (as noted in §3.2.2, Clarke raises these same objections against Leibniz)? In order to meet this challenge, many relationists have adopted, as we have seen, a modal or dispositional account of spatial structures, such as metrical structure, which extends beyond the existing material bodies to cover all "physically possible configurations". Since this modal capacity is often taken as a primitive notion, the substantivalist can thus claim that the relationists have illegitimately utilized an absolute structure that transcends the actual material relations. It has long been recognized that a physics limited to the mere relational motion of bodies, based on a strictly eliminativist brand of spatial relationism, faces serious difficulties in trying to capture the full content of modern physical theories. In GR, furthermore, the formalism of the theory makes it meaningful to determine if a lone body rotates in an otherwise empty universe, or whether the whole material content of the universe rotates in unison, but such possibilities are excluded on a strict eliminativist relationism since there are no material bodies or frameworks from which to measure that single body's (or all bodies') rotation. A potential relationist rejoinder is to reject the strict version of relationism, and simply hold that the spatial structures needed to make sense of motion and its effects do not supervene on some underlying, independent entity called "substantival space (spacetime)". The rejection of a strictly eliminativist brand of spatial relationism, i.e., sophisticated relationism, allows the relationist to freely adopt any spatial structure required to explicate dynamical behavior, e.g., affine structure, ∇, just as long as it is acknowledged that these structures are instantiated by material bodies or fields in some fashion. Sophisticated relationism, as noted in Chap. 1, includes all forms of relationism other than strict eliminative relationism, such as modal relationism. Yet, by embracing these richer structures, sophisticated relationism is open to the charge of being an "instrumentalist rip-off" of substantivalism, the same allegation leveled at Sklar's property theory, as noted in §5.1.3.

In the context of GR, however, the most plausible form of sophisticated relationism is "metric field relationism". As revealed above, Hoefer views the metric as representing substantival space, but Einstein judged the metric field, which is also the gravitational field, as more closely resembling Descartes' theory of space—and Descartes' conception of space is normally categorized as relationist:

> If we imagine the gravitational field, *i.e.* the functions g_{ik}, to be removed, there does not remain a space...but absolutely *nothing*....There is no such thing as an empty space, *i.e.* a space without field. Space-time does not claim existence on its own, but only as a structural quality of the field.
>
> Thus Descartes was not so far from the truth when he believed he must exclude the existence of an empty space....It requires the idea of the field as the representative of reality, in combination with the general principle of relativity, to show the true kernel of Descartes' idea; there exists no space "empty of field". (Einstein 1961, 155–156)

As first discussed in our examination of the property theory, by holding that all fields, including the metric/gravitational field, g_{ik}, are physical fields, the vacuum solutions to GR no longer correspond to empty spacetimes, thus eliminating a major relationist obstacle. According to the standard matter-based conception of relationism (or property theory), a spacetime entirely void of matter corresponds to either a meaningless state-of-affairs or a universe without space (spacetime). Yet, the vacuum solutions to the field equations in GR (where the stress-energy field, T, is 0) are a meaningful, as well as possible, state-of-affairs. Hence, if a relationist deems the metric field to be a physical field, then this maneuver automatically renders the vacuum solutions of GR to be relationally meaningful since there is no spacetime, to use Einstein's phrase, "empty of [metric] field" (even in a matter-less vacuum): on Dieks' succinct explanation of this version of relationism, "[t]he metric field, a physical system of the same kind as [particle physics], is the bearer of the geometrical space-time properties" (Dieks 2001a, 14; besides Einstein, see also Rovelli 1997; Auyang 2001).

A substantivalist might reject this interpretation of the metric field, g, and strive to formulate some ontological criterion for differentiating substantival and non-substantival aspects of GR. Unfortunately, given the peculiar complexities of the field concept in physics, as well as the argument that this particular field can produce measurable physical effects, it becomes difficult to imagine how such a criterion could avoid arbitrary ontological stipulations that beg the question against the relationist. As put forth in Earman and Norton (1987), the gravitational waves that can propagate through the empty spacetime solutions of GR are associated with the physical/material content of the spacetime, and not the underlying substantival space, since "in principle [the wave's] energy could be collected and converted into other types of energy, such as heat or light energy or even massive particles" (1987, 519).

Hoefer (2000) challenges Earman and Norton's conclusion by insisting that it depends on a well-defined conception of the stress-energy carried by the gravitational field, which, he adds, is not actually sanctioned by GR. Specifically, the problem arises because the term that represents the stress-energy of the gravitational field, t^{ab}, is a pseudo-tensor, where "its non-tensorial nature means that there is no well-defined, intrinsic 'amount of stuff' present at any given point" (2000, 193). Yet, Hoefer's response is not without its own set of difficulties, for it seems to entail that the energy lost by a gravity wave source is a real loss, and is not simply somewhere else in space; likewise, the energy gained by a gravity wave detector is a real gain, in apparent *ex nihilo* fashion. As Hoefer admits, "such a perspective seems to strain our general cause-effect intuitions by positing a cause-effect relationship without an intermediary carrier" (196).[15] In short, Hoefer's interpretation would forfeit

---

[15] See Slowik (2005b), for an analysis of other aspects of Hoefer's arguments. For the committed relationist, the vacuum solutions to GR without gravity waves might seem to undermine the suggestion made by Earman and Norton, since the absence of gravity waves means that there is no energy to convert into particles, etc. However, since the spacetime would still possess the capacity to generate waves, it is unclear that there is any real difference in this case for those who side with Einstein's hypothesis that the g field is a physical field. A relationist could also accept Harré's suggestion that the vacuum solutions to the field equations of GR have "no reasonable physical interpretations" (Harré 1986, 131), but this seems a rather unmotivated and unwarranted strategy since, as explained above, the vacuum solutions are mathematically meaningful.

energy-momentum conservation in GR, which is counter-intuitive given the long-standing presumption of this conservation law as a cornerstone of modern physics. On these grounds, if one is forced to choose between the rejection of the conservation of energy-momentum in GR versus a non-localizable conception of the stress-energy associated with the gravitational field, then the latter seems a much more preferable alternative. Friedman has likewise strived to uphold a substantivalist interpretation of the metric field, g, arguing that "it accords with the general-relativistic practice of not counting the gravitational energy induced by g as a component of the total energy, and it allows us to preserve a measure of continuity between general relativity and our previous space-time theories" (Friedman 1983, 223). But the practice of not counting the gravitational energy does not undermine the argument put forward by Earman and Norton above, i.e., that gravitational energy can be converted into other forms of energy, and is thus real. Likewise, the alleged continuity between GR's metric field and our previous theory of the large scale structure of space, Newtonian gravitation theory, is precisely the point at issue, and so one cannot simply assume it. While the basic metric and manifold structure of g is, of course, closely related to the Euclidean geometry of Newtonian theory, the gravitational field aspect of g is decidedly not. Following Einstein's lead, Rovelli argues that g is much closer to matter than a spatiotemporal backdrop: "In general relativity, the metric/gravitational field has acquired most, if not all, the attributes that have characterized matter (as opposed to spacetime) from Descartes to Feynman: it satisfies differential equations, it carries energy and momentum, and, in Leibnizian terms, *it can act and also be acted upon*, and so on" (1997, 193).[16]

A substantivalist could also argue that the proper or most defensible conception of relationism in the context of GR is the Machian account, which in its most robust form, dubbed "Mach-heavy" in Huggett and Hoefer (2015), dictates that the spatial geometry and inertial structure of the spacetime, g, must be determined by the material content of that spacetime, i.e., the stress-energy, T.[17] Yet, Mach-heavy is just one interpretation of relationism in a GR setting; and, if Einstein's views are deemed a deciding factor, then he had long abandoned the Mach-heavy conception by the time he posited his Cartesian interpretation of the theory (as opposed to the Newton interpretation) in the quotation provided above. Consequently, the proper relationist account of GR is as indeterminate as the proper relationist interpretation of Newtonian theory.

---

[16] Another issue that involves the gravitational field specifically pertains to the pseudo-tensor, t^{ab}; namely, whether only tensors should be granted the status of a real quantity. On the whole, it remains unclear that the ontological classification of a quantity like t^{ab} turns on a tensor formulation alone: as long as there exists at least one reference frame that measures a non-zero t^{ab}, that may be sufficient (albeit awkward) for any relationist or property theory interpretation. Likewise, the pseudo-tensorial status of t^{ab} may be regarded as simply an artifact of its non-localizable nature, with no further ontological implications warranted. On the general issue of vanishing fields, Weyl reasons that if, say, the electric field, E, is 0 in some part of space, then it is not absent in that part of space, but merely in a "state of rest" in that part (1949, 172).

[17] "Mach-lite" is the standard anti-substantivalist rejection of substantival space for a relationally acceptable alternative, such as the fixed stars, or, better yet, the center-of-mass reference frame of the world, which Mach stipulates must not accelerate (SM 287). The most significant problem for Mach-heavy is the fact that the boundary conditions of GR's field equations are not totally determined by T, but have to be specified with respect to a choice of g as well.

On the whole, the predicament posed by these conflicting substantivalist and relationist interpretations of *g* has led some to question the relevance of the standard dichotomy (substantivalism versus relationism) in the context of GR: if *g* can be coherently viewed as supporting either relationism or substantivalism, then what remains of the standard dichotomy that is useful in analyzing the ontology of GR (see Rynasiewicz 1996)? In subsequent chapters, the conflicting ontological interpretations of *g* will remain a major theme, yet, before leaving this subject, there is an advantage that a property theory interpretation of GR's metric field enjoys over its relationist opponent that should be addressed. On the general relationist ethos, a synchronous or uniform motion of the universe's material content should be prohibited; however, not only does *g* encode inertial structure, but a unison rotation of the sort just described is a possible scenario, e.g., the Gödel metric. The metric field relationist will claim that this scenario is nonetheless relationally consistent by insisting that the unison rotation of matter is determined against the backdrop of the metric field, which is itself a physical field, and hence the rotation is a relation among physical things, a physically-conceived metric and matter. On the other hand, since a unison rotation of the entire material universe is, in general, an exceedingly non-relational state-of-affairs, the property theorist will interpret these scenarios as favoring their view. That is, since a property theory interpretation of the metric field is not bound to any form of relationism, in particular, relational motion, and can accept inertial structure or any other spacetime structures that are normally associated with the absolutist side of the standard dichotomy, the apparent "absoluteness" of the world's rotation does not constitute the embarrassing complication that is does pose for the metric field relationist. Indeed, recalling Einstein's claim that *g* is "a structural quality of the field", it would seem that a property theory reading, as in "structural *quality*", is a more natural ontological assessment than metric field relationism. Finally, a similar criticism can be raised for all non-field sophisticated relationist strategies, such as Teller (1991), that admit inertial structure and thereby reject a strict relational account of motion. To summarize: a property theory seems to be a better ontological classification than relationism for an interpretation that accepts both the possible rotation of a lone body in an empty universe or if the material contents of the universe undergo a uniform rotation.

5.2.2   The Structural Realist Alternative

An additional option in the spacetime ontology dispute is represented by "structural realism" (SR), a form of scientific realism that departs from a straightforward realist commitment to theoretical entities. Since a structural realist holds that what is preserved in successive theory change is abstract mathematical or structural content, it is these mathematical structures that receive the ontological commitment. As will be explained in more detail in Chap. 8, some of the benefits that can be gained from an SR approach include an explanation of the progressive and convergent empirical success of the scientific process, while also accommodating the fact that the specific entities incorporated by these distinct, evolving theories often differ quite radically. If space or spacetime is conceived along SR lines, the result is "spacetime structural realism", or "SSR", a designation coined by Dorato (2000, 2008). Nevertheless, in order to limit the aggregation of terminological differences, and since our investigation is exclusively concerned with space and spacetime, we will continue to use "SR" to refer to spacetime structural realism (SSR).

The SR viewpoint insists, therefore, that there remains a core geometric structure underlying our best spacetime theories, presumably from Aristotle all the way to GR and beyond. An invariance of geometric structure across specific versions Newtonian theory has been carefully documented in the work of Friedman (1983), and may lend credence to the SR belief that the continuing evolution of spacetime theories will manifest an underlying geometric structure that underwrites the success of all spacetime theories. In moving from the classical Newtonian theory (with a spacetime rigging to define absolute place) to the more complex neo-Newtonian formulation (which eliminates both absolute place and absolute velocity, and incorporates a geometrized gravitational potential), three geometric structures are invariant features in all versions of the theory: an affine connection, ∇, which provides the inertial trajectories (and may be either absolute or dynamic); a co-vector field, dt, which represents absolute time; and the metric tensor, h, which secures a Euclidean measure of distance (Friedman 1983, 71–124; here, a fourth structure, the point manifold, M, should also be included). In all versions of the Newtonian theory, these structures play an integral role in the spacetime structure, thus they cannot be removed without undermining the overall Newtonian theory and its various physical laws (such as the laws of motion and the gravitational law). In moving to the spacetime of GR, the geometric structures shrink to just (M, g), since the affine and temporal structures of the Newtonian theories, ∇ and dt, are now provided by the semi-Riemannian metric tensor g (which also replaces the Euclidean metric, h; 177–215).

Once again, a more elaborate examination will be given in Chap. 8, but recent SR interpretations of the ontology of GR parallel Freidman's investigation by focusing upon the metric/gravitational field, g (with or without, M, as explained above) as the key mathematical structure. A spacetime structural realist construal of the spacetime in GR has been put forward by, e.g., Dorato (2000, 2008), Slowik (2005a), Rickles (2008a), and Esfeld and Lam (2008), to name only a few. Yet, the ontological content of the SR conception appears indistinguishable from the brand of relationism (disclosed above) that regards the metric field as a physical field. In fact, Dorato's (2000) account would also seem to resemble a property theory: he claims that, since spatiotemporal structure is "a mind-independent *property* of certain physical systems, structural realists about spacetime may rely on the fact that *properties* (spatiotemporal ones included) *simply are,* in any respectable metaphysical theories, *the causal powers of the entities having them*"; and, "[w]hat causes the deflection of the orbit of a massive body is the gravitational field, a thoroughly *physical* field, *via* its geometrically, causally active relational properties" (1616, original italics). Since SR reckons the metric to be a physical field, how does the SR conception of spatial ontology, in a GR setting, differ from the sophisticated version of relationism, or, for that matter, substantivalism and the property theory? Overall, and while acknowledging the difficulty in determining this issue, a structural realist could make the claim that the metric is a unique physical field, a field that is a sort of hybrid of the relationist's physical metric field and the substantivalist's metric field substance. Dorato comments that the metric field "performs the typical individuating functions of *classical space and time* regarded as *principia individuationis,* and is therefore *not* a matter field like any other" (1611). On the other hand, it could be argued that there is merely a conventional difference among these ontological positions (namely, SR, metric field relationism, metric field substantivalism, metric field version of the property theory), since they all accept (a) that the metric provides both the spatiotemporal individuation of the manifold points and the spatiotemporal location of all other objects/fields, and (b) that it is a dynamical field that affects, and is affected in turn, by other objects/fields (more on this in Chap. 7). On the positive side, a structural realist ontology does have an advantage over the sophisticated versions of both relationism and substantivalism, since (a) is more on the substantival side, whereas (b) is more on the relational side, and so GR's metric does indeed seem to be a unique combination of traditional substantivalism and relationism. Then again, since the sophisticated versions of both substantivalism and relationism, as well as the property theory, can likewise claim that they provide a unique combination of traditional substantivalism and relationism, the impasse over these competing ontological interpretations appears impossible to resolve— although, as discussed at the end of §5.2.1, a metric field property theory does retain a decided advantage over metric field relationism if certain universal states of motion are allowed, such as the rotation of the universe. Likewise, as noted above, structure is often conceived along the lines of a property of some entity, an outlook that favors the property theory over its rivals.

The main reason that our investigation of an SR conception of GR is practically indistinguishable from its rivals is due to the fact that we have thus far focused on the ontological side of structural realism as opposed to the epistemic side. While a detailed analysis will be taken up in Chap. 8, a brief description of the basic idea behind the ontic and epistemic branches of SR can be provided at this stage in our investigation. One view, dubbed "epistemic structural realism" (ESR), regards structure from a purely epistemological perspective, insisting that the mathematical structures that turn up in our best scientific theories do not provide any information on the actual entities that underlie the observed structural relationships. In contrast, "ontic structural realism" (OSR), does claim that these mathematical structures reveal facts about the underlying ontology—and, in fact, may *be* the underlying ontology. Hence, on the ontic version of SR (OSR), the metric field becomes a physical field, which leads Dorato (2008) to admit that "if we think that the dispute between substantivalists and relationists is genuine", then "*structural spacetime realism is a form of relationism*" (2008, 24; i.e., OSR is identical to sophisticated metric field relationism). Dorato's use of the term "relationism" in this context, it should be noted, seems to denote the more general concept of "materially-grounded", and is thus consistent with the definitions of sophisticated relationism and the property theory of space surveyed in our investigation. Hence, both OSR and ESR are consistent with the property theory, although the main emphasis in this chapter will devoted to ESR.

Turning to the epistemic version of SR (ESR), it might seem at first glance to hold little advantage for the assessment of rival spatial ontologies, such as substantivalism and relationism. Yet, as will be argued throughout the remainder of our investigation, ESR can resolve a number of problems that beset its better known competitors. One of those problems, revealed above, involves the apparently arbitrary decision to foist either a substantivalist, relationist, or property theory interpretation on GR's metric field. Since ESR exclusively concerns epistemic facts or truths about mathematical structure, and not ontology, the difference (if there is truly a difference at all) between the opposing substantivalist, relationist, and property theory camps is no longer a factor.

The analysis of the GR substantival/relational debate in §5.2.1 nicely demonstrates how two divergent ontological interpretations can nevertheless agree on the necessity of a common structure: for both Einstein and Hoefer, the metric field g is the structure identified with spacetime, whether as its "real representor" (Hoefer), or where spacetime is thought to be a "structural quality of the [g] field" (Einstein). In either case, if you remove g, then you remove spacetime. On an ESR construal of the debate, consequently, both Hoefer's substantivalist and Einstein's relationist or property version of GR would appear to constitute different ontological interpretations of the very same underlying physical theory, since the key mathematical structure equated with the nature of spacetime is identical in both cases. In fact, a host of diverse ontological evaluations of GR (or Newtonian physics, etc.) could fall under the same ESR category for the same reason, i.e., if they all straightforwardly accept the theory's spacetime structure regardless of their different ontological appraisals of that structure, whether substantivalist, relationist, or even the property theory. For instance, the various proposals that posit spatiotemporal structure as a type of property (Sklar 1974; Teller 1987; Dieks 2001a), or as a form of sophisticated modal relationism (Manders 1982; Teller 1991; Hinckfuss 1975), presumably do not differ on the mathematical structure and predictive scope of the relevant spacetime theory.[18]

---

[18] An objection to grouping similarly structured substantival and relationist theories into a single ESR category might draw on Maudlin (1993), where a classificational scheme is presented that attempts to formulate both relationist and substantivalist versions of specific spacetime structures: e.g., Newtonian substantivalism, Newtonian relationism, etc. For many sophisticated relationist theories, however, both the structures advocated, *and* the physical implications of those structures, seem identical to those espoused by the substantivalists. A case in point is Teller's later (1991) "liberalized" relationism, which not only employs the same inertial structure as the substantivalist (1991, 381), but may only require the existence of one object or event (396) to ground those structures—since Teller rejects a strict eliminativist relationism, his liberalized relationism can agree on the meaningful possibility of, say, a lone rotating body in an empty universe, a state-of-affairs that is completely unacceptable under strict relationism. The vacuum solutions to GR need not be seen as violating this "one object/event" rule, either, since one can simply side with Einstein's hypothesis that the metric field is a physical field, and is thus a physical "object" that instantiates the mathematical structure g. Maudlin's (1993) conception of substantivalist and relationist spacetime structures will be critically examined in §9.5.3.

In brief, the rationale for the ESR approach stems from the apparently irresolvable ontological dispute between substantivalists and relationists: the ESR theorist maintains that all we can ever know about spacetime is its structure, and not the competing claims that, say, the spacetime of GR is either a unique non-material substance (sophisticated substantivalism) or a unique physical/material substance (sophisticated relationism) or a unique physical/material property (property theory). More carefully, any ontological interpretation of a spacetime theory that puts forward the same mathematical structure constitutes the same ESR spacetime theory. So, given the fact that most sophisticated relationists and sophisticated substantivalists, as well as property theorists, (1) accept the standard formalism of the relevant spacetime structure, such as M and g from the set (M,g,T) in GR, as well as (2) accept the implications of these structures (e.g., our lone rotating body), it follows that ESR must regard these apparently different theories as identical.[19] The essential criterion for an ESR approach to spacetime is the structure actually utilized in the theory, and neither the ontological ranking of those structures nor the attempt to prove that some structures are more privileged from a mathematical perspective is relevant to the theory's ESR classification.[20] Consequently, even those ontological interpretations of GR that strive to identify the substance of spacetime with either M or g will separately fall under the same ESR category that endorses the joint M and g structure. As mentioned previously, these interpretations do not eliminate, but rather still employ, both M and g (e.g., as discussed above in §5.2.1, Hoefer admits that M cannot be dropped altogether since "it represents the continuity of spacetime and the global topology"; 1998, 24).

A key feature of ESR, first pioneered in Worrall's (Worrall 1989) adaptation of structuralism for a scientific realist audience, is that it can be helpful in explicating our evolving understanding of the importance of structure considered over the course of theory change. In the same manner that the theoretical success of both Fresnel's elastic sold ether and Maxwell's electromagnetic field can be explained as due to a common structure (i.e., Maxwell's field equations) despite their different ontologies, so the evolution of spacetime theories can demonstrates how conflicting spacetime commitments may, or may not, incorporate the same necessary struc-

---

[19] This is a fairly informal presentation of the identity of structuralist theories. Formulating an identity criterion for structural realism is a work-in-progress, as noted in Da Costa and French (2003, 122), but one could utilize their method of partial isomorphisms among the sub-structures of models (48–52). Shapiro (1997, 91–93) also describes several means of capturing "sameness of (mathematical) structure" across different systems through the use of a full, or partial, isomorphisms of the objects and relations of the compared structures.

[20] Some of the different mathematical formulations of spacetime theories can be regarded in this manner; i.e., they may employ the same structures, but simply disagree on which structure is more fundamental, a point raised above with respect to the Quine-Carnap debate. The necessity of the mathematical structure to the function of the theory is the key idea, here, regardless of its primary or derived status. In Chap. 8, we will return to this topic, with the competing mathematical formulations of a spacetime theory (such as GR) subsumed under the more general problem of underdeterminism, so that there are many different theoretical combinations of "geometry coordinated to physical processes" that save the phenomena.

tures.[21] For instance, whereas Newton's conception of space and time can be faulted for postulating an unnecessarily rigid structure (absolute position and velocity), Descartes' competing conception lacks the necessary structure required to make sense of his own laws of motion. It was only much later that the spacetime structure mandated by Newtonian mechanics came to be fully recognized—and, not coincidentally, the newly discovered structure can be seen as combining facets of relationism (in the symmetry group that eliminate absolute position and velocity) alongside the more absolutist insights of the substantivalists (as in the affine structure). Likewise, in the wake of the modern hole argument, the striking resemblance among many sophisticated substantivalist and sophisticated relationist hypotheses can be seen as a further manifestation of this structuralist evolutionary tendency. As the analysis of spacetime theories progresses (the insufficiency of strict relationism, the hole argument, etc.), the structures put forward by the competing ontologies draws ever more closer, and may have reached a point where there is no longer any significant difference. It is this capacity of ESR to reduce the seemingly irresolvable ontological conflicts, and focus on the crucial role of structure, that marks its true advantage in the spacetime debates.

5.3   Conclusion

In this chapter, we have examined the so-called third-way interpretations of the ontology of space and spacetime, with special emphasis placed on the property theory and spacetime structural realism. One of the goals has been to determine if there is a contemporary form of the property theory that matches the type of spatial ontology put forth by Newton and Leibniz, since both of these seventeenth century natural philosophers posited, respectively, an hypothesis that matches key components of a property theory of space (see Chaps. 2 and 3, in particular). As we have seen, the traditional version of the property theory, where space is a property of matter, fails to correspond with Leibniz' conception, let alone Newton's. And, leaving aside the historical comparisons with ancient and contemporary hypotheses, not only does a matter-based property theory hold little advantage over the standard substantivalist/relationist dichotomy, but the vacuum solutions in GR represent a grave threat to its plausibility as well. However, if the spatial property is associated with the metric field in GR, and not matter (i.e., the stress-energy tensor), then the property theory not only eliminates the vacuum solution problem, but it also gains credibility via its similarity with metric field versions of substantivalism, relationism, and spacetime structural realism. Likewise, a property theory of the metric field does not need to explain away the seemingly non-relational state-of-affairs

---

[21] "Fresnel's equations are taken over completely intact into the superseding theory [Maxwell's]—reappearing there newly interpreted but, as mathematical equations, entirely unchanged" (Worrall 1989, 120). Worrall's use of Fresnel's equations as an example is problematic for various reasons, at least for the spacetime theorist, as will be explained in Chap. 8.

embodied in a uniform rotation of the universe, a scenario that raises awkward interpretational problems for any form of relationism.

As regards the spacetime structural realist hypothesis, SR, we have argued that the epistemic brand of SR also holds an advantage over the other metric field interpretations for a single, if contentious, reason: by representing the ontologically neutral mathematical facts of spacetime structure, ESR treats all of the other contenders in the ontological dispute, such as sophisticated substantivalism and sophisticated relationism, as the very same type of SR spacetime theory, thus eliminating a great deal of fruitless quarreling over which ontological label to affix to the metric. To recall our previous query: Is there really any meaningful difference between calling GR's metric field either a unique substance (substantivalism), a unique physical entity (relationism), or even a unique property (property theory)? An ESR theorist (as well as an OSR theorist, for that matter) would interpret all of these different ontologies as the very same ontology, although, as often mentioned, a plausible case could be made that the property theory has much more in common with ESR than sophisticated substantivalism or relationism, since a structure is normally conceived as a structure *of* some thing, and hence it fits the general category of a property as opposed to conceiving the metric field as a substance or relation. Yet, since ESR involves mathematical structure, it has the potential to fill in the details of the property theory by specifying the relevant structures at issue in a physical theory. As for the drawbacks of the ESR proposal, it can be argued that the move towards a structural realist philosophy only replaces one mystery, whether spacetime is a substance, relation, or property, with another mystery, how ESR, with its emphasis on epistemology, relates to ontology in general. Given that our overall project is centered upon the ontology of space, more details will need to offered on this aspect of ESR, as well as on the interrelated issues of causation, the possible underdetermination of mathematical structures, and how structuralism and ontological disputes in the philosophy of mathematics factor into a structural realist conception of space. In Chaps. 7 and 8, we will take up these issues, along with an in depth examination of OSR, but the next chapter will explore a different structuralist topic, namely, Newton's argument on the identity of spatial points.

Chapter 6
Newton's Immobility Arguments and the Holism of Spatial Ontology

For those philosophers of space and time inclined towards a structuralist, third-way approach to ontology, Newton's defense of the immobility of the parts of space is often regarded as a powerful historical precedent, prompting a host of structuralist-leaning commentary over the past several decades. Newton's arguments, which appear in his early *De grav* and the *Principia*'s scholium on space and time, have been analyzed by Stein, DiSalle, Healey, Torretti, and many others committed to a more nuanced spatial ontology than traditional substantivalism and relationism offer. Recently, however, there have appeared two important assessments, by Nerlich and Huggett, that question whether Newton's structuralist or holistic conception of the identity of spatial parts ultimately undermines his overall conception of space, a problem that, interestingly, does not appear to be a connected with his espoused absolutism or alleged substantivalism. Since Newton bases the identity of the parts of space on their structural relationships, and since all the parts of his infinite Euclidean space manifest the same structural relationships with one another, do these parts thereby lack the necessary identity criterion for a coherent theory of space? In order to better grasp Newton's arguments and his general conception of these issues, this chapter will explore the background of, and the possible sources of influence on, Newton's theory of the identity of spatial parts, as well as critique several important interpretations and arguments put forward by commentators. Yet, this chapter is not limited to an historical examination of seventeenth century theories alone, since a contemporary analogue of the problems associated with Newton's treatment of the identity of spatial parts finds a home in contemporary spacetime debates. The goal of this chapter, consequently, is two-fold: first, we will rebut the problems raised by both Nerlich and Huggett by means of a more intricate historical and philosophical analysis of the spatial holism intrinsic to Newton's theory; second, we will argue that modern debates on the ontology of spacetime, some of which have been motivated by similar puzzles, have either unwittingly followed, or could benefit from, Newton's holistic conception of spatial ontology.

As for the chapter outline: §6.1 will survey Newton's "immobility arguments", as they are called, whereas §6.2 will be devoted to the historical context of Newton's

© Springer International Publishing Switzerland 2016

E. Slowik, *The Deep Metaphysics of Space*, European Studies in Philosophy of Science, DOI 10.1007/978-3-319-44868-8_6

arguments. Specifically, §6.2 will analyze the holistic character of Newton's conception of space, i.e., its simplicity and oneness, a holism that, from the modern perspective, amounts to the interconnection of metrical and topological structure. The important historical work of McGuire and Tamny will figure prominently in this part of our inquiry. In §6.3, the lessons gathered from our investigation will be juxtaposed with similar disputes and stratagems in contemporary spacetime debates, thus demonstrating the continuing relevance of these issues for the contemporary exploration of spacetime ontology.

6.1   Newton's Immobility Arguments

In Newton's *De gravitatione* (probably early 1680s), the important unpublished tract first discussed in Chap. 2, a seemingly innocuous argument is put forward for the immobility of space that, ultimately, would spawn much debate among Newton scholars and philosophers of space and time. Following Huggett's (2008) valuable contribution to these issues, the apparent change of emphasis in this passage will be labeled (Ai) and (Aii) respectively:

> [(Ai)] The parts of space are motionless. If they moved, it would have to be said either that the motion of each part is a translation from the vicinity of other contiguous parts, as Descartes defined the motion of bodies, and it has been sufficiently demonstrated that this is absurd; or that it is a translation out of space into space, that is out of itself, unless perhaps it is said that two spaces everywhere coincide, a moving one and a motionless one. [(Aii)] Moreover, the immobility of space will be best exemplified by duration. For just as the parts of duration are individuated by their order, so that (for example) if yesterday could change places with today and become the later of the two, it would lose its individuality and would no longer be yesterday, but today; so the parts of space are individuated by their positions (*positiones*), so that if any two could change their positions, they would change their individuality at the same time and each would be converted numerically (*numerice*) into the other. The parts of duration and space are understood to be the same as they really are only because of their mutual order and position (*ordinem et positiones inter se partes*); nor do they have any principle of individuation apart from that order and position, which consequently cannot be altered. (N 25)

An argument similar to (Ai) also turns up in the scholium on space and time from the first edition of the *Principia* (1687), a passage we will identify as (B).

> [(B)] Just as the order of the parts of time is unchangeable, so, too, is the order of the parts of space. Let the parts of space move from their places, and they will move (so to speak) from themselves. For times and spaces are, as it were, the places of themselves and of all things. All things are placed in time with reference to order of succession and in space with reference to order of situation (*situs*). It is of the essence of spaces to be places, and for primary places to move is absurd. They are therefore absolute places, and it is only changes of position (*translationes*) from these places that are absolute motions. (N 66)

In brief, (Ai) and (B) argue that the parts of space cannot move since that would entail the allegedly contradictory or impossible state-of-affairs that a part could move "out of itself", (Ai), or "from themselves", (B). (In what follows, "parts" and "points" will be used interchangeably.)

We will return to (Ai) and (B) in §6.2, but a more in-depth examination of the "identity" argument, (Aii), as we will call it, is in order: since the parts of space are understood to be the same due to their "the mutual order and position", and since any interchange of parts preserves the same mutual order, thus there can be no interchange of parts, and thus the parts cannot really move/interchange. The trouble with (Aii), as Huggett succinctly puts it, is that "if any two parts of space are indistinguishable with respect to their metrical relations to the other parts of space, then they are strictly identical" (2008, 396–397). More carefully, Newton claims that points have no "principle of individuation apart from [their] position", where Newton's phrase is taken by Huggett as pertaining to the metrical relations *between* points; thus, given the symmetries of (infinite) Euclidean space (where every point has the same metrical relations to every other point), and given that there are no previously identified points relative to which others can be identified, it follows that the points of Newton's (Euclidean) space are really the *same* point—henceforth we will refer to this dilemma as the "collapse" problem or argument. In order to establish this obstacle for Newton's theory of space, something like Leibniz' "principle of the identity of the indiscernibles", PII, must be in play: if two things have identical properties (so that neither has a property different than the other), then they are the same thing (i.e., one thing and not two). The upshot of the collapse problem, where the points of Newton's (Euclidean) space collapse into a single point, not only raises a contradiction for Newton's conception of absolute (substantival) space, but it would apparently undermine Euclidean geometry as well. Nerlich's diagnosis of the (Aii) collapse problem is similar, although his overall interpretation of the philosophical import of Newton's arguments differs considerably from Huggett's: "Every point in Euclidean space satisfies [Newton's (Aii)]"; and "order and situation without some hint of individuality independent of that order is powerless to identify—to distinguish any point from any other" (Nerlich 2005, 123). As will become evident later, a key issue is whether or not Newton's phrase, "mutual order and position", really pertains to metrical (distance) relations. In §6.2, this will be demonstrated to be a correct assessment, although it needs to be qualified because Newton considers order and position to be attributes of space, which is a quantity. Specifically, we will need to examine the metaphysical background to Newton's conception of space, a framework that relies on such notions as attribute, quantity, quality, and their interrelationships.

6.2 The Historical Background to Newton's Immobility Arguments

From a geometrical perspective, it is the symmetric and homogeneous nature of Newton's Euclidean space that generates the collapse problem, although other symmetric spaces of constant curvature, such as a spherical or hyperbolic space, would suffer the same fate, but not most variably curved and dynamical spaces (see, §6.3, and Wüthrich 2009, for an (Aii)-like criticism of spacetime structuralism). Huggett strives to block the consequences of the collapse problem, and thereby evade the unpalatable conclusion that all the points in Newton's absolute space are the very same point, by utilizing a sophisticated form of *de re* representation of points across states or worlds. (Huggett's attempted solution is not the subject of this chapter, but see footnote 13.) Nevertheless, as will be argued, an in-depth historical analysis of Newton's conception of space, and the "immobility" arguments in general, i.e., (Ai), (Aii), and (B), can side step the collapse problem without the need for these more complex modern strategies.

The first issue that requires attention is the intended purpose of Newton's immobility arguments. As is clear from the title of his article, "Can the Parts of Space Move?", Nerlich sees the immobility arguments primarily as a metaphysical effort to counter that very possibility. Huggett's understanding of these arguments is similar, since he eventually judges, with respect to (Ai), that "it is not clear how this argument secures Newton against the motion of the parts of space relative to one another"; and, "although [A(ii)] does demonstrate the relative immobility of the parts of space, since Newton cannot consistently hold it, he has no demonstration at all" (2008, 394). But, are Newton's immobility arguments solely intended to preclude the *motion* of the parts of space? This section will strive to show that Newton actually had different objectives in mind; specifically, to refute the notion that space has real, divisible parts, and (at least potentially) to deny that space itself requires a space or place. It should be noted, finally, that Newton's immobility arguments were prompted in part by Descartes' hypotheses of place, motion, and body, which Newton inferred led to problems associated with the mobility of place, but the connection with Descartes will only assume a small part of our discussion.

6.2.1   Oneness, Indiscerpibility and Simplicity

Starting probably with Zeno's paradox of place, one of the traditional difficulties with this concept is the potential regress that ensues given the stipulation that all things, with the usual exception of God, require a place: If place required a place, then the "place of place" would need a place, too, etc. These problems were well known by the late Medieval period, as is evident in the writings of Albert of Saxony and John Buridan (see, Grant 1981, 18).

The Cambridge Neoplatonist, Henry More, may have tackled the regress issue, albeit indirectly, in his *Enchiridium Metaphysicum* (1671), a work that almost certainly influenced the content of Newton's *De grav*, a point discussed at length in Chap. 2.[1] In an elaborate metaphysical examination of the properties of space, More's insistence that infinite spatial extension is "one" would seem to encompass concerns about a multiplicity of places: "infinite extension distinct from matter…is one to the extent that it is absolutely impossible that to that one there be many, or that it make many, since it has no physical parts from which they can be combined

---

[1] On Newton's indebtedness to his predecessors (e.g., Charleton, More, Barrow, Wallis), see, e.g., Hall (2002), McGuire and Tamny (1983), and McGuire and Slowik (2012).

and into which they can be truly and physically divided" (EM 58). On More's estimation, this oneness of space is inextricably linked with its "simplicity" (that space is without parts) and "indiscerpibility" (a word coined by More that denies the actual or real divisibility of space, i.e., physically or metaphysically, by a process of tearing or cutting; see, once again, Holden 2004 for an analysis of the different conceptions of divisibility in the Early Modern period). In the *Enchiridium*, More relates these features of infinite spatial extension to his conception of immobility. His aim is to explain "*in what way that infinite immobile extension distinct from matter is one, simple, and immobile*", after which he defines oneness and simplicity: "[infinite extension] is aptly called simple, seeing that it has, as I have said, no physical parts" (EM 58). More continues:

> And this simplicity, however, is easily understood of its immobility. For, no infinite extension which is not combined from parts, nor is condensed or thickened in some way, can be moved, either from part to part, since the whole is simple and indiscerpible, nor can the whole at the same time, since it is infinite, be contracted into less space, since it is not condensed anywhere nor can it leave its place, since this infinite is the intimate place of all things, within or beyond which there is nothing. (58)

In short, (1) the immobility of the parts of space is a direct consequence of the oneness, simplicity, and indiscerpibility of infinite extension, and (2) the whole of space cannot move since there is nothing (e.g., a second place/space) relative to which it can move. While (2) does provide a rationale for (1), the simplicity and oneness of space does not necessarily rule out a regress of one, simple spaces. Besides the sheer unintelligibility of a regress of space, however, More (and Newton, as will be argued below) appealed to ontological concerns of a deeper sort—namely, God—to justify their acceptance of (2). More claims that immobility "is celebrated as the most excellent attribute of First Being in Aristotle" (58), and contends that space is God's attribute (57), thus securing space's immobility as a whole by brute ontological fiat. That is, if God is immobile, then God's properties (attributes), such as space, acquire the same immobility as well.

In the wake of McGuire's groundbreaking historical work, the claim that Newton's immobility arguments, (Ai), (Aii), and (B), are predicated on a set of beliefs similar to More's is, to put it bluntly, practically indisputable. In an unpublished piece from the early 1690s brought to light by McGuire, labeled "Tempus et Locus", Newton follows More's lead by forthrightly asserting that "space itself has no parts which can be separated from one another,.... For it is a single being, most simple, and most perfect in its kind" (TeL 117). The likely incentive for both Newton and More's viewpoints is the troubling prospect that discerpibility may also be ascribed to the ontological foundation of space, namely, the omnipresent God. In *De grav*, Newton cautions that "lest anyone should…imagine God to be like a body, extended and made of divisible parts, it should be known that spaces themselves are not actually divisible" (N 26). By denying that space is comprised of separable parts, Newton thus blocks any ploy to attribute parthood to God via spatial divisibility, e.g., Leibniz' well-known line of attack in the Leibniz-Clarke correspondence (L.V.42). Given the assistance that Newton likely rendered to Clarke, it is therefore not surprising that many of the themes of oneness, simplicity, and indiscerpibility, both for space and God, figure prominently in Clarke's detailed replies:

For infinite space is one, absolutely and essentially indivisible, and to suppose it parted is a contradiction in terms, because there must be space in the partition itself, which is to suppose it parted and yet not parted at the same time. The immensity or omnipresence of God is no more a dividing of his substance into parts than his duration or continuance of existing is a dividing of his existence into parts. (C.III.3)

[I]nfinite space, though it may be...conceived as composed of parts, yet since those parts (improperly so called) are essentially [indiscerpible][2] and immovable from each other and not able to be parted without an express contradiction in terms..., space consequently is in itself essentially one, and absolutely indivisible. (C.IV.11-12)

With respect to Newton's immobility arguments, Clarke's contention that "there would be space in the partition" is the likely analogue of Newton's earlier claims that a part of space would move "out of itself" in (Ai) or "from themselves" in (B). The contradiction that Newton had tried to articulate is, arguably, expressed more successfully by Clarke: since "there would be space in the partition", this "is to suppose it parted and not yet parted". This is a somewhat cleaner formulation than Newton's assertion that spaces are the "places of themselves" in (B).

Consequently, it is important to bear in mind that the immobility arguments are not solely intended to establish a thesis prohibiting the motion of the parts of space: rather, their specific purpose is to offer reasons for denying that the parts of space can be *really divided* or separated from one another. It is preserving the oneness and simplicity of space that is the intended target of More and Newton's immobility arguments.

Before leaving this topic, it is worth delving further into the relationship between, on the one hand, the regress argument, and on the other, Newton's views on the oneness, simplicity, and indiscerpibility of both God and space. As noted above, the oneness, simplicity, and indiscerpibility of space does not necessarily stop a regress of spaces (places), but, since Newton declares that the ontological foundation of space is an infinite God, there is accordingly a unique irreducible "object" that grounds the existence of space together with its immobility and infinity (as is also the case with More). There are abundant passages from which to choose, many previously examined in Chap. 2: *De grav*, "space is eternal in duration and immutable in nature because it is the emanative effect of an eternal and immutable being" (N 26); General Scholium, *Principia* (1713), "He [God] endures forever, and is everywhere present; and by existing always and everywhere, he constitutes (*constituit*) duration and space" (91). This last conjecture, that God "constitutes" space, is particularly important for it would seem to rule out the possibility that any regress of spaces (places), or a motion, is applicable to this entity (against Khamara 2006, 111–112). To be specific, it is only meaningful to entertain a regress of the space of God if it is possible to meaningfully discuss a space apart from God—but, since space is "brought about" by God, there can be no spatial framework, or second place/space, upon which to establish the meaningfulness of this entity's regress or motion. This explanation gains support from a passage in TeL where Newton asserts

---

[2] As Koyré and Cohen point out (1962, 91), most modern translations incorrectly use the term "indiscernible" in place of "indiscerpible", the latter being the term actually used in Clarke's original reply but mistranslated in the published versions of the correspondence.

that God contains "all other substances in Him as their underlying principle and place" (TeL 132). Whether or not this from of response constitutes a successful resolution of the regress problem is unclear, needless to say, but it apparently forms an instance of a conception popular from the early Medieval period up through Newton's own time: namely, that God was "in Himself" prior to the creation of the world—and there are seventeenth century precedents for equating God's being "in Himself" with not being in a place (see, Grant 1981, 330, n.57).[3] All of these themes, incidentally, support the ontological dependence conception of the property theory of space, P(O-dep), first introduced in Chap. 1, over a substantivalist or relationist interpretation.

6.2.2 Simplicity and Spatial Holism

At this juncture in our historical examination of the immobility arguments, it is essential to assess the background of the identity argument, (Aii), for it is the origin of many current disputes concerning the part-whole relationship in Newton's conception of space. The problem, to recap, is that if the identity of spatial parts is established by their mutual metrical relationships, which roughly correlates with Newton's "mutual order and position", then this leads, purportedly, to the parts of space being identical since their mutual order relationships are identical, i.e., the collapse problem.

One route out of this difficulty is to embrace a form of spatial holism or monism, so that the parts (points) of space are no longer viewed as independent elements that directly form or construct the whole of space.[4] Rather, the dependency relationship goes the other way, with the whole of space comprising the basic ontological entity, and the parts derived from, or supervening on, the whole; hence each part upholds the PII and the collapse problem is avoided. Nerlich, drawing on Healey (1995),

---

[3] It is also interesting that the main published works that link the ontology of God and space do not appear until the General Scholium of the second edition of the *Principia* (1713), although this would seem consistent with Newton's general avoidance of God's role in his published natural philosophy prior to his later years (post 1700). Incidentally, another means of establishing God's immobility in the Scholastic period is to emphasize the connection with God's infinity (omnipresence): Vasquez reasons that, if God is infinite, then there is literally no place for God to move (see, Grant 1981, 369, n.125), a view that Locke accepts in the *Essay* as well (E II.xxiii.21).

[4] Monism and holism, as used in this monograph, will refer to an entity's "oneness", and is thus equivalent to the notion of oneness used by More and Newton. Therefore, to employ the term "monism" or "holism" with respect to, say, the metric field of GR is to claim that its topology or other structures are not independent of the metric, rather they are all parts of a single entity, g. Presumably, there is a difference between holism and monism, since monism implies the holism of that entity, but it would seem that an entity could be holistic but retain an ontological independence of its parts. In what follows, consequently, our use of holism will reject the ontological independence of the parts of an entity (in keeping with the demand for the oneness and simple nature of space advocated by both More and Newton).

ultimately favors this interpretation of Newton's spatial theory: "Assume that space is real, but it is not *made up* of its parts, nor yet *analyzable into* parts with any kind of ontic *independence*. Perhaps, even, that spatial parts and their relations are, ontologically, *supervenient* on the structure of space" (Nerlich 2005, 131). On this model, since infinite (Euclidean) space is *itself* the "structure-instantiating entity", the identity of the parts/points is determined relative to the whole, and not via their relationships to one another. In accordance with the PII, each part of space retains a unique identity, although these parts are now understood to be supervenient features of the whole of space. The collapse argument, in contrast, is based on the premise that the relations among the parts determines each part's identity, an assumption that leads to the collapse of space into a single part via the PII (because these mutual relations are identical for all parts/points).

What are the ontological implications of this form of spatial holism? While there are a number of ways to construct a holistic account of space, some of the main contenders might be the following: (i) invoke the supervenience of parts on the whole of space, which includes all spatial structures and does not single out any one in particular; (ii) limit the supervenience to just the distance (metrical) relations, so that metric relations are primary and the parts/points are derived; or claim (iii) that spatial relations are internal relations of each point, and not external relations (although this last strategy might be deemed only indirectly holistic or structuralist).[5] On the holistic interpretation in general, it is clearly true that the homogenous nature of infinite Euclidean space makes it difficult to distinguish spatial parts, but this is an empirical problem far removed from the troubling ontological worries associated with Huggett's reading of (Aii), which, as noted previously, apparently conceives space as derived from a ground floor of points and "their metrical relations to one another" (2008, 397).[6] There are objections which might be raised against this holis-

---

[5] If one interprets our holistic maneuver as the view (ii), that metrical relations are primary, with the identity of points dependent on these relation, what then accounts for the identity of metrical relations? This new version of the collapse problem (for metric relations) fails, however: given any two metric relations among points, say, g^1 and g^2, their identity will be secured via a larger metrical relation, g^3, which includes both g^1 and g^2 within its scope, and so on for any extent of space (to infinity). Internal relations are employed in (iii), where internal relations are sometimes described as the relational equivalent of an essential or monadic property: i.e., the relation, R, that a point, p, bears with another point, q, is viewed as an internal relation of p if p bears R to q in all possible worlds. Unlike an external relation among points, therefore, an internal relation R incorporates the identity of the point p, and thereby does not violate the PII and is not subject to the collapse argument. (This strategy was suggested by a referee with respect to an earlier version of this chapter.) Since (iii) refers to individual points and their properties, it exhibits a somewhat non-holistic appearance, but it leads to the same interconnected holism of space as in (i) and (ii). A further investigation of strategies (i), (ii), and (iii), and all of the other possible constructions, is clearly required, however.

[6] Some material thing would seem to be required to serve as a coordinating basis to resolve these epistemological worries. Moreover, references to the "whole" of space include, unless otherwise noted, all structures in space along with lesser (non-three) dimensional structures and the R^3 point manifold.

tic interpretation that draw upon the division of labor, topological and metrical, in modern differential geometry, but we will postpone that discussion until §6.3.

This holistic strategy for interpreting Newton's (Aii), which has been advanced by Howard Stein as well (e.g., Stein 2002, 272),[7] also bears a certain resemblance with structuralism in the philosophy of mathematics, as Healey's remarks in connection with the immobility arguments would seem to imply: "[I]t is its place in a certain relational structure that makes p the spacetime point that it is. In this respect spacetime points are analogous to mathematical objects. It is its position in the natural numbers which makes 3 the number that it is" (Healey 1995, 303).[8] Ironically, the holistic conception of space that both Healey and Nerlich posit can be best described using the Neoplatonic terminology utilized by both More and Newton; namely, that space—including the metric (which is roughly akin to Newton's order of situation/position of spatial parts)—is one, simple, and indiscerpible! Although Healey, Nerlich, and Stein do not provide historical support for their respective interpretations, the discussions above do indeed corroborate a reading of Newton's spatial theory that is consistent with a holistic/monistic interpretation of the parts of space—an interpretation, moreover, that has much in common with the philosophy underlying contemporary spacetime structuralism, the sophisticated brands of both substantivalism and relationism, and the property theory (see §6.3).

The intent of (Aii), put simply, is to demonstrate that space is a non-aggregate, partless whole, whereby the very individuality of spatial parts is derived from the whole. The (Aii) argument, hence, provides a more detailed explanation why the motion of the parts of space, critiqued initially in (Ai), is not possible. In a previously quoted passage from TeL, space's non-aggregate structure, which is both single and simple, is defended using the same arguments, in the (Ai) and (Aii) vein, about the immobility of spatial parts:

> But neither does Place argue the divisibility of a thing or the multitude of its parts,…, since space itself has no parts which can be separated from one another, or be moved among themselves, or be distinguished from one another by any inherent marks. Space is not compounded of aggregated parts since there is no least in it, no small or great or greatest, nor are there more parts in the totality of space than there are in any place which the very least body of all occupies. In each of its points it is like itself and uniform nor does it truly have parts other than mathematical points, that is everywhere infinite in number and nothing in magnitude. For it is a single being, most simple, and most perfect in its kind. To be bounded in time and in place, or to be changeable does argue imperfection, but to be the same always and everywhere is supreme perfection. (TeL 117)

---

[7] Stein provides the following comment on (Aii): "This can be taken, in rather modern terms, as saying that space is a *structure,* or "relational system", which can be conceived of independently of anything else [i.e., it is not simply a relation among existents, contra strict relationism]; its constituents are individuated just be *their relations to one another, as elements of this relational system*" (Stein 2002, 272).

[8] In correspondence, Nick Huggett has pointed out that a more adequate analogy would be to a set with the ordinal properties of the integers alone, without labels (such as 3), since every member of this set bears the same relation to some other member, and so this relationship would be preserved under the mapping.

Among the many revelations in this passage, it is worth drawing specific attention to Newton's assertion that space only has parts in the sense of "mathematical points, that is everywhere infinite in number and nothing in magnitude", and his following claim that, "nor are there more parts in the totality of space than there are in any place which the very least body of all occupies." In other words, his conception of the part-whole constitution of space adhere to what we may call the classical or Aristotelian-Euclidean view of geometry, wherein a line of any length can be conceptually decomposed into an infinity of points, although the line itself is not actually constructed by a process of adding points (since they have no magnitude). This aspect of Newton's theory clearly has holistic overtones, but the truly non-reducible character of the spatial metric, and its relationship with (Aii), will only become evident after exegesis of Newton's other tracts on spatial ontology.

6.2.3 The Order of Position of Spatial Parts

Overall, the classical geometric inclinations that run through a host of mature works, such as *De grav*, *Principia*, and TeL, can be traced back to one of Newton's earliest investigations, namely, the Trinity Notebooks from 1664 to 1665 (*Questiones*). The similarities between the *Questiones* and these later writings can be described, roughly, as pertaining to the individuation of points and the continuity of space, two aspects of Newton's treatment that are intimately linked to the question of space's holistic, or simple, nature.

In the *Metaphysics*, Aristotle puts forward an hypothesis on the difference between points and units that can be seen, in retrospect, as one of the principal ideas motivating Newton's (Aii): "that which is indivisible in quantity is called a unit if it is not divisible in any dimension and is without position, a point if it is not divisible in any dimension, and has position" (CWA V.6.1016b24-27). In short, points are without dimension, but they have position, unlike units/numbers. As McGuire and Tamny explain, in the Aristotelian-Euclidean tradition, "the point itself lacks existence independent of the line, but it can be distinguished by its position relative to another point, or with respect to the line itself" (McGuire and Tamny 1983, 62). The incentive behind the use of position as a means of identification likely resides in the unique complications associated with points and the definition of continuity (see, *Physics*, W VI.1.231a21-231b18). Since points are partless, points cannot touch without completely overlapping and losing their individuality. Put differently, if two points were in contact they would possess common extremities; but two points that possess common extremities are continuous and one, since they occupy the same place. This interrelationship between place and continuity is echoed in Newton's *Questiones:* "Extension is related to places, as time to days, years, etc. Place is the *principium individuationis* of straight lines and of equal and like figures; the surfaces of two bodies becoming but one when they are contiguous, because in but one place" (CPQ 351). Likewise, "if you say then that [a point] might touch one of the other points that makes the line, I say then that that point is in the same place with the point that it touches" (421). As is equally the case in (Aii), the geometric elements (points, parts) are individuated by way of the overall spatial backdrop (places, order of situation of spatial parts), since the peculiar character of the geometric elements on the Aristotelian-Euclidean scheme renders them incapable of securing their own individuation (due to the continuity problem). For our purposes, the implications of this form of reasoning is that it undermines any attempt to construct the metric of space from the relationships among *independently* established parts—indeed, if the actual identity of the parts is dependent on the whole of space, which would include space's metrical structure, then Nerlich's claim that the parts supervene on the overall structure of space would appear to be vindicated.

As a rejoinder, the critic might contend that Newton's appeal to place as a means of individuating geometrical elements only commits him to the weaker (topological) notions of coincidence/non-coincidence, and not a metric, the latter approximate to Newton's "order of situation". Newton's explanation that "extension is related to places, as time to days, years, etc." would seem to undermine this line of response, however. Since a day or year is a part of duration and has a particular finite duration, it follows that place can possesses a particular finite extension as well—hence it is very difficult to tie Newton's use of "place" exclusively to a non-metrical, topological conception. This last inference is also supported by arguments put forth in the Leibniz-Clarke correspondence, where space and time are categorized as "quantities, which situation and order [are] not" (C.V.54). A criticism of Leibniz' definition of space, as the "order of coexistences" (L.III.4), prompts Clarke's explanation of the importance of quantity as opposed to situation or order: "the distance, interval, or quantity of time or space...is entirely a distinct thing from the situation or order and does not constitute any quantity of situation of order; the situation or order may be the same when the quantity of time or space intervening is very different" (C.V.54). (Nevertheless, as discussed in Chap. 3, Leibniz insists that his conception of the "order of coexistence", which is a ratio or proportion, has quantity, contra Clarke; L.V.54). For Clarke and Newton, situation and order are likened to ratios and proportions, which "are not quantities but the proportion of quantities" (C.V.54). Overall, Clarke's explanation nicely demonstrates that Newtonian natural philosophy presumes the metric (distance) to be a basic quantitative feature of space—and this, of course, imparts a metrical significance to all of its constitutive parts, whether points, lines, surfaces or volumes. The scholium on space and time makes the same point in a passage we shall label (C):

[(C)] Place is the part of space that a body occupies, and it is, depending on the space, either absolute or relative. I say the part of space, not the situation [*situs*] of the body or its outer surface. For the places of equal solids are always equal, while their surfaces are for the most part unequal because of the dissimilarity of shapes; and situations, properly speaking, do not have quantity and are not so much places as attributes of places [*quam affectiones locorum*]. (N 65)

As with Clarke's account, situations "properly speaking" do not have a quantity, unlike space/place—indeed, situation is an *attribute* (property) of the quantity place, and hence space, given the oneness of space discussed above.

McGuire and Tamny interpret this last passage, (C), as a reversal of (Aii), however. The parts of space do not obtain their identity from the order of position (situation), as in (Aii); rather, since positions are the attributes of places, it follows that places, as the parts of space, are prior to the order of position. They conclude that "this indicates that situations derive their character from the parts of space on which they depend, and not the converse as *De gravitatione* states" (McGuire and Tamny 1983, 73). But, this would seem to confuse the order of position *per se*, namely, as in Clarke's examples of pure ratios or proportions, which do not have quantity, with the order of position of *spatial parts*, which does have quantity (hence the rationale behind our use of the latter designation throughout this chapter). That is, Newton's immobility arguments always associate order or situation with space, which is a quantity; e.g., "the order of the parts of space" in (B), and "the parts of space are individuated by their positions" (Aii). So, there is little evidence to support the idea that Newton changed his conception of the identity of spatial part from *De grav* to the *Principia*: both assume space is a quantity. On the other hand, McGuire and Tamny's are correct in pointing out that situation alone does not account for the identity of Newton's spatial parts, for quantity, as an attribute, is also required.

Returning to (C), since Newton specifically mentions "the situation of the body or its outer surface", it would seem that, like Clarke, his goal in this passage is to criticize the general relationist strategy of determining place by means of the mutual situations of bodies, as well as dismiss the Scholastic/Cartesian idea that place is the boundary of the contained/containing bodies (both aspects included within Descartes' conception of place; Pr II 13–15; see also §1.3.2). Moreover, as Rynasiewicz (1995, 141) has noted, Newton's explanation that "the places of equal solids are always equal, while their surfaces are for the most part unequal because of the dissimilarity of shapes" is a reference to the intrinsic volume of place (as opposed to the surface area of the body's boundary or the non-quantity order/situation of bodies), and volume is, of course, a metric measure. Therefore, when (C) is added to the holistic, simple characterization of space in *De grav*, TeL, etc., explored above, the basic metrical nature of space, as a quantity with the attribute of order/situation, becomes readily apparent. Moreover, although (C) does not refer to points and other geometrical elements, it would be a mistake to single out only place or volume as the key components in Newton's spatial ontology. Belkind (2007) attempts to make a case for volume as central to Newton's defense of absolute space, based largely on the scholium on space and time, and employing Newton's anti-relationist argument that "whole and absolute motions can be determined only by means of unmoving places" (N 67). Yet, as we have seen, non-dimensionless points are held to be as much a part of space as place (volume), e.g., TeL 117, and points are also distinguished by the metric, as in the *Questiones* extract (CPQ 421). Despite the fact that Newton's examples utilize moving bodies, which have volume, the appeal to the compound motions of the constitutive parts of bodies could have employed bodily points as easily as bodily volumes to attack Descartes, along with using the points of absolute space to determine absolute motions (as opposed to relative motions).[9]

---

[9] Concerning other aspects of Belkind's (2007) innovative analysis: that a Cartesian body's quantity of motion involves volume, or internal place (actually, the volume of its second and third ele-

Before leaving this section, it would be useful to examine a passage from *De grav* that further validates Newton's adherence to the Aristotelian-Euclidean conception of the continuity of geometry/space, a passage which also highlights the relationship between parts and points:

> In all directions, space can be distinguished into parts whose common boundaries we usually call surfaces; and these surfaces can be distinguished in all directions into parts whose common boundaries we usually call lines; and again these lines can be distinguished in all directions into parts which we call points. And hence surfaces do not have depth, nor lines breadth, nor points dimension, unless you say that coterminous spaces penetrate each other as far as the depth of the surface between them, namely what I have said to be the boundary of both or the common limit; and the same applies to lines and points. Furthermore, spaces are everywhere contiguous to spaces, and extension is everywhere placed next to extension, and so there are everywhere common boundaries of contiguous parts;.... (N 22)

This explanation provides important details on the geometrical composition of Newton's spatial ontology: points, lines, surfaces, and thus volumes, are all elements of Newton's one, simple, and indiscerpible space (much as TeL describes space's parts as mathematical points). Indeed, as is also the case with Leibniz' holistic conception of geometry (see Chap. 3), lines are the "common boundaries" of surfaces, and points the common boundaries of lines: contra the collapse argument, points are *not* free standing or independent geometric entities that form relations (distance) among other independent points, where these relations supervene on the points. If supervenience is involved, it is points that supervene on lines, and lines that supervene on surfaces, etc., which is in keeping with the holism examined above. This realization thus undermines any attempt to foist the distinctions of modern differential geometry on Newton's conception of space, since the clear division between a topological manifold and an overlaying metric in the modern theory finds scant support in Newton's classical approach.

A possible objection to the holistic supervenience ontology surveyed above is that is simply unintuitive in some fashion, for instance, in that it fails to conform to our common experience of parts retaining their identity apart from the whole. While

---

ments; see, Slowik 2002, chapter 4) is not undermined by the fact that the constitutive parts of that body may have their own motions, and thus their own quantities of motion linked to their own volumes. This is no more a problem for Descartes than for Newton, especially given the latter's Corollary 5 (the principle of Galilean relativity). Newton's ship example is, ironically, Descartes' own part-whole illustration: "on a ship, all motions are the same with respect to one another whether the ship is at rest or is moving uniformly straight forward" (N 78; cf. Descartes, Pr II 13). That is, if the scholium's part-whole critique (above) is devised "to support the concept of momentum" (2007, 288), as opposed to a conceptual criticism of Cartesian motion, then Corollary 5 would undermine Newton's own mechanics as well (since the true motion of the ship would need to be determined in order to calculate the momentum of any interactions on the ship). Furthermore, the rotating bucket and globes examples are probably best viewed as an inference to the best explanation in support of absolute space, since the non-inertial effects of rotation are not correlated with the relative motions of the bodies, contra relationism, and this is the only legitimate grounds that Newton can offer in support absolutism *if* the choice is confined to just absolutism and relationism—although Belkind's point (290) may be that this constitutes a false dichotomy.

the philosophical foundation of the holistic notion of space is indeed a difficult topic, it must be conceded that the ontology of all spatial theories suffer from a similar defect, but in different ways (see also §5.1.3). For example, as regards the layered set of structures employed by differential geometry (and assumed in the modern spacetime debates on ontology), it is hard to grasp the reality of a bare topological point manifold bereft of metric structure, nor is it easy to imagine a finite space, etc. So, given the inherent difficulties in gauging the intuitive coherence of all, or most, spatial structures and ideas, the holistic supervenience concept fairs no worse, and may even be more intuitive, relative to its non-holistic competitors. More, Newton and Clarke, it should be recalled, considered the potential separation of the parts of space to be so counter-intuitive that they used that scenario as the basis of a reductio argument in order to establish the opposite case, i.e., a holistic spatial hypothesis (oneness and simplicity).

6.2.4   The Least Distance Hypothesis

The inference that space has an essential metric structure is corroborated elsewhere in the Trinity notebook, where Newton explores, and ultimately rejects, the possibility that spatial lengths can be comprised, bottom-up, from a least unit of distance that resembles something akin to atoms of space, but is also linked to the topology of its constitutive mathematical points. The "least distance" hypothesis seems motivated by the Epicurean atomist idea that there exists a minimal indivisible quantity of matter, so that the minimal distances become "the basis of all other extensions and the mould of atoms" (CPQ 423). Newton uses a cipher method of marking off the points on a line, with the stipulation that the ciphers "resist being the same" (421), that is, they retain a power of non-coincidence (cf. Huggett 2008, 398) that generates the least unit of spatial distance. The collection of ciphers thereby characterizes the units of least distances among the points, *partes extra partes*, along the line. Given a point, if "there be another point with which it refuses to be joined,… then there is distance between the two, though indivisible, and the least that can be" (CPQ 423). Unfortunately, Newton's assumption that these least distances are indivisible runs into the obvious difficulty that, at least conceptually, "the least extension is infinitely larger than a point and therefore can contain it and be divided by it" (425). This prompts the reply, "I confess it is so", along with an abortive effort to establish that, although a least distances "has no inside, no midst, nor center", it must be the case that the infinite number of points in that least extension "must be all in the borders or sides and outward superficies of it, and that cannot make out a place for division" (425). For our purposes, it is important to note that Newton ultimately crossed out these notebook pages, i.e., the pages that elaborate his least distance thought experiment, likely due to the untenability of his defense of their indivisibility.

Another contributing factor in the demise of Newton's least distance hypothesis is the inevitable implication that there must exist a direct correlation between the length of a line (figure) and its likely *finite* number of constitutive points. Central to Newton's hypothesis, as we have seen, is the notion that the points "are imbued with such a power as that they could not touch or be in one place", which leads to the following conclusion: "add these [points] as close in a line as they can stand together. *Every point added* must make *some extension* to the length, because it cannot sink into the former's place or touch it" (343, emphasis added). This conclusion not only conflicts with other views advanced in the notebooks (e.g., "points added between points infinitely are equivalent to a finite line"; 345), but it is clearly alien to the Aristotelian-Euclidean direction that Newton's mathematical thought would increasingly take after 1665. Recalling Newton's claims, in TeL, that "space is not compounded of aggregated parts", and his denial that there are "more parts in the totality of space than there are in any place", it would appear that space acquired its simple, holistic structure fairly early in his philosophical development, since his non-simple, non-holistic conception of an atomic least distance is absent from all later works subsequent to these (deleted) pages of the Trinity notebook.[10] In short, given his failure to construct a metric from a topology of points that possess an elemental power of non-conjunction, and since the subsequent portions of the notebooks accept that his geometrical elements are individuated by a metrically-influenced concept of place (e.g., CPQ 351), it is thus not surprising that his later utilization of the order of position/situation of spatial parts, in (Aii) and (B), is similarly imbued with a metrical significance.

Lastly, based on their analysis of the cipher construction in the *Questiones*, McGuire and Tamny offer a prescient observation with respect to *De grav*'s (Aii) which foreshadows the difficulties later developed by Nerlich and Huggett. They observe that "positions are positions of parts, and they depend for their character on the parts themselves", rather than on the points of the earlier cipher method (McGuire and Tamny 1983, 72). But, the infinity of space necessitates that "one position, any one, be nameable independent of the others" which "cannot be done" (72). However, repeating our earlier critique, McGuire and Tamny err by overlooking the simplicity and oneness of space: to claim that the positions of parts "depend for their character on the parts themselves" is tantamount to asking for a criterion of the individuality of the parts independent of the whole, which raises a host of problems for Newton's other non-reductive, holistic pronouncements on the parts of space (as examined above)—e.g., "nor are there more parts in the totality of space than there are in any place which the very least body of all occupies" (TeL 117). How does one make sense of this passage on McGuire and Tamny's suggestion? On a more positive note, they construe these difficulties "from an epistemic perspective" (McGuire and Tamny 1983, 72), which marks a notable improvement over the more troubling ontological allegations submitted by both Nerlich and Huggett.

---

[10] McGuire (1982) and Koslow (1976, 254) attempt to make a case for a least spatial unit in Newton's post-*Questiones* natural philosophy, or that Newton's spatial ontology at least does not countenance dimensionless points. But, the passage quoted from TeL above (TeL 117) utterly refutes these readings, and, in fact, McGuire ultimately rejects the least distance interpretation in an endnote added later to his essay (McGuire 1982, 185).

Part of the motivation for McGuire and Tamny's criticism of (Aii) may lie in their assertion that one of the ideas that might have survived Newton's abandonment of the least distance hypothesis, at least in some form, is the cipher method for demarcating the units of distance, i.e., a metric: "every cipher…being different or distant from all the former by the quantity of a unit" (CPQ 423). While it may be true one can draw an analogy between the cipher method and Newton's later discussion of the spatial order of position in *De grav* and later works, the generality of the concept of a spatial order hardly requires the earlier precedent, although the relational structure of the arrayed ciphers does bear a vague resemblance with the identity argument in (Aii). That is, given a collection of 11 ciphers, demarcating 10 units of distance, the collected ciphers represent a non-aggregate, simple metrical structure wherein one cannot add another cipher "into the midst of them, as between [ciphers] five and six" (423). Yet, unlike (Aii), the relational structure of the ciphers is generated from "a nature and quality that they will resist being the same"—in other words, each cipher retains enough *individuality* that it can resist becoming identical to another cipher, and from these individual "powers" the metric is constructed. (Aii) stipulates, in contrast, that the individuality of spatial parts is determined by the metric (order of position of spatial parts) alone, with no hint of any individual traits apart from, or prior to, the metric of the whole (simple, one, indiscerpible) space.

6.3 Contemporary Spacetime Ontology and the Immobility Arguments

To briefly summarize §6.2, we have demonstrated that Newton posits a holistic conception of space such that the identity of the parts supervenes on the whole, thereby avoiding the collapse problem. Not only is this conclusion supported by an analysis of the texts on the oneness and simplicity of space which are directly linked to, and the basis of, the immobility arguments themselves (especially, TeL 117), but this conclusion is likewise bolstered through his abandonment of the non-holistic least distance hypothesis in the Trinity notebooks. In this section, we will continue our examination of the more prominent contemporary analysis pertinent for understanding Newton's views, but the emphasis will progressively shift to the relevance of Newton's arguments and hypotheses for current debates on the ontology and structure of space (spacetime).

6.3.1 Spatial Transformations and Leibniz Shifts

Returning to *De grav*'s version of the immobility arguments, Torretti has proposed that (Aii) can be interpreted as postulating a criterion of the identity of points "but only up to isomorphism" (Torretti 1999, 55; where "isomorphism" is defined as "a structure-preserving one-to-one mappings" of the points of space):

Newtonian—that is, Euclidean—space admits an infinity of distinct internal isomorphisms.... In particular, if we designate one of these copies be *E* and we represent by the
vector **v** a translation of each point of *E* in the direction of **v** by a distance equal to **v**'s
length, then,..., the translation *t***v** yield the successive positions of a frame *E***v** moving
through *E* with a constant velocity **v**. (56)

Torretti draws the conclusion that, based on this reading of (Aii), "all inertial frames
are equivalent" (56), and hints that this Newton-inspired approach can also help to
resolve Einstein's hole argument: i.e., the hole argument "forgets the fact, so clearly
set forth by Newton, that points in a structured manifold have no individuality apart
from their structural relations" (297; see Chap. 5 on the hole argument). While
Torretti is correct in his overall holistic conclusions as regards the lack of primitive
identity for Newton's points, the implications that he draws for other aspects of
Newton's ontology of space are quite problematic. Nerlich (2005, 129) rightly criticizes Torretti's analysis as incompatible with the last sentence of (B), which posits
motionless absolute places, so that "changes of position from these places...are
absolute motions". Specifically, while Newton's *Principia* draws a distinction
between absolute and relative space (with the latter being inertially related copies of
absolute space), the true rest frame of the material world is absolute space, and thus
not all inertial frames are *ontologically* equivalent (more on this below).

Yet, it is more instructive to examine Torretti's reading against the backdrop of
De grav's immobility arguments, since he employs these passages to support his
interpretation of Newton's spatial ontology, and not the *Principia*'s (B). While not a
mathematical mapping of the parts of space per se, (Ai) does give two reasons for
rejecting the motion of spatial parts. First, Newton states that the motion of a part of
space might be "a translation from the vicinity of other contiguous parts, as
Descartes defined the motion of bodies, and it has been sufficiently demonstrated
that this is absurd" (N 25). In a preceding section of *De grav*, Newton offers a number of arguments against Descartes' conception of external place, as the boundary of
the contained and containing bodies, and its corresponding definition of motion as
change of place (Pr II 25). Newton reckons that "after the completion of some
motion the position of the surrounding bodies no longer stays the same" (N 19),
and, since these contiguous plenum bodies must fill the vacancy left after the body
moves, this reordering thereby eliminates the original material boundary, external
place, required to determine the motion.[11] Newton's accusation of the absurdity of
moving spatial parts in (Ai) would therefore seem to be based on a premise similar
to his line of criticism of Cartesian motion, namely, that the motion of a part of
space/extension would bring about a corresponding reshuffling of the remaining
spatial parts, so that the motion of the part would be likewise indeterminate. Overall,
it difficult to grasp how this peculiar model of the motion of the parts *through* space,
i.e., the strange consequences that stem from this plenum-based conception of mov-

---

[11] At length, Newton argues: "Now since it is impossible to pick out the place in which a motion
began,...for this place no longer exists after the motion is completed, that traversed space, having
no beginning, can have no length; and since velocity depends upon the length of the space passed
over in a given time, it follows that the moving body can have no velocity" (N 20).

ing spatial parts, could in any way qualify as the equivalent of a modern mathematical transformation.

Newton's second (Ai) criticism of the idea that spatial parts can move incorporates more familiar themes: "or...it is translation out of space into space, that is out of itself, unless perhaps it is said that two spaces everywhere coincide, a moving one and a motionless one". We have already examined the alleged contradiction in claiming that space can move "out of itself" (see §6.2.1); namely, using Clarke's explanation, that it is "to suppose it parted and yet not parted". In the second half of this sentence, the phrase "unless perhaps it is said" is of particular interest, for it apparently signifies that it is an exception to the idea that the part moves "out of space into space", and this is consistent with the remainder of the sentence: specifically, Newton imagines that the supposed moving part does not actually leave its space, but merely occupies two spaces simultaneously, the original motionless space and a moving space that "everywhere coincides" with it. If this interpretation is correct, then this brief aside may constitute the closest approximation to a geometric transformation concept in Newton's natural philosophy of space, and it shows that he, at least temporarily, entertained the idea of multiple spaces. Yet, the type of transformation that Newton envisages in this passage is not an active transformation, "a one-one mapping of spacetime onto itself" (labeled a "point transformation" in Torretti 1999, 263), since this implies the "out of space into space" type of mapping which Newton rejects. Nerlich also finds Torretti's exegeses a violation of the "out of space into space" prohibition (Nerlich 2005, 128), but he fails to take the transformation analogy a bit further. The form of mapping that best correlates with Newton's (Ai) explanation would seem to fall within the category of a passive (or coordinate) transformations, where the geometric objects *remain fixed* under a substitution of coordinates — in the (Ai) case, it would be a transformation of a coordinate frame x at a point p, to a another coordinate frame y also at p, where y is related to x by a velocity boost $\mathbf{v}$, rather than as an active mapping h from p to its image under the mapping, h(p) (see, e.g., Friedman 1983, 51–53; Torretti 1999, 263–264). Consequently, if any proposed resolutions of Einstein's hole argument were to necessitate an active (point) transformation, as Torretti seemingly maintains (1999, 297), then his citing Newton's (Ai) as an historical precedent is wide of the mark.[12]

---

[12] Then again, if the hole argument is conceived employing a passive (coordinate) transformation, so that the original and mapped geometric structures (for instance, g and $\breve{g}$, see Chap. 5) are merely alternative representations of the *same* reality, then maybe Newton's (Ai) can indeed be seen as resolving this issue. Unlike active transformations, which purportedly describe a troubling physical underdeterminism involving two distinct states (or worlds), passive transformations do not pose any epistemological or ontological mysteries since they are trivial (coordinate) redescriptions of the same state (or world). The modern version of the hole argument is predicated on the active reading of transformations, of course; if not, it would fail to represent a problem for substantivalism or any other theory. Yet, since many text books on differential geometry and GR move happily between the active and passive use and interpretation of transformations, the hole argument is open to the objection that it is foisting a specific interpretation of geometric practice on substantivalists without argument. That is, while the active reading is the preferred basis for understanding many structures and models in contemporary physics, whether or not the hole argument poses a valid

A further impediment for Torretti's "active transformation" interpretation of Newton's immobility arguments stems from the drastically different conceptions of geometry employed in the seventeenth century and modern spatiotemporal theories, a point raised earlier. Modern differential geometry, with its many component geometric structures (manifold, affine, metric, etc.), is utterly foreign to the classical, Euclidean conception shared by both Newton and Leibniz, where each component geometrical structure is regarded as a limit or boundary of the next higher dimensional component: e.g., "space can be distinguished into parts whose common boundaries we usually call surfaces; and these surfaces can be distinguished in all directions into parts whose common boundaries we usually call lines; and again these lines can be distinguished in all directions into parts which we call points" (N 22). Likewise, Leibniz often asserts that points "strictly speaking, are extremities of extension, and not in any way, the constitutive parts of things; geometry shows this sufficiently" (AG 228). On this form of geometric holism or monism, the kinds of isomorphic mappings that generate the modern hole argument are simply inapplicable, since points cannot be separated from the larger geometric structures if they are mere boundaries: i.e., there is no point manifold, M, and thus no diffeomorphisms, $h(g)=\check{g}$, $h(T)=\check{T}$, that can generate the (observationally indistinguishable) models $(M,\check{g},\check{T})$, from (M,g,T), that plague the modern manifold substantivalist (see, Chap. 5 on the modern form of hole argument against substantivalism). Put simply, the independence of the various geometric structures in modern differential geometry generate the familiar underdetermination worries, such as the hole argument.

Lastly, the identity argument (Aii) is also relevant to the Leibniz shift scenarios, a topic first presented in Chap. 3. As previously mentioned, Huggett strives to avoid the implications of the identity argument by developing a representational account of points: "if two points in different worlds or states have the same metrical relations to other points then they represent the same points" (Huggett 2008, 401). Yet, given the homogeneity of Newton's Euclidean space, it therefore follows that the material world cannot differ as regards its position in absolute space (static shift), nor possess a different state of absolute motion (kinematic shift):

> Since representation *de re* supervenes on the metrical relations between points, one state cannot unequivocally represent any point as standing in any different relation to bodies from the other. But bodies can only move with respect to space—i.e., absolutely—if they can occupy different parts of space at different times; and that is impossible if matter isn't represented as occupying different parts of space in the possible states in the world. (2008, 405)

---

physics problem is precisely the issue at hand. Defenders of the passive interpretation of the hole argument can insist that isomorphic models, such as $\check{g}$, are the side effect of the *geometric* redundancies inherent in modern differential geometry, and hence not a true case of physical underdetermination at all. In fact, maybe a criterion should be invoked that confines legitimate ontological worries to only those underdetermination cases that arise under *both* the active and passive interpretations. At any rate, this assessment does not deny that the gauge-invariant interpretations of spacetime theories raise a host conceptual difficulties, such as the problem of frozen time and the status of "observables" in GR and quantum gravity (see, Belot and Earman 2001)—rather, our critique applies only to the issue of (unobservable) spacetime points apropos substantivalism. For an active interpretation of Einstein's Hole argument for spacetime theories, see, Earman and Norton 1987.

DiSalle's provides a different interpretation of (Aii), but one which has similar consequences, for he claims that Newton "is expressly denying that the points of space…are 'irreducible objects of first order predication' (cf. Earman 1989) and therefore denying by implication that the material universe would be intrinsically different if it existed at different spatial and temporal points" (DiSalle 1994, 267). But, as Huggett correctly points out (2008, 404–405), Clarke admits that Leibniz shifts are distinct, possible states of the world (C.V.1-20), a stance that Newton apparently found unobjectionable in his review of the Leibniz-Clarke correspondence for Des Maizeaux (see, Koyré and Cohen 1962). Indeed, one of the main goals of the scholium is to demonstrate that "absolute and relative rest and motion are distinguished from each other" (N 66). It is thus not surprising that Huggett finds the demise of Leibniz shifts to be a major obstacle for his own *de re* representation strategy.[13]

To recap the various themes discussed thus far, it is important to bear in mind that Newton provides a fairly body-centered exegesis of absolute and relative place/space in the scholium: "relative space is any movable measure or dimension of… absolute space; such a measure or dimension is determined by our senses from the situation (*situm*) of the space with respect to bodies and is popularly used for immovable space" (N 64). This manner of depicting the absolute/relative dichotomy could be seen as upholding a form of coordinate transformation, and it would naturally align with Corollary 5 of the *Principia,* i.e., the principle of Galilean relativity. Nevertheless, it is not a transformation of the sort expressed in passage (Aii) from *De grav*, where the transformations are conceived metaphysically or conceptually and *only* involve the parts of space. Rather, since the *Principia*'s Galilean transformations are defined, so to speak, materially or "operationally" by using bodies (N 64–67),[14] it follows that these active transformations operate at the purely phenomenal level, so that one, and only one, of the potentially infinite set of transformations corresponds to Newton's immobile absolute place/space. And, despite Newton's valiant efforts to distinguish absolute from relative space by way of absolute motions and their effects, the scholium's thought experiments that strive to determine the absolute motion of bodies (i.e., the rotating bucket and rotating

---

[13] It should be added, here, that Huggett's analysis, which examines the possibility of different spatial structures, is a very worthwhile exercise in its own right. His project is based on the idea that Newton's theory allows counterfactual situations that require an account of the identity of points across such models. Overall, a modern theory based loosely on Newton's views, and employing modern differential geometry, can benefit greatly from Huggett's analysis. The intention of this chapter, however, is to make the historical case that Newton's ideas do not, in fact, support the reality or possibility of these counterfactual states, as well as to make the philosophical point that modern spacetime debates prompted by contemporary versions of the identity argument can actually benefit from following Newton's conceptions more closely.

[14] For example: "place is the part of space that a body occupies" (N 65); and "we define all places on the basis of the positions and distances of things from some body that we regard as immovable, and then we reckon all motions with respect to these places" (66). He adds that "absolute motions can be determined only by means of unmoving places,…and relative motions to movable places" (67). As for absolute place/space: "the only places that are unmoving are those that all keep given positions in relation to one another from infinity to infinity" (67).

globes) are incapable of breaking the symmetry of the Galilean transformations and thereby reveal the true rest state of space (i.e., absolute space). The reading that both Torretti and DiSalle favor, accordingly, does not capture the intended meaning of Newton's (Aii), which concerns the metaphysics of the parts of space, and not the symmetries of material inertial systems. DiSalle's attempt to deny the Leibniz shift scenarios, furthermore, suggests that there is no ontological basis for absolute rest, i.e., Newton's God, a stance that would seem to indorse the strong third-way reading of Newton first introduced in Chap. 2.

6.3.2 Conclusion: Modern Spacetime Ontologies and Newton's Immobility Arguments

As a means of wrapping up our analysis, it would be useful to briefly contrast Newton's views on space, both in their ontological and holistic aspects, with analogous debates in the contemporary philosophy of spacetime, in particular, as regards the main contenders examined in Chap. 5: metric field substantivalism, sophisticated (metric field) relationism, structural realism, and the property theory. To summarize our earlier findings, the most basic distinction among these ontologies is that, unlike substantivalism, sophisticated relationism, structural realism, and the property theory reject the existence of space (spacetime) in the absence of physical objects or fields. Furthermore, it is important to recall that analogues of the collapse argument can be advanced against modern spacetime ontologies if they appeal to a structuralist conception of parts like Newton's (Aii) argument (more on this below).

Overall, a plausible case can be made that many of the current crop of sophisticated spacetime theories, whether sophisticated (metric field) substantivalism, sophisticated (metric field) relationism, structural realism, and the property theory, are consistent with the broad outlines of Newton's (Aii) conception. First, these theories all endorse a holistic approach similar to Newton's by emphasizing the crucial role of the metric (approximate to Newton's "order of position of spatial parts") in securing the identity of the points of the manifold: to be precise, whereas manifold substantivalism accepts the primitive identity of manifold points, with Field (1980) being a prime example, structural realism, the property theory, sophisticated metric field relationism, and sophisticated substantivalism all reject this primitive identity, and strive instead to place both metrical and topological structure on at least an even footing. Indeed, the holistic/monistic nature of spacetime structure is a recurrent theme in many of these contemporary interpretations of classical gravitation theories, such as Newtonian theory or general relativity (GR), or other field theories, like quantum field theory (see, e.g., Auyang 1995, Esfeld and Lam 2008, and the other references below). Of course, there are other reasons for preferring some of these modern ontologies as better representatives of Newton's spatial theory (see below, and Part III), but, as argued in §5.2.2, there is little or no difference between sophisticated substantivalism, metric field relationism, structural realism, and the property theory, hence they all represent Newton's (Aii) argument equally well.

Second, in the context of GR, the rationale for claiming that these modern ontologies are consistent with Newton's spatial holism is that all four predicate their holistic spacetime structure on a pre-given "entity" of sorts, namely, the metric field (or metric plus manifold, the latter without primitive identity of points, of course).[15] Specifically: the sophisticated substantivalist deems the metric field to be a unique substance, dubbed "spacetime", that can nonetheless interact with other fields (e.g., Hoefer 1996); the sophisticated metric field relationist hypothesis views the metric, through its tie to the gravitational field connection, to be just another material/physical field (e.g., Rovelli 1997); the structural realist judges the metric to be a unique physical field that is a kind of hybrid of the relationist's material/physical field and the substantivalist's spacetime substance view (e.g., Dorato 2000); and the property theory would put forward a similar, albeit property-oriented, interpretation.[16] As discussed in §6.2.1 and Chap. 2, Newton likewise grounds space on a unique entity: God "is everywhere present; and by existing always and everywhere, he constitutes duration and space" (N 91). So, given that both Newton and these modern spacetime ontologies posit a holistic entity to undergird spatial (spatiotemporal) properties, respectively, God or metric field, which is not merely mathematically or conceptually but actually holistic, it follows that the collapse argument simply begs the question since its mathematical/conceptual machinery is being applied to a domain that is, by stipulation, already holistic in the substantival/physical/(supernatural) sense. Put differently, the collapse argument must rely on something like the following premise in order to gain purchase: the properties of the parts of a whole (here, identity) need to be fixed prior to examining the properties of the whole. But, the non-local character of GR's metric and other physical fields (and possibly theological entities) would seem to stand as a direct counter-example to this line of reasoning—so why should a structuralist accept the collapse arguments' basic premise? To be precise, GR's metric/gravitational field, g, exhibits a form of holism (i.e., non-locality) because the term that represents the energy of the gravitational field, t^{ab}, is a pseudo-tensor, where "its non-tensorial nature means that there is no well-defined, intrinsic 'amount of stuff' present at any given point" (Hoefer 2000, 193; see, also, Lam 2011). Consequently, since the energy of the gravitational field cannot be pinned down to specific points, but can only be attributed to the whole field or extended regions, there is a well-established holistic precedent upon which to base a holistic interpretation of other aspects of g, contra the collapse argument and its non-holistic assumption of locality.

Finally, an additional consequence of the ontological grounding issue is that, like Newton's theory, all four of our modern spacetime ontologies allow vacuum solu-

---

[15] In what follows, similar conclusions can be reached for quantum gravity hypotheses, although that discussion will be the topic of Chap. 10. Hence our analysis in this chapter will remain confined to GR.

[16] As noted in Chap. 5, all references to GR's metric incorporate its unique relationship with the gravitational field, via the Christoffel symbols of the metric. As Cao explains, "although the spatio-temporal relations are constituted by the chrono-geometrical structure (the metric), the latter itself is constituted, or ontologically supported, by the inertio-gravitational field (the connection)" (Cao 2006, 45).

tions since GR's metric is never absent in any region of spacetime, even in a spatial region empty of matter (stress-energy), and this is quite unlike the more traditional non-sophisticated, non-metric field relationisms (see §1.1.1). However, while Newton's God is both extended and present in space, and can act on matter, God is beyond a reciprocal influence in the same way that matter acts upon GR's metric, and thus there is a dissimilarity in the case of GR between Newton and these modern holistic spacetime ontologies (see Chap. 2 as well).

As regards specific examples, there are a host of recent structuralist hypotheses that can be viewed as following the general outlines of Newton's (Aii) argument. Cao's structuralism "takes the metric and connection as holistic structures that enjoy ontological priority over their components" and thus "takes the ultimate reality of spacetime as being field-theoretical in nature" (Cao 2006, 46; and footnote 16). Stachel's structuralism draws upon a distinction between quiddity, which refers to a classification employing natural kinds, and haecceity, which pertains to individuality or "primitive thisness":

> The points of spacetime [in GR] have quiddity as such, but only gain haecceity (to the extent that they do) from the properties they inherit from the metrical or other physical relations imposed on them. In particular, the points can obtain haecceity from the inertio-gravitational field associated with the metric tensor: For example, the non-vanishing invariants of the Riemann tensor in an empty spacetime can be used to individuate these points in the generic case. (Stachel 2006, 57).

Another holistic-leaning interpretation is Hoefer's metric field substantivalism, which aims "to strip primitive identity from space-time points", so that "the focus of this view is on the metric tensor as the real representor of space-time" (Hoefer 1996, 24; see Chap. 5). A holistic interpretation might also find support in Maudlin's work, especially if his earlier "metric essentialism" thesis—that the "parts of space bear their metrical relations essentially" (1988, 86), put forward in the context of (Aii)—is regarded as akin to an internal relation of each point (see strategy (iii) in §6.2.2, and footnote 5). Yet, leaving aside the question of the relationship between holism and internal relations, since Maudlin (1988) apparently relies on the standard separation of manifold and metric structures in differential geometry, metric essentialism likely amounts to a bottom-up approach, from points to the whole of space. The essential metrical qualities of the parts of Newton's space, as we have seen, are secured by its oneness and simplicity, a top-down approach that it is likely the converse of metric essentialism's scheme. As a result, Maudlin's later espousal of a fibre bundle strategy for characterizing spatial length, wherein the base space of points is closely linked to the fibres and other higher structures, would seem to more accurately capture the spirit of Newton's spatial holism (see, also, the quotient space fibre bundle formulation in Stachel 2002, 234–235). On Maudlin's assessment of fibre bundles, path lengths in space are primary, whereas the external relationships between points are derived (2007a, 87). Since "all points related by distance to one another must be parts of a single, common, connected space" (89), fibre bundles could therefore be seen as a contemporary analogue of Newton's holistic "points as boundaries" conception. Indeed, Maudlin refers to the ontology implicit in his account of fibre bundles as "Spinozistic" (102), a description that nicely demonstrates its close allegiance to the holistic/monistic account of spatial geometry advanced in this chapter. On the other hand, Auyang's more detailed investigation reveals that substantivalist and structuralist conceptions can be discerned among competing fibre bundles interpretations as well. If one adopts a bottom-up stance, viewing the base space as the foundation upon which all other structures are established, then a fibre bundle version of manifold substantivalism can be defended. The top-down, structuralist approach, in contrast, would seem to be the proper holistic fibre bundle equivalent of Newton's conception. On this top-down conception, the base space is "neither a substratum nor an entity that can stand on its own. Rather, it is an arching structure of the physical gauge field as a whole" (Auyang 2000, 492). This form of top-down structuralist strategy, whether for fibre bundles or other structures, would seem ideally suited for defeating the modern versions of the collapse problem, e.g., Wüthrich 2009, for the identity of the parts is determined by the whole physical field. That is, one need not posit an autonomous identity for the constituent points of a physical field *prior* to ascertaining the larger, global properties of that physical field—that general assumption, once again, underlies the collapse argument's would-be trap of the unsuspecting structuralist. Rather, as in the case of Newton explored in §6.2.2, the structuralist can insist that the whole field is prior to its supervening points.

Naturally, there are components of Newton's theory of space that will not be congenial to the modern holistic strategies surveyed above, especially given the contemporary theoretical context. We have already touched on a few, in particular, the lack of a reciprocal influence between matter and Newton's space-instantiating entity, God, which is unlike the dynamical relationship between matter (stress-energy) and GR's space-instantiating entity, the metric/gravitational field. This difference is, of course, a consequence of Newton's static conception of physical geometry, as opposed to GR's dynamic standpoint, but the philosophical basis of the former likely stems from the intersection of Newton's beliefs concerning theology and geometry. Much like the unchanging circular motions that comprise a main component of ancient celestial hypotheses, Newton's God-grounded ontology of space provides a key insight into the motivations underlying his infinite, unchanging spatial geometry. In the passage from TeL quoted in §6.2.1, which refers to space, but could equally describe his theological ontology, he states: "to be changeable does argue imperfection, but to be the same always and everywhere is supreme perfection" (TeL 117). In short, the metrical properties of Newton's space can never change because they are the direct result of a deeper unchanging entity (see Chap. 2). This realization thereby calls into question the alternative, dynamic scenarios of his spatial metric envisaged by both Nerlich (2005, 131) and Huggett (2008, 403). And, just as God is really extended in space, geometric structures would also seem to be really in space. The structure of physical space is, in fact, practically identified with Euclidean geometry, as the last quote's ensuing discussion reveals:

> For the delineation of any material figure is not a new production of that figure with respect to space, but only a corporeal representation of it, so that what was formerly insensible in space now appears before the senses....We firmly believe that the space was spherical before the sphere occupied it, so that it could contain the sphere; and hence as there are everywhere spaces that can adequately contain any material sphere, it is clear that space is everywhere spherical. And so of other figures. (N 22)

By declaring that bodies merely expose the geometric forms that are actually present in space, Newton's geometric conception of space is quite literal, and not a mere rhetorical flourish. So, while it is true that many of the modern structuralist ontologies also draw inspiration from a realism about mathematical structures, they would almost certainly recoil from embracing the type of pseudo or virtual platonism that is central to Newton's spatial geometry; i.e., since space would apparently lack existence absent Newton's instantiating entity, God, space does not qualify as platonist in the full sense (see Chap. 7 for more on these issues). Unlike the contemporary spacetime scene, furthermore, Newton's *Euclidean* spatial realism benefits from the absence of an underdetermination of alternative geometric formalisms (twistors, Einstein algebras, etc.) or alternative physical constructions (a la Poincaré) that complicate the modern picture (see Chap. 8 on these general concerns). As argued above, *De grav*'s (Aii) is not the seventeenth century equivalent of a transformation argument, nor is the clean delineation of geometric structures (metric, manifold, affine, etc.) assumed by practitioners of modern differential geometry a part of Newton's brand of Euclidean spatial realism.

In conclusion, the classical holistic conception of geometry implicit in Newton's handling of spatial ontology has long been a neglected aspect of his natural philosophy. This oversight might be rooted, at least for philosophers of physics, in the standard conception of differential geometry often utilized to interpret past spatial theories—a geometric scheme that presupposes a seemingly self-sufficient topological manifold upon which higher structures are placed (e.g., the tensors on manifold method that has been the basis of philosophical reconstructions of pre-twentieth century spatial hypotheses by contemporary philosophers of space and time). Given the obstacles that such a layered approach can create (e.g., the collapse problem, as well as the hole argument), there has been a growing awareness of the philosophical benefits that can be obtained from either a more holistic interpretation of these standard methods (e.g., sophisticated substantivalism, sophisticated relationism, structural realism, and the property theory), or for adopting different geometric techniques that more naturally lend themselves to a holistic interpretation (e.g., fibre bundles). That the modern "holistic turn" in physical geometry has a rough counterpart in the classical geometric assumptions of the seventeenth century is, therefore, both a topic ripe for further investigation and a cautionary tale of the potential historical bias that our modern mathematical methodologies can unwittingly impose.

Chapter 7
The 'Space' at the Intersection of Physics, Metaphysics, and Mathematics

As developed in Chap. 5, the attempts to capture the ontology of space and space-time using the conceptual apparatus of substantivalism and relationism—that space is, respectively, either an independently existing entity or a relation among material substances—have proved to be quite problematic since the sophisticated versions of both substantivalism and relationism seem to be identical in the context of general relativity (GR), our best theory on the large scale structure of space. But, whether space is a substance or a relation is just one of the many conceptual distinctions that have entered the modern debate on the nature of space. In addition to substantivalism and relationism, one also finds references to realism versus anti-realism, platonism versus nominalism, and background independence versus background dependence. In this chapter, we will begin the examination of these new dichotomies, their impact on the traditional substantival/relational distinction, and their relationship to one of the least discussed, but quite central, components of the spatial ontology debate, namely, how the philosophy of mathematics factors into the philosophy of space and time.

But, can these additional resources, especially platonism and nominalism, shed light on the ontological disputes concerning space (spacetime), or, more generally, the realism/anti-realism debate as regards the geometric structures used in scientific theories? Specifically, this chapter will examine two interrelated questions: (i) the relationship, if any, between platonism/nominalism and the traditional substantival/relational distinction, as well as (ii) how the spatial/geometric structures utilized by spacetime theorists and scientific realists can benefit from the introduction of platonism/nominalism. Against the backdrop of the spatial ontology controversy, both historically (Leibniz, Newton) and the modern setting (GR, quantum mechanics, quantum gravity), our investigation will demonstrate that a more nuanced version of the traditional platonism/nominalism dichotomy—one which incorporates a distinction, adapted from Pincock (2012), between a fictionalist nominalism and a liberal truth-based nominalism—provides distinctive and important information on various assumptions that undergird the spatial ontology debate, as well as helps to elucidate the scientific realist's commitment to spatial structures. Truth-based

© Springer International Publishing Switzerland 2016   175
E. Slowik, *The Deep Metaphysics of Space*, European Studies in Philosophy of Science, DOI 10.1007/978-3-319-44868-8_7

nominalism, moreover, will be shown to be the most versatile and successful hypotheses among the three platonist/nominalist categories in the context of theories of space. Although a resolution of the ontology debate by means of platonist/ nominalist notions is unlikely, the underlying assumptions that motivate substantivalism and relationism gain clarity and context when viewed from the mathematical and metaphysical perspective of platonism and nominalism, thus the introduction of these additional tools for assessment is both warranted and constructive. In particular, it will be demonstrated that this more refined platonist/nominalist scheme not only accounts for the details of both Newton and Leibniz' spatial ontologies, but also incorporates the best features of some of the other contending dichotomies, such as realism/anti-realism and background dependence/independence.

The outline of the chapter is as follows. The analysis of platonism/nominalism will begin in §7.1 by focusing on its role in the philosophy of mathematics, in particular, contemporary forms of mathematical structuralism. On this basis, our three-part distinction between immanent realism, fictionalist nominalism, and truth-based nominalism will be introduced and assessed alongside other strategies. While §7.2 will delve into the role that causation plays in both a general mathematical setting as well as for spacetime theories, §7.3 will demonstrate that our new platonist/nominalist categorization can deliver a quite successful explanation of the various complicating factors that shape Newton and Leibniz' spatial ontologies. Finally, in §7.4, some of the other applications of platonism/nominalism will be critically examined, both within the spacetime ontology debate (Field, Arntzenius) and as regards recent forms of scientific realism (Psillos).

7.1 Platonism, Nominalism, and Structuralism in Spacetime Theories and Mathematics

In examining the potential stratagems for an interpretation of spacetime theories founded on the platonist/nominalist divide, it will be beneficial to analyze the parallel developments towards a platonist/nominalist account of mathematics, especially those forms compatible with the structuralist trend in contemporary philosophy of science and spacetime theories. Although other strategies could be employed, since mathematical structuralism both encompasses the platonist/nominalist dichotomy and has been discussed in previous works on the ontology of space and time (e.g., Pooley 2006), it is ideally suited to represent the general issues that pertain to platonism and nominalism. Likewise, mathematical structuralism has been discussed in Chap. 5, and will be a major factor in Chap. 8, so using a structuralist-oriented version of the platonist/nominalist dispute has that advantage as well.

Finally, while platonism and nominalism can be formulated in countless ways, the conception developed in this chapter will focus primarily on the acceptance/ rejection of the existence of *non-spatiotemporal* abstract objects (or universals), a choice that will be explained below. Consequently, as employed throughout the investigation, the distinction between platonism and nominalism is linked to the acceptance of non-spatial abstract objects by the former and their rejection by the latter; or, put in structuralist terms, whether mathematical structures are independent of (platonism), or dependent on (nominalism), the entities that instantiate those structures. If forced to provide a metaphysical perspective, however, then the nominalism that we will ultimately adopt is consistent with both: (a) traditional nominalism, that only particulars exist, and thus any discussion of a body's properties (as predicates of that particular) are conceptual distinctions or linguistic conventions; and (b), trope theory, that only particulars exist, although a body's properties (as predicates of that particular) are further particulars, but not abstract objects.[1]

7.1.1 The Spatial Context of Platonism and Nominalism

As just disclosed, an initial dilemma that faces the application of platonism and nominalism to the ontology of space concerns the, for lack of a better term, ontological domain of the spatiotemporal structures themselves. Do they exist as a sort of platonic form or universal, independent of all physical objects or events in spacetime, or are they dependent on matter/events for their very existence or instantiation? This problem arises for a philosopher of mathematics in an analogous fashion, of course, since she also needs to explicate the origins of mathematical structures, whether set theory, arithmetic, etc. For instance, mathematical structuralism can be classified according to whether the structures are regarded as independent or dependent on their instantiation in systems—*ante rem* and *in re* structuralism, respectively—where a "system" is defined as a collection of objects and their interrelationships. *Ante rem* structuralism, as favored by Resnik (1997) and Shapiro (1997, 2000), is thus akin to the traditional absolute or substantival conception of spacetime, for a structure is held to "exist independent of any systems that exemplify it" (Shapiro 2000, 263). *In re* structuralism, on the other hand, limits the existence of structures to their exemplification in physical systems, thereby mimicking relationism.

Yet, the claim that the platonist/nominalist dispute is similar to, or can assist in the philosophical analysis of, the substantival/relational distinction is problematic

---

[1] See, Chakravartty 2011, who raises problems for using the traditional mind-based, or linguistic-conventional, type of nominalism in conjunction with scientific realism (SR): i.e., if a body's properties are mind-dependent, then scientific realism is undermined (since SR claims that a mind-independent reality explains the success of our best scientific theories). Chakravartty's seems to advocate a version of a trope theory to overcome this difficulty and uphold SR; i.e., he claims that a thing's properties can allow different mind-based interpretations without undermining the reality of the property (2011, 170–171). However, the traditional nominalist could simply accept that certain interpretations of an object's properties are true, and others false (which thereby upholds SR as well), but without reifying the object's properties as real particular things (a la trope theory). In fact, Chakravartty's view may constitute a trope-based version of the type of (neo-Meinongian) nominalism advanced in this chapter. See, also, footnotes 2 and 3.

in various ways. First, it is extremely difficult to apply the standard conception of the platonist/nominalist dispute in mathematics and metaphysics, including its contemporary analogue, universals/particulars, to the ontology of space and time. As noted above, platonism is traditionally characterized as an hypothesis that assumes the existence of non-spatial and non-temporal entities, properties, or facts of some sort; i.e., platonists believe that "there are uninstantiated properties, kinds, and relations" (Loux 2006, 41; see also Lowe 2002, chap. 19, for a discussion of the complex issues involved in the spatiotemporality of universals and particulars). *Ante rem* structuralism, for example, holds that structures can exist apart from the systems that exemplify them—but what is the nature of those existing, but uninstantiated, structures? Are they akin to ideal or immaterial entities, such as Plato's Forms? And, where do they exist (in Plato's heaven)? If universals and uninstantiated *ante rem* structures do not exist in space and time, then the platonic/nominalist distinction would appear to be of little relevance to both substantivalism and relationism, since one could accept either platonism or nominalism independently of one's stance on the spacetime ontology dichotomy. For instance, given the setting of GR's curved spacetime (and leaving aside metric conventionalist worries; see Chap. 8), both the sophisticated substantivalist and sophisticated relationist could hold that, say, a flat Newtonian spacetime structure is a platonic uninstantiated entity or property, or both could opt for the nominalist view that Euclidean spacetime structure only exists in the entity (material field for the relationist, unique substance for the substantivalist) that exemplifies that Euclidean structure. Presumably, some philosophers of mathematics might be comfortable with either of these proposed applications of the platonist/nominalist dichotomy, but it is doubtful that scientific realists would embrace platonic uninstantiated entities or properties over a nominalist view that can conveniently forsake these extravagant ontological items (but see §7.4.3 below on Psillos' view). In short, if platonism requires the non-spatiality of abstract mathematical structures, then platonism will fail to be of service to the spatial ontology dispute.

Our second criticism of the platonist/nominalist approach to spatial ontology is closely related to the first. As first argued in Slowik (2005a), given a platonism that assumes the existence of uninstantiated structures, the mathematical structures contained in *all* spacetime theories would, as a direct result, fall within a nominalist classification. That is, since "system" and "object" must be given a broad reading, without any ontological assumptions associated with the basis of the proposed structure, it would seem to follow that substantivalism does not qualify as *ante rem* structuralism. The spacetime structures posited by a substantivalist are structures *in* a substance, namely, a unique substance/entity called "spacetime" that exemplifies the structure, whereas an *ante rem* structure exists independent of any and all systems that exemplify the structure. Accordingly, if, as the nominalists insist, mathematical structures are grounded on the prior existence of some sort of entity, then both the substantivalists and relationists would appear to sanction mathematical nominalism, whether that entity is conceived as a unique non-material substance (substantivalism) or as a physical field/object/event (relationism, of either the eliminative, modal, or metric field variety). In Slowik (2005a), this default nominalist classification of all spatial ontologies is presented as a virtue on the grounds that it prevents the platonism/nominalism dichotomy from reverting to a simple reinstatement, but under a different guise, of the substantivalist/relationist dispute. Nevertheless, as will be argued below, there are alternative resources for constructing a platonist-like conception of spatial ontology that might help to shed light on the differences between substantivalism and relationism (and Newton and Leibniz). So, before declaring a nominalist victory, it is imperative to address these alternative conceptions of spatial ontology that retain important features of platonism.

First, the substantivalist might try to avoid the default nominalist classification just described by declaring that their spacetime structures are actually closer in spirit to a pure absolutism, without need of any underlying entity/substance to instantiate or house the structures; hence, it could be claimed that "substantivalism" is simply an unfortunate label. Nevertheless, as revealed in Chap. 2, Newton's own conception of absolute space does not support this contention (since an entity, God, is the required foundation for space and time), nor it is very convincing in the context of GR, especially as regards the sophisticated substantivalist theories described in Chap. 5. Given the reciprocal relationship between the metric and matter fields, it becomes quite mysterious how an alleged non-substantival, absolute structure, g, can be affected by, and in turn affect, the matter field T. For the *ante rem* structuralist, abstract mathematical structures do not enter into these sorts of quasi-causal interrelationships with physical things; rather, things "exemplify" structures (see also footnotes 9 and 10). Accordingly, one of the initial advantages of examining spacetime structures from within the context of mathematical ontology is that it drives a wedge between an absolutism about quantitative structure and the metaphysics of substantivalism, although the two are typically, and mistakenly, treated as identical. This verdict could change, of course, if a non-substance absolutist conception of spacetime becomes popular in GR, but this seems highly unlikely given the dual nature of g as both the metric and gravitational field. (However, in a non-GR setting, such as quantum mechanics or quantum gravity hypotheses, a non-substance absolutist interpretation of spacetime structure may be slightly more plausible given the absence of the type of reciprocal causal influence evident in GR; see §7.2.2 below).

A second, and much better, option for a platonist-inspired conception of space is to drop the requirement that abstract objects are non-spatial. Henceforth, our analysis will refer to "immanent realism" as the hypothesis that abstract objects, such as mathematical structures or universals, only exist in entities, broadly construed, whether that entity is a substance, a physical field/object/event, or some other type of entity altogether, e.g., a seventeenth century conception of God.[2] Essentially, this

---

[2] Immanent realism differs from a trope theory in that the former accepts abstract objects (so that, e.g., all square objects instantiate "squareness", the latter being the very same abstract object in all square bodies), whereas trope theorists reject abstract objects (hence, all square bodies have a property of squareness, the latter being a particular that stands in relations of "resemblance" with other square particulars in other square material objects). In addition, Sepkoski (2007, 16) suggests a three part division of mathematical ontology that is close to the three part division offered in our

option turns immanent realism into a form of nominalism, since we have defined nominalism as the hypothesis that there are no non-spatiotemporal abstract objects—and there are precedents for interpreting immanent realism in this manner, i.e., as an ontological position that rivals traditional platonism and thus amounts to a species of nominalism (see, e.g., Balaguer 2009). Yet, since both platonism and immanent realism accept the existence of abstract objects, such as mathematical structures, but only differ on where they exist, there is still a close relationship between these views. Indeed, since all of the remaining contenders in the realm of mathematical ontology reject the existence of abstract (or mathematical) objects, there is a good justification for accepting immanent realism as a surrogate for traditional platonism in the assessment of spatial ontology.

Yet, there are tricky conceptual difficulties connected with immanent realism that may ultimately render it an unattractive option in the spatial ontology debate. In particular, whereas it seems (relatively) unproblematic to claim that the abstract object "sphere" is located in a particular spherical body in spacetime, how does one make sense of the claim that spacetime *itself* is located or exemplified in a substance (substantivalism) or a physical body/field (relationism)? Take, for instance, sophisticated substantivalism and its identification of the metric field g with spacetime: since an abstract object provides the defining essence of a substance, the abstract object g makes a particular substance, call it x, the spacetime substance g. But, since x lacks all spatial properties or attributes prior to its exemplification of the abstract object g, how can it situate or locate the abstract object g at all? More generally, how can a non-spatial substance, like x, exemplify or instantiate an abstract object that is spatial? Presumably, one can invoke the fact that x is a substance, and thus must possesses some form of haecceity, or "bare numerical difference", that can serve as the grounds of the exemplification relationship. But this tactic, which is employed by Armstrong (1997, 109) to avoid similar worries, only raises the question whether "bare numerical difference" is itself a spatial/topological notion, and hence inadmissible prior to the exemplification of g (here, it is important to recall that sophisticated substantivalism holds that g provides the local topology; see, §5.2.1). There may be strategies available to the immanent realist to forestall these difficulties, such as a non-spatial or non-topological notion of "bare numerical difference". Yet, it is interesting to note that Armstrong, one of the chief advocates of the platonist interpretation of immanent realism, ultimately supports a position that takes spatiotemporal relations, conceived as abstract objects or universals, as constitutive of spacetime but not themselves located in spacetime:

> Where are external spatiotemporal relations located? Here, it seems to me, we could cheerfully concede, if we wanted to, that they are not located, yet not place them 'outside space and time'. For it is part of the essence of space and time that they involve such spatiotemporal relations, whether these be conceived as relations between things [relationism], or between particular places and times [substantivalism]. So, if they help to constitute spacetime, then it is no objection to their spatiotemporality that they are not located in space-time. (Armstrong 1988, 111–112; see also Armstrong 1997, 136–138; and Magalhães 2006)

---

investigation, although he equates the view that we have labeled "immanent realism" with Aristotle's position as regards the existence of substantial forms.

Interestingly, the difficulties involved with understanding the process by which an abstract object or universal brings about spacetime, but is itself not located in spacetime, forms a counterpart to the relationship between Leibniz' extended bodies and his non-spatial monads and/or non-situated God (see Chap. 4).

7.1.2   Fictionalist Nominalism Versus Truth-Based Nominalism

Just as the ontological domain of abstract objects justifies a multifaceted conception of platonism, a complex subdivision of nominalist interpretations will likewise be of use in our investigation. We have already discussed one type of nominalism, namely, immanent realism, but the most familiar and viable versions do not embrace abstract objects. For instance, an eliminativist approach to nominalism is evident in Field's (1980) attempt to treat mathematical objects and structures as entirely dispensable or fictional. Field postulates a continuum of spacetime points, conceived as entities in the manner of the manifold substantivalist, in his effort to rewrite Newtonian gravitation theory along mathematically anti-realist lines. This form of nominalism is dubbed "fictionalist nominalism", or simply "fictionalism", since the fictional mathematical ontology is eliminated in favor of an isomorphic physical/substantival ontology, with the latter providing the meaning of the former. Fictionalism is thus a close analogue of eliminativist and super-eliminativist relationism, for both replace structure, mathematical and spacetime, respectively, with existing physical entities—that is, just as eliminative relationism requires a congruence of spatiotemporal relations and extant physical relations, a fictionalist nominalism requires a congruence of mathematical structure and physical or substantival structure. In Field (1980), a work which brought the relationship between spacetime ontology and platonism/nominalism to the forefront for many philosophers of science, an infinity of substantival spacetime points are posited, isomorphic to R^4, in order to capture the content of the real numbers essential for the mathematics of classical gravitation theories, hence there is a congruence of the structure of the underlying ontology (spacetime points that exemplify the real number structure) and the higher mathematical structures employed in Newtonian theory which rely on the real numbers. In Chap. 1, we briefly examined the details of several eliminativist conceptions of space, such as by Descartes and Barbour, that would comprise the relationist equivalent of fictionalist nominalism. Sophisticated modal relationists, like Teller (1991), would not constitute the spacetime analogue of Field's program, accordingly, since this form of relationism sanctions modal spacetime structures that can transcend the structures exhibited by the actually existing physical objects (see Chap. 5): e.g., the affine structure, ∇, instantiated by a lone rotating body. Unlike the strict eliminative relationist, the sophisticated relationist can allow modal or other types of structures to serve this function, thus releasing the physical ontology of this heavy ontological burden.

The mathematical equivalent of sophisticated relationism is, rather, any of the less stringent nominalist theories that reject Field's fictionalism; e.g., Chihara (2004), Azzouni (2004), Lewis' (1993) mereological approach, or Hellman's (1989) *in re* structuralism. Much like the sophisticated relationist theories surveyed previously, the "truth-based nominalists", as we will dub this view, do not allow structures to exist independently of the systems that exemplify those structures, yet they do not believe that those structures can be eliminated in favor of an underlying physical ontology that is isomorphic to the real numbers or some other mathematical structure.[3] Hellman's *in re* structuralism, for instance, employs "possible structures" as a means of avoiding a commitment to an infinite physical ontology of spacetime points, a feature that helps to explain its common nominalist classification (see, Hellman 1989, 47–52, and Chihara 2004, 107–115). The truth-based nominalist theories often differ on how the mathematical structures are constructed from their basic ontology, as well as how to precisely construe the meaning of statements about mathematical structures, but most of these nominalisms accept, in one way or another, a realism about the truth-values of mathematical statements (e.g., Azzouni limits the truth-values of mathematical statements to only applied mathematics; 2004, 30–48). Specifically, what links these similar truth-based nominalist strategies is (i) the rejection of fictionalist nominalism (which replaces the mathematical ontology with an isomorphic physical ontology), (ii) a firm reliance on some form of modality or possibilia, (iii) the rejection of the existence of abstract objects, whether spatiotemporal or non-spatiotemporal, and (iv) a realism about the truth-values of mathematical statements. Newstead and Franklin (2012), for example, follow Aristotle's suggestion and construe points utilizing the possible divisions of an extended entity: "the truthmaker for many statements of real analysis could be a merely *possible* mathematical continuum. There is no need to be wedded to the view that there is a (physical) continuum in space-time" (2012, 95).[4]

Given our analysis of platonism and nominalism above, the philosophy of mathematics thus offers a three-part distinction: immanent realism, fictionalist nominalism, and truth-based nominalism. A similar three-part division is advanced by Pincock (2012), who explains that fictionalists, like Field, "are not truth-value realists, while the standard nominalists remain realists about truth-values, even if they are not realists about abstract objects. Platonists are realists about both truth-values and abstract objects" (15). However, Pincock's three-part distinction among

---

[3] We have employed the term "truth-based nominalism", which is adapted from Pincock's (2012) concept, "truth-value nominalism", so that Pincock's own usage is kept separate from our interpretation. Alternatively, truth-based nominalism can be called "liberal nominalism", a designation used in Slowik (2005a). In Balaguer (2009), the view that we have defined as truth-based nominalism is categorized as "neo-Meinongianism", and he includes a number of the authors discussed in this chapter as among its defenders, e.g., Salmon 1998, Azzouni 2004, Priest 2005, and Bueno 2005.

[4] However, Newstead and Franklin's view seems to be a mixture of our truth-based nominalism and immanent realism (2012, 83–84). Overall, there are many options for constructing an hypothesis that lies between the extremes of fictionalism and traditional platonism (where abstract objects are non-spatial in Plato's sense). Examining all of these potential views, and how they might differ from one another, is beyond the scope of this chapter.

platonist and nominalist categories differs in various ways from our usage, most conspicuously in that our analysis replaces platonism with immanent realism (which is nominalist in orientation), but there are other important differences as well that pertain to how to interpret a truth-based form of nominalism (see Balaguer et al. 2013, 271–272).[5] Unlike the strategy advanced in Slowik (2005a), this new three-part conception allows a view closely related to platonism to be a live option in an assessment of spatial ontology derived from the platonist/nominalist distinction (although the broad nominalist conclusion of that earlier article remains intact since immanent realism counts as a nominalist hypothesis under our new scheme).

Furthermore, returning to the larger question of the relationship between the substantival/relational and platonism/nominalism debates, it is arguably the case that the philosophy of mathematics is a more proper arena for assessing the structures employed by spacetime theories, at least as opposed to the indistinct and underdeveloped metaphysics associated with contemporary substantivalism and relationism. Not only has the traditional substantival/relational dichotomy failed to explain how these mathematical structures arise from their basic ontology, but, as discussed in Chap. 5, the mathematical structures advocated by the sophisticated versions of both substantivalism and relationism are identical. Whether that foundational entity is called a substance or a physical object/field is irrelevant, and probably a conventional stipulation, since the real work, as judged from the mathematical perspective, concerns the nature of the mathematical structures themselves, that is, mathematical ontology—and the competing claims of substance or physical existent do not come to grips with this question about mathematical ontology. Leaving aside the relational motion issue, the only apparent difference between the sophisticated substantivalism and sophisticated relationism is *where* those mathematical structures are located: either *internal* to the substance or field (for the substantivalist and metric field relationist, respectively), or external to bodies/events (for non-field formulations of sophisticated relationism, such as Teller 1991). Needless to say, this internal/external distinction does not provide much information on mathematical ontology; rather, it reveals the pervasive influence of the age-old substance/property dichotomy on contemporary philosophers of space and time, an unfortunate legacy that has done little to advance the debate on spatial ontology.[6]

---

[5] In particular, Pincock (in Balaguer et al. 2013, 271–272) has doubts concerning the hypothesis entitled, "heavy duty platonism" (see below), although our analysis treats this concept as equivalent to truth-based nominalism. The problem probably resides in the difficulty in determining just what Field intends by this concept, as will be further explained below.

[6] The view of structure advocated in Brading and Landry (2006), called "minimal structuralism", would seem to be in accord with both truth-based nominalism and the ESR view advocated in this monograph (and the ESR-L version that will be developed in Chap. 8). "What we call minimal structuralism is committed only to the claim that the kinds of objects that a theory talks about are presented through the shared structure of its theoretical models and that the theory applies to the phenomena just in case the theoretical models and the data models share the same kind of structure. No ontological commitment—nothing about the nature, individuality or modality of particular objects—is entailed" (2006, 577). If one interprets "objects" as including substantival space, bodies, fields, etc., then this characterization of structure could apply to the conception of spacetime structure advocated in our investigation.

Finally, since the competing truth-based nominalist constructions are not judged solely from a mathematical perspective, but from a scientific and empirical standpoint as well, a few words are in order on the relevance of empirical evidence in assessing spacetime structure. This issue will be addressed further in Chap. 8, but, in brief, it is unlikely that any physical evidence could provide strong confirmation of any one of the competing truth-based nominalist theories described above. In effect, these nominalist constructions are only being utilized to explain the grounding of the spacetime structures, such as M or g, that appear in our best physical theories, with the important qualification that these nominalist constructions do not commit the physical theory to any problematic or meaningless physical outcomes. As for the sophisticated brands of both substantivalism and relationism, all of the truth-based nominalist construction are apparently identical as regards their implications for possible spacetime scenarios and meaningful physical states. Unless other reasons are brought forward, the choice among the competing truth-based nominalist constructions could thus be viewed as conventional, since it is difficult to conceive how empirical evidence could reach the deep mathematical levels where the differences in their nominalist construction of spacetime structures might come into play. These last observations are not meant to downplay the importance of the ongoing research in the philosophy of mathematics on the origin of structures, however, for it is always possible that substantial problems will arise for some of these truth-based nominalist theories, thus eliminating them from contention.

7.2   Causation, Explanation, and Instrumentalist Rip-Offs

This section will present a case for the similarity of mathematical structuralism and spacetime structuralism by way of their ambiguous relationship with causal explanation, as well as introduce a counter-argument against the claim that relationism reverts to instrumentalism. These issues will greatly assist our evaluation of the role that the platonism/nominalism dichotomy plays in spatial ontology as we move forward in the investigation of third-way theories.

7.2.1   Spacetime Structure and Non-inertial Forces

One of the benefits of the various structuralist approaches in the philosophy of mathematics is that it allows the spacetime philosopher to use (or construct) higher mathematical structures without obligating one to posit the existence of other, more problematic, mathematical objects, such as numbers—one can simply accept that, say, numbers exist as places in a larger structure, such as the natural number system, rather than as independently existing, transcendent entities of some sort (which raise a host of other problems, see, Shapiro 1997, 77–81). "The structuralist vigorously rejects any sort of ontological independence among the natural numbers. The essence of a natural number is its *relations* to other natural numbers" (Shapiro 2000, 258). Accordingly, a variation on a well-known mathematical structure, such as exchanging the natural numbers 3 and 7, does not create a new structure, but merely gives the same structure relabeled (with 7 now playing the role of 3, and visa-verse). This structuralist tactic is familiar to spacetime theorists, as discussed in Chap. 5, for not only has it been adopted by sophisticated substantivalists to undermine an ontological commitment to the independent existence of the manifold points of *M*, but it is tacitly contained in all relational theories, since they would count the initial embeddings of all material objects and their relations in a spacetime as isomorphic (as in Leibniz's rejection of absolute place).

A critical question remains, however: since spacetime structure is geometric structure, how does an approach to spacetime modeled on mathematical structuralism, of either the platonist or nominalist variety, differ in general from mathematical structuralism? Is a spacetime structuralist theory just the application of mathematical structuralism to spatial geometry (or, more accurately, differential geometry), rather than arithmetic or the natural number series? While it may sound counter-intuitive, the structuralist should answer this question in the affirmative—the reason being, quite simply, that the puzzle of how mathematical spacetime structures apply to reality, or are exemplified in the real world, is identical to the problem of how all mathematical structures are exemplified in the real world. Philosophical theories of mathematics, especially nominalist theories, commonly take as their starting point the fact that certain mathematical structures are exemplified in our common experience, while others are excluded. To take a simple example, a large collection of coins can exemplify the finite structure of the natural numbers (without zero) up to the finite number of coins, but not, say, real numbers or rational numbers (unless some elaborate system of representing irrational numbers or fractions is devised using the coins). In short, not all mathematical structures find real-world exemplars (although the truth-based nominalists will hold that these structures can be given a modal construction). The same holds for spacetime theories: while empirical evidence currently favors the mathematical structures utilized in GR, so that the physical world exemplifies the metric, g, a host of other geometric structures, such as the flat Newtonian metric, h, are not exemplified.[7]

The critic will likely respond that there is substantial difference between the mathematical structures that appear in physical theories and the mathematical structures relevant to everyday experience. For the former, and not the latter, the mathematical structures will vary along with the postulated physical forces and laws; and this explains why there are a number of competing spacetime theories, and thus different mathematical structures, compatible with the same evidence: in Poincaré

---

[7] Another question concerns what counts as a spacetime theory. If any use of geometric relations were to qualify as the type of structuralist spacetime theory considered above, then nearly all physical theories would trivially count given an actual or potential geometric formulation. A case in point is Fresnel's wave optics, cited in Chap. 5, which utilizes a trigonometric relationship to relate the intensities of reflected and refracted light that pass through mediums of different optical density (Worrall 1989, 119). One could classify such theories as spacetime theories, of course, but the designation seems more apt for the most basic dynamical theories that correlate inertial motion and force against a particular spacetime backdrop (e.g., Newtonian, Leibnizian, etc.).

fashion, Newtonian rivals to GR can still employ h as long as special distorting forces are introduced (we will return to this topic at length in Chap. 8). Yet, under-determination can plague even the most simple arithmetical experience, a fact well known in the philosophy of mathematics and in measurement theory. In Chihara (2004), for example, an assessment of the empiricist interpretation of mathematics prompts the following conclusion: "the fact that adding 5 gallons of alcohol to 2 gallons of water does not yield 7 gallons of liquid does not refute any law of logic or arithmetic ["$5+2=7$"] but only a mistaken physical assumption about the conser-vation of liquids when mixed" (237). While Chihara's conclusion is obviously cor-rect, he could have also mentioned that, in order to capture our common-sense intuitions about mathematics, the application of the mathematical structure in such cases requires coordination with a physical measuring convention that preserves the identity of each individual entity, or unit, both before and after the mixing. In the mixing experiment, perhaps atoms should have served as the objects coordinated to the natural numbers, since the stability of individual atoms would have prevented the, as it were, "blurring together" of the original individual units ("gallon of liq-uid") responsible for the arithmetically deviant results. By choosing a different basis for the coordination, the mixing experiment can thus be judged to uphold, or exemplify, the statement "$5+2=7$". What all of this helps to demonstrate is that mathematics, for both complex geometrical spacetime structures and simple non-geometrical structures, cannot be empirically applied without stipulating *physical* hypotheses and/or conventions about the objects that model the mathematics. Consequently, as regards real world applications, there is no difference in kind between the mathematical structures that are exemplified in spacetime physics and in our everyday experience; rather, they only differ in their degree of abstractness and the sophistication of the physical hypotheses or conventions required for their application. Both in the simple mathematical case and in the spacetime case, more-over, the decision to adopt a particular convention or hypothesis is normally based on a judgment of its overall viability and consistency with our total scientific view (i.e., the scientific method): we do not countenance a world where macroscopic objects can, against the known laws of physics, lose their identity by blending into one another (as in the liquid addition example above), nor do we sanction otherwise undetectable universal forces simply for the sake of saving a cherished metric.

Another significant shared feature of spacetime and mathematical structure is the apparent absence of causal powers or effects, even though the relevant structures seem to play some kind of explanatory role in physical phenomena. For instance, if one were to ask for an explanation of the event, "adding five coins to another seven", and why it resulted in twelve, one could simply respond by stating, "$5+7=12$", which is an explanation of sorts, although not in the scientific sense. On the whole, philosophers have found it difficult to offer a satisfactory account of the relationship between general mathematical structures (arithmetic/$5+7=12$) and the physical manifestations of those structures (the outcome of the coin adding). As succinctly put by M. Liston: "Why should appeals to mathematical objects [numbers, etc.] whose very nature is non-physical make any contribution to sound inferences whose conclusions apply to physical objects?" (Liston 2000, 191). One response to the question can be comfortably dismissed, however: mathematical structures did not *cause* the outcome of the coin adding, for this would seem to imply that numbers (in this case, $5+7=12$) somehow had a mysterious influence over the course of material affairs.[8]

In the context of the spacetime ontology debate, there has been a corresponding reluctance on the part of both substantivalists and relationists to explain how space and time differentiate the inertial and non-inertial motions of bodies; and, in particular, what role spacetime plays in the origins of non-inertial force effects. Returning once more to our universe with a single rotating body, and assuming that no other forces or causes, it would be somewhat peculiar to claim that the causal agent responsible for the observed force effects of the motion is either substantival spacetime or the relative motions of bodies (or, more accurately, the motion of bodies relative to a privileged reference frame, or possible trajectories, etc.).[9] Yet, since it is the motion of the body relative to either substantival space, other bodies/fields, privileged frames, possible trajectories, etc., that explains (or identifies, defines) the presence of the non-inertial force effects of the acceleration of the lone rotating body, both theories are therefore in need of an explanation of the relationship between space and these force effects.[10] Traditionally, most substantivalists have followed Newton's precedent by simply pointing to the fact that the non-inertial forces of bodies do not align with the presence or absence of relational motion among bodies (e.g., the rotating bucket experiment in the *Principia*). Unfortunately, this tactic, while devastating for strict eliminative relationist definitions of motion, does not secure any form of positive explanation of the substantivalists' own use of substantival spacetime as an alternative basis for identifying the non-inertial motions via their force effects. To be fair, the sophisticated relationists face an equally unsavory alternative as regards this issue, for how can the relative motion of bodies (or, as noted above, of bodies relative to a privileged reference frame or possible

---

[8] The Quine-Putnam Indispensability thesis is the best known philosophical debate related to the interface of mathematics and empirical reality, although the thesis does not really concern the interaction between the two realms; rather, the contention is that one should be a realist about mathematics if one is also a realist about scientific laws, objects, etc.

[9] It should be noted that the problem of incongruent counterparts (Kant's handed objects) could stand as a counter-argument against the claim that spacetime lacks causal powers. On the other hand, one could always invoke a maneuver similar to that advocated in Sklar (1985, 234–248), which accepts an "intrinsic" property of handedness (i.e., a primitive property defined via continuous rigid motions) to avoid the commitment to substantival space. On the whole, there are a variety of strategies that the relationist can offer to undermine the inference that space "causes" the change in handedness. That is, while the background spatial structure is obviously relevant to *determining* an object's handedness, the claim that space *causes* a change in handedness is as dubious as the claim that inertial structure is the cause of the observed non-inertial forces of a (non-uniformly) accelerating body.

[10] See, Teller (1991), Sklar (1990), Bricker (1990), and Azzouni (2004, 196–212), for similar arguments about the causal irrelevance of spacetime structures for explaining accelerated motions and effects. An important early critique is Einstein (1923, 112–113), who labels absolute space as "a fictitious cause" since rotation with respect to absolute space is not "an observable fact of experience". Furthermore, with respect to the interrelationship between the metric and matter fields in GR, this interrelationship does not explain why non-inertial forces are associated with accelerated motion; rather, the interrelationship is relevant to explaining the metric curvature.

trajectories) explain the origin of the non-inertial forces? The strict eliminative (and super-eliminativist) relationists also face a different, if not less daunting, task; for they must reinterpret the standard formulations of, say, Newtonian theory in such a way that the rotation of our lone body in empty space, or the rotation of the entire universe, is not possible. To accomplish this goal, the eliminative relationist must draw upon different mathematical resources and adopt various physical assumptions that may, or may not, ultimately conflict with empirical evidence: for example, they must stipulate that the angular momentum of the entire universe is 0 (e.g., Barbour 1982).

To sum up, all participants in the spacetime ontology debate are confronted with the nagging puzzle of understanding the relationship between, on the one hand, the empirical facts of non-inertial bodily forces, and, on the other hand, the apparently non-empirical, *mathematical* properties of the spacetime structure that are somehow inextricably involved in any adequate explanation of those non-inertial forces— namely, for the sophisticated versions of substantivalism and relationism, the affine structure ∇ that lays down the geodesic paths of inertially moving bodies. The task of explaining this connection between the empirical and abstract mathematical or quantitative aspects of spacetime theories is thus identical to elucidating the mathematical problem of how numbers relate to our common experience (e.g., how "5+7=12" figures in our experience of adding coins). Likewise, there is a parallel situation in the fact that substantivalists and relationists seem to shy away from positing a direct causal connection between material bodies and space (or privileged frames, possible trajectories, etc.) in order to account for non-inertial force effects, just as a mathematical platonist would recoil from ascribing causal powers to numbers so as to explain our common experience of adding and subtracting material objects.

7.2.2 Instrumentalism and Mathematical Structure

An insight into the non-causal, mathematical role of spacetime structures can also assist the sophisticated relationist in defending against the charge of instrumentalism, as, for instance, in deflecting Earman's criticisms of Sklar's "absolute acceleration" hypothesis. As explained in Chap. 5, Sklar's monadic property of absolute acceleration rejects the common understanding of acceleration as a species of relative motion, whether that motion is relative to substantival space, bodies, or privileged reference frames (Sklar 1974, 225–234). Earman's objection to this maneuver centers upon the utilization of spacetime structures to describe the primitive acceleration property: "it remains magic that the representative [of Sklar's absolute acceleration] is neo-Newtonian acceleration $(d^2x^i/dt^2)+\Gamma^i_{jk}(dx_j/dt)(dx_k/dt)$ [i.e., ∇]" (Earman 1989, 127–128). Earman's critique of Sklar's brand of relationism would seem to cut against all sophisticated relationist hypotheses, for he regards the exercise of these richer spacetime structures, like ∇, as tacitly endorsing the absolute/ substantivalist side of the dispute: "the Newtonian apparatus can be used to make the predictions and afterwards discarded as a convenient fiction, but this ploy is hardly distinguishable from instrumentalism, which, taken to its logical conclusion, trivializes the absolute-relationist debate" (128).

Leaving aside the inherent difficulties with Sklar's maneuver, the drawbacks of Earman's own argument is more readily discernable in the wake of our lengthy examination of the parallel function of mathematical and spacetime structures— since, to put it bluntly, does the equivalent use of mathematical structures, such as "$5 + 7 = 12$", compel the mathematician to accept the same realist commitment, in this case, to a platonism about numbers and arithmetic (so that these structures also exist independently of all exemplifying systems)? Most substantivalists would likely demur, and insist that the platonism/nominalism question as it pertains to basic mathematical structures is different than the substantivalism/relationism dispute over higher spacetime structures. Yet, how can a substantivalist justify this difference? If the straightforward employment of basic mathematics does not entail either a platonist or nominalist theory of mathematics (as most mathematicians would likely agree), then why must the equivalent use of the mathematical structures of spacetime physics, such as ∇, require a substantivalist conception of ∇ as opposed to a sophisticated relationist conception of ∇ (with platonism/nominalism comparable to, respectively, substantivalism/relationism)? Put differently, does a substantivalist commitment to $(d^2x^i/dt^2) + \Gamma^i_{jk}(dx_j/dt)(dx_k/dt)$, whose overall function is to determine the spacetime's straight-line inertial trajectories, also necessitate a substantivalist commitment to its components, such as the vector, d/dt, along with its limiting process and mapping into R, i.e., so that d/dt and R qualify as substantival in the same way that the larger equation $(d^2x^i/dt^2) + \Gamma^i_{jk}(dx_j/dt)(dx_k/dt)$ qualifies as substantival? In short, how can one adopt an ontological realist commitment to higher mathematical structures (usually, of second derivative order or higher) when those structures are grounded on, and derived from, lower mathematical structures whose reality is denied or called into question. A non-instrumentalist interpretation of some component of the theory's quantitative structure is often justified if that component can be given a plausible causal role (as in particle physics)—but, as noted above, ∇ does not appear to cause anything in spacetime theories.[11]

In addition, Earman's argument may prove too much, for if we accept his conclusion at face value, then the introduction of any mathematical or quantitative device that is useful in describing or measuring physical events would saddle the ontology with a bizarre type of entity. For instance, if a group of people came to the realization that their difficult domestic circumstances were due to their income falling below the poverty line, i.e., that the mathematical number representing a minimum subsistence level *explains* their failure to advance in economic terms, does this admission similarly imply the existence of a "poverty line" substance (alongside such other independently existing substances as "gross national product, "average household family size", etc.)? Now, the substantivalist might interject at this point that a concept like "gross national product" does not correspond to an independently existing entity, but is simply an abstraction from other "real" material entities and

---

[11] Since truth-based nominalists reject platonism, they do not claim, of course, that mathematical entities and their relationship enter into spacetime theories. But, spacetime structures are not just the *physical relationships* between the physical objects, either, since (as argued above) causation is not a spacetime relationship (whereas causation is probably the most prevalent physical relationship among physical objects). Spacetime structures, like all other mathematical structures, are unique in this regard: systems can exemplify the structures, but the structures are neither identical to the systems, nor (for the nominalists) can they exists in the absence of systems.

processes (such as factory goods, people's salaries, etc.); but, of course, this is exactly the line of reasoning put forward by the relationists, who maintain that, say, privileged reference frames, like ∇, are also useful non-causal explanatory abstractions from real physical entities and processes. As for an instance of a geometric structure that provides a similarly useful non-causal explanatory function, but whose substantive existence we would be inclined to reject as well, Dieks' offers the case of a three-dimensional color solid: "Different colours and their shades can be represented in various ways; one way is as points on a 3-dimensional colour solid. But the proposal to regard this 'colour space' as something substantive, needed to ground the concept of colour, would be absurd" (Dieks 2001b, 230). In conclusion, unless the substantivalists can submit a consistent and persuasive argument for why the non-causal explanatory role of spacetime structures requires a substantival commitment, while simultaneously demonstrating why other non-causal explanatory devices fail to obligate the same commitment (e.g., gross national product or colour space), there is little reason to accept Earman's charge of instrumentalism.

Among the important contributions to the philosophy of space and time that have strongly emphasized the non-causal role of spatiotemporal structure, the work of Howard Stein (1977b), and Robert DiSalle (1994, 1995) comes immediately to mind. Focusing on DiSalle, he suggests that the non-causal explanatory significance of mathematical structures, in this case, Euclidean geometry, is analogous to the similarly non-causal role of spacetime structure in explicating the dynamical behavior of bodies:

> To claim that space is Euclidean *only means* that measurements agree with the Euclidean metric; Euclidean geometry, if true, can't *causally explain* those measurements, because it only expresses the constraints to which those measurements will conform....To claim that [the] formal structure [of Euclidean geometry] is *really* the structure of actual space is not to posit an underlying cause of the appearances. It is only to claim that, *modulo* the initial coordination [of spatial measurement with a basic physical process, such as motion of rigid bodies], the appearances conform to the laws of that structure. This claim is no less a form of realism than the supposed causal postulate. But it is a form of realism that captures much more clearly the relationship between geometry and experience. (DiSalle 1995, 324)

Finally, it is worth commenting on a strategy for interpreting a scientific theory's mathematical structure that takes a different stance on the causation issue. "Background-independence" is often defined, in simplest terms, as the denial of a fixed geometric structure for all models of a particular physical theory, but, drawing on the work of, e.g., Brown (2005, 140) and Brown and Pooley (2006, 71), Alexander Bird has argued that background independence might also incorporate the "action-reaction principle". In the case of space, this principle stipulates that the geometric (i.e., spatial) component of a theory's structure must be able to act and be acted upon, a state of affairs that is manifest in the (oft-mentioned) reciprocal influence between GR's metric and (non-gravitational) stress-energy. This leads to the following definition of background independence (which Bird dubs "background-free"): "In a true theory, any structure appearing in the laws of that theory is subject to being affected by changes elsewhere" (Bird 2007, 165).

A few observations are in order concerning this strategy. First, as argued in Chaps. 2 and 3, space cannot act upon things for either Newton or Leibniz (a point also noted by Brown and Pooley 2006, 71), thus modern spatial ontologies based on the action-reaction principle cannot appeal to the traditional substantival/relational dichotomy for support. Nor does the action-reaction principle apply to the foundational entity in that period, whether God (e.g., "space is eternal in duration and immutable in nature because it is the emanative effect of an eternal and immutable being", N 26) or monads (since a monad is "a certain world of its own, having no connections of dependency except with God", AG 199). Rather, it is only corporeal substance (matter) and finite immaterial substances (souls/angels) that fit the criterion of the action-reaction principle in the seventeenth century. Second, if the action-reaction principle is, in fact, used as the basis of a new conception of spatial ontology, then it also marks a departure from the platonism/nominalism distinction, since platonism and nominalism only pertain to the existence/non-existence of mathematical structures in the absence of matter; i.e., causal powers are excluded. Overall, the action-reaction principle would seem to be an attempt to put space within the category of physical entities, an approach that, as noted above, is quite different than the standard conceptions of physical geometry and substantivalism that eschew talk of space's causal influence. As a result, this strategy raises troubling questions as regards the extent to which the background structures are, so to speak, "physicalized". Unlike GR, all of the existing QM and QG (quantum gravity) hypotheses would seem to accept some form of background structure, whether metric, topological, Hilbert space, etc., that is not subject to action-reaction, so there is little justification within QM and QG for this interpretation of background independence as regards these fixed structures. Given the common belief that GR will eventually be supplanted by a QM-based quantum gravity theory, the justification for invoking the action-reaction principle is, therefore, significantly weakened. Moreover, since Bird declares that "any structure appearing in the laws of that theory" must be subject to being affected, does this imply, once again, that arithmetic or logic must be subject to the action-reaction principle as well? Since numbers and arithmetic appear in all theories, whether GR, QM, or QG, it must therefore partake in action-reaction affects. And, if not, then why is the mathematical field of geometry subject to that principle and not the mathematical field of number theory (or logic, etc.)? As with Earman's charge of instrumentalism, the advocates of the action-reaction principle have only succeeded in revealing the bias that exempts lower-level mathematical structures, such as numbers or arithmetic, from their strategy while simultaneously demanding the inclusion of higher geometric structures, such as affine or metric structure (even though the latter depend on the former).

7.3 Platonism/Nominalism in Newton and Leibniz' Spatial Ontologies

At this point, we will turn the focus of our investigation of the role of platonism and nominalism in spatial ontology to the seventeenth century's foremost (alleged) representatives of substantivalism and relationism, Newton and Leibniz. This analysis, which will culminate in the conclusions reached by the end of Chap. 10, will reveal the importance of platonist and nominalist conceptions of space, together with the different ontological levels of geometric structure, with respect to the content of both Newton and Leibniz' theories. Most of the material discussed in this section was first introduced in Chaps. 2 and 3, but we will now begin the process of integrating our results within a more comprehensive framework for understanding spatial ontology, ultimately leading to the new taxonomy presented in Part III. In contradistinction to the failure of the substantivalist/relationist dichotomy in elucidating the spatial hypotheses of Newton and Leibniz, it will be argued that different versions of nominalism best capture the content of, respectively, Newton's alleged substantivalism and Leibniz' assumed relationist view of space. Building on these conclusions, the final section of the chapter will then demonstrate the benefits that platonism/nominalism hold for current debates in spacetime philosophy and scientific realism, and conclude with an assessment of alternative approaches to nominalism.

7.3.1   Newton

As first discussed in Chap. 2, Newton reckons that space is not a substance because "[i]t is not absolute in itself, but is as it were an emanative effect of God...; on the other hand, because it is not among the proper affections that denote substance, namely actions" (N 21). Yet, space is it not an accident (property) of body either, "since we can clearly conceive extension existing without any subject, as when we imagine spaces outside the world or places empty of any body whatsoever" (22). Nonetheless, given his claim that space is not absolute but God's emanative effect, space still seems to fit the general category of property, i.e., as a unique property of God along the lines of the P(O-dep) concept explored in Chap. 1. Newton often claims, moreover, that space is dependent on that entity: God contains "all other substances in Him as their underlying principle and place", and he denies "that a dwarf-god should fill only a tiny part of infinite space with this visible world created by him" (TeL 123).

Can a case be made for a platonist-inspired interpretation of Newton's spatial ontology, especially immanent realism? (It is important to bear in mind that, as used in our investigation, immanent realism is a nominalist position, although it appeals to abstract objects, that exist in space and time, in a decidedly platonist fashion.) In *De grav*, the presence in space of Euclidean geometry is revealed in his claim that "the delineation of any material figure is not a new production of that figure with respect to space, but only a corporeal representation of it, so that what was formerly insensible in space now appears before the senses" (N 22). This claim appears consistent with immanent realism, but it also rules out Armstrong's brand of constitutive platonism, where the universals that constitute space are not themselves in space. Yet, to qualify as a version of immanent realism, abstract objects or universals would need to be responsible for the very structure of space. There is no evidence to corroborate such a view, however, even granting that one might, somewhat implausibly, take *De grav*'s sole reference to space as a "being" (N 20–21) as indirectly signifying an abstract object. Rather, because it is God's very existence or being that brings about space, abstract objects would seem to be a superfluous addition: e.g., from the 1713 *Principia,* "He endures forever, and is everywhere present; and by existing always and everywhere, he constitutes duration and space" (N 91). In the passage from *De grav* cited above (N 22), bodies reveal the geometric figures that are instantiated or exemplified in God's being or substance, since God is actually present in space. The "determined quantities of extension" thesis, where bodies are depicted as movable properties in God's infinite spatial extension, is the most straightforward presentation of this God-space relationship (see Chap. 2); and, while possible, abstract objects appear to play no role in this spatial ontology.

In short, the evidence of the texts supports a general nominalist stance whereby space is instantiated by God's own being—we will dub this view, "incorporeal nominalism", a category that potentially includes all varieties of nominalism within its scope (including immanent realism). Nevertheless, given the theological problems connected with the identification of space with God's being *per se* in the seventeenth century, fictionalism is almost certainly not the correct form of nominalism. That is, fictionalist nominalism would demand the elimination or replacement of the spatial ontology with God's ontology, but Clarke, who was likely advised by Newton in his correspondence with Leibniz, straightforwardly rejects this view: "Space is not a being, an eternal and infinite being, but a property or a consequence of the existence of an infinite and eternal being. Infinite space is immensity, but immensity is not God; and therefore infinite space is not God" (C.III.3). Newton's assertions follow this line of reasoning as well: e.g., space is "an emanative effect of God" (N 21), and, in the Des Maizeaux drafts, space and time are "unbounded *modes* & consequences of the existence of a substance that which is really necessary & substantially Omnipresent & Eternal" (Koyré and Cohen 1962, 96–97). That is, space is an "effect", "mode", "consequence", etc., of God,—and all of these descriptions seem to put some ontological room between God's being and space, so that God brings about space but is not literally identical to space.[12] On a fictionalist construal of this particular ontology, conversely, it really would be true to say that space is God, since space would constitute a fictional order/relation that can be replaced by that ontology (God)—this parallels Field's scheme, where mathematical structures are really the substantivalist's point manifold and only a fictional order/relation that can be eliminated in favor of that ontology. Therefore, given the obstacles that face the immanent realist and fictionalist construals of Newton's spatial ontology, truth-based nominalism is the most plausible, as well as the only remaining, option; i.e.,

---

[12] Some of Newton's descriptions might challenge this conclusion, such as his contention that God "constitutes" space (N 91), if one takes that term as synonymous with "is". Likewise, one may strive to read such terms as "effects", "affection", "attribute", "modes", and "consequences", as Newton's somewhat clumsy way of describing a fictionalist elimination of space in favor of God's being. Yet, Clarke's straightforward denial that "space is consequence of God" equates with "space is God" poses a significant challenge, as does the tenor of Newton's Des Maizeaux drafts, which rejects Clarke's use of the term "property" in favor of "modes" and "consequences"; likewise, he does not object to Clarke's claim that "space is not God". In short, a fictionalist interpretation faces severe obstacles given this evidence.

God's being instantiates the truths of Euclidean geometry alongside space itself, but spatial geometry is neither an abstract object nor identical with God.

Lastly, as briefly mentioned in Chap. 6, since space is independent of matter at the level of material bodies, one could ascribe a sort of "virtual-platonism" to Newton's spatial ontology, where this term designates the simple fact that all aspects of spatial geometry, but especially its instantiation by an entity, are independent of matter due to his espousal of incorporeal nominalism. One reading of the term "virtual" is through the definition "being something in practice", which is a good description of the way Newton's incorporeal nominalism functions at the level of material bodies as regards abstract objects; i.e., incorporeal nominalism function like traditional platonism in that space's abstract structures are independent of matter, even though Newton's nominalism obviates the need for these structures to be explained by way of abstract objects. In Chaps. 9 and 10, virtual-platonism will figure prominently in the analysis of the different levels of spatial ontology.

7.3.2 Leibniz

Like Newton, Leibniz reckons that space cannot act upon things (NE II.xiii.17), nor does it inhere in individual beings in the traditional Scholastic sense as an accident (L.V.47). In accordance with the substance/accident metaphysics, he insists that there can be no *ontologically unsupported* spatial properties: "Now extension must be the affection of something extended. But if that space is empty, it will be an attribute without a subject, an extension without anything extended" (L.IV.9). Yet, even granting that extension is not an individual bodily property that can transfer from one subject to another, Leibniz does not rule out the possibility that the world's spatial extension could still function as something like a unique holistic property of the entire material world that is ultimately grounded in God. As argued in Chaps. 3 and 4, the global, or holistic, basis for space is linked to the dynamical interconnections among all bodies, but these interconnections are ultimately derived from the monadic realm. Leibniz goes on to argue that extended quantity and position "are mere results, which do not constitute any intrinsic denominations *per se,* and so they are merely relations which demand a foundation derived from the category of quality [i.e., the action or primitive force of monads], that is, from an intrinsic accidental denomination" (MP 133–134).

Leibniz further contends that space's "truth and reality are grounded in God, like all eternal truths", and that "space is an order [of situations] but that God is the source" (NE II.xiii.17). Unlike Newton, however, God is not present in space: "God is not present to things by situation but by essence; his presence is manifested by his immediate operation" (L.III.12). At greater length, he follows Descartes by positing that only God's operations can be assigned a spatial location, a view also known in as the "extension of power" doctrine (EP):

> Where space is in question, we must attribute immensity to God, and this also gives parts and order to his immediate operations. He is the source of possibilities and of existents alike, the one by his essence and the other by his will. So that space like time derives its reality only from him, and he can fill up the void whenever he pleases. It is in this way that he is omnipresent. (NE II.xv.2)

To summarize, space (and time) obtain their "reality only from [God]" even though God is not situated in space; furthermore, God's "essence" is responsible for the *possibility* of anything existing in space. Relations receive a similar treatment: "relations and orderings are to some extent 'beings of reason', although they have their foundations in things; for one can say that their reality, like that of eternal truths and of possibilities, comes from the Supreme Reason" (NE II.xxv.1). Finally, as regards the basic constituents of Leibniz' world, monads, God plays a similar foundational role: "a monad, like a soul, is, as it were, a certain world of its own, having no connections of dependency except with God" (AG 199). There is a parallel to Leibniz' God-space relationship in his monad-matter relationship, moreover, for matter "results from" monads and their primitive forces; e.g., "properly speaking, matter is not composed of constitutive unities [monads], but results from them" (AG 179), and monads are not situated in space (AG 201; more on this in Chap. 9). Therefore, to return to the issues first raised in Part I, the non-spatial monads underwrite the possibilities of extended matter along the lines of the P(O-dep) conception in the same way that his non-spatial God underwrites the possibilities tied to spatial geometry by means of P(O-dep).

Turning to the platonism/nominalism issue, since "there is no space where there is no matter and that space in itself is not an absolute reality" (L.V.62), Leibniz opts for a nominalist account of space at the level of material bodies.[13] Yet, although space is not "an absolute being" (L.III.5), it still represents "real truth" (L.V.47), even in the absence of matter: "[t]ime and space are of the nature of eternal truths, which equally concern the possible and the actual" (NE II.xiv.26). But, since "truth and reality are grounded in God, like all eternal truths" (II.xiii.17), these God-grounded spatial "truths" also determine the range of possibilities for the arrangement of bodies in space. On the numerous ways that the world could be filled with matter, Leibniz states that "there would be as much as there possibly can be, given the capacity of time and space (that is, the capacity of the order of possible existence); in a word, it is just like tiles laid down so as to contain as many as possible in a given area" (AG 151). Truth-based nominalism best describes Leibniz' approach, moreover, especially when the difficulties associated with a fictionalist interpretation of the space-God relationship are taken into account. As with Newton's spatial ontology, fictionalist nominalism in this case would require that space can be eliminated and replaced by an analogous isomorphic structure in God,

---

[13] One of the most straightforward declarations of nominalism can be found in the *New Essays*, where numbers and extension are compared (see also Chap. 3): "[I]n conceiving several things at once one conceives something in addition to the number, namely the things numbered; and yet there are not two pluralities, one of them abstract (for the number) and the other concrete (for the things numbered). In the same way, there is no need to postulate two extensions, one abstract (for space) and the other concrete (for body)" (NE II.iv.5).

but that would imply, also in keeping with the fictionalism, that the ontology of space is really God, although called by a different name. Yet, as we have seen, Leibniz rebuffs any attempt to link matter/space to God in this manner, preferring to state that "space is an order but that God is the source" while simultaneously rejecting the claim that "space is God or that it is only an order or relation" (NE II.xiii.17). On a fictionalist interpretation, as noted above, the structure of space is, in fact, God's structure. Leibniz does claim that God has the property of "immensity" (NE II.xv.2), but he explicitly denies that God's immensity is spatial: "the property of God is immensity but…space (which is often commensurate with bodies) and God's immensity are not the same things" (L.V.36). Likewise, since "space cannot be in God because it has parts" (L.V.51), it is rather difficult to envisage a structure preserving isomorphism between space, which is extended and has parts, and God, who is neither extended nor has parts. To recap, Leibniz' appeal to the truths of spatial geometry appears to be inconsistent with fictionalist nominalism, nor is the replacement of space with God consistent with the mature Leibniz' denial that God is present in space (although early Leibniz texts may have supported that notion; see Chap. 3).

The one major difference between the modern conception of truth-based nominalism in the philosophy of mathematics and Leibniz' version as regards space is that the latter utilizes God as the grounds of spatial truths while the former uses matter, but this parallels Newton's incorporeal nominalism, where God's being, and not matter, is the instantiating entity. Hence, there are interesting differences between Newton and Leibniz' versions of spatial nominalism: for Newton, God directly instantiates both spatial extension and either the truth-values or abstract objects pertaining to spatial geometry; for Leibniz, God only instantiates the truth-values associated with spatial geometry, while matter instantiates spatial extension *per se* but not the mathematical/geometric abstract objects associated with those truth-values (since they are ideal and do not exist in reality). On this rather slender, but hugely important, issue—Newton's incorporeal nominalism versus Leibniz' unique God-matter formulation of truth-based nominalism, where God is the foundation of the spatial geometric truths that govern the behavior of bodies—one can trace most of the specific differences among their respective conceptions of space. In Chap. 9, an examination of the two ontological levels, foundational (God for Newton, God and monads for Leibniz) and material, will allow us to track the differences between Newton and Leibniz' views in further detail.

At this point, a few words are in order regarding Leibniz' remarks on the possibility of a vacuum, and how it relates to his spatial ontology and the eternal truths of space. In brief, as explained in Chap. 3, Leibniz often admits that space is not necessarily a plenum, whether outside the confines of a finite material world, or within the material world: concerning the former, he states that "[a]bsolutely speaking, it appears that God can make the material universe finite in extension" (L.V.30); and, regarding the latter, he judges that "we can refute someone who says that if there is a vacuum between two bodies then they touch, since two opposite poles within an empty sphere cannot touch—geometry forbids it" (NE II.xv.11). There are even hints that a complete vacuum state is possible, for Leibniz explains that God could prevent a monad's primitive force from bringing about extended matter, and thus potentially all of the monads could be effected in this way. In a letter from 1692, he states that primitive force is "a higher principle of action and resistance, from which extension and impenetrability emanate when God does not prevent it by a superior order" (A.I.vii.249; Adams 1994, 351).[14] Furthermore, since "there is no space where there is no matter" (L.V.62), and "extension must be the affection of something extended" (L.IV.9), the Leibnizian void lacks extension. As E. Sylla (2002) explains, there were similar theories put forward in the late Medieval period, in particular, by the fourteenth century natural philosopher, Nicole Oresme, who accepted a nominalist account of space that is strikingly similar to Leibniz' view on these very points. Like Leibniz, Oresme contemplates a spherical vacuum that lacks extension, but which nonetheless can be attributed a sort of indirect extension by means of the possible bodies that God could bring about to fill that void (Sylla 2002, 269–271). In the *New Essays,* Leibniz adds that "[i]f there were a vacuum in space (for instance, if a sphere were empty inside), one could establish its size" (NE II.xv.11). As for a temporal vacuum, on the other hand, "it would be impossible to establish its length"; likewise, "if space were only a line, and if bodies were immobile, it would also be impossible to establish the length of the vacuum between two bodies" (NE II.xv.11). Consequently, three-dimensional space allows the surrounding bodies to completely determine the size of a vacuum, unlike the contrasting one-dimensional cases he considers. This would appear to explain the intended meaning behind Leibniz' claim, quoted above, about the predetermined capacity of space to be filled with bodies (AG 151): although unextended, these void spaces are set within a fixed geometric structure secured via God's immensity (i.e., God's essence grounds the possibilities of space).

7.4 Platonism/Nominalism and Contemporary Spacetime Philosophies

The three-part distinction advanced above—immanent realism, truth-based nominalism, and fictionalist nominalism—can also be of service in evaluating some of the other assessments of platonism and nominalism that have appeared in the literature on the philosophy of physics (Field, Arntzenius) and scientific realism (Psillos, French). As will be demonstrated, given the conceptual resources of our tripartite classification, assumptions that underlie a number of spatial ontologies and arguments can be seen in a different light, often revealing a mixture of mathematical and physical notions that goes beyond the resources of the traditional substantival/relational dichotomy or the realism/anti-realism dispute.

---

[14] In more detail, although a monad's extended secondary matter would not arise if God prevented it, with secondary matter equating with bodily extension at the macrolevel, the unextended primary matter would remain at the microlevel. See Chap. 4 for more on the primary/secondary, and primitive/derivative force, distinctions.

7.4.1   Nominalist Substantivalism

Interestingly, the claim that the platonism/nominalism divide in mathematics can benefit the analysis of substantivalism and relationism was briefly mentioned in Sklar (1974, 165), one of the seminal works in the modern approach to the philosophy of space and time. Sklar regarded nominalism as an exclusively relationist standpoint, a verdict that is in keeping with relationism's historical disdain for absolute space, the latter suggesting, perhaps mistakenly, a platonist's commitment to abstract objects. Nevertheless (and leaving aside Slowik 2005a), the combination of nominalism and substantivalism is a central element in Field's program (1980), and has recently been advocated by Arntzenius, who favors a fictionalist version of nominalism in line with Field's viewpoint (Arntzenius 2012, 213–268). Field's rejects relationist nominalism, moreover:

> According to the substantivalist view, which I accept, space-time points (and/or space-time regions) are entities that exist in their own right. In contrast to this are two forms of relationalist view. According to the first (*reductive relationalism*), points and regions of space-time are some sort of set-theoretic construction out of physical objects and their parts; according to the second (*eliminative relationalism*), it is illegitimate to quantify over points and regions of space-time at all. It is clear that reductive relationalism is unavailable to the nominalist: for according to this form of relationalism, points and regions of space-time are mathematical entities, and hence entities that the nominalist has to reject. (Field 1980, 34)

In short, Field reasons that spacetime physics requires spacetime points, which relationism cannot tolerate, so substantivalism is the only option remaining for a nominalist. The two forms of relationism put forward by Field match the relationist categories utilized in this work (see Chap. 1): Field's "eliminative relationalism" is identical to our super-eliminative and (probably) eliminative brands of relationism, i.e., space and time are eliminated in favor of a purely physical account of those structures; whereas Field's "reductive relationalism" corresponds to our sophisticated relationism, i.e., spatiotemporal structures are grounded in, or can be reduced to, physical entities, although these structures cannot be eliminated given their important function within the overall theory (e.g., the use of inertial structure, ∇, by sophisticated relationists like Sklar 1974 or Teller 1991).

Field's reference to spacetime points as "mathematical entities" prompts a number of questions, in particular, whether truth-values, abstract objects, or both, are incorporated into his conception. As explained in §7.1.2, a sophisticated relationist can insist that the employment of mathematical structures is consistent with nominalism as long as that commitment is limited to the truth-values associated with these structures. In his later (1989) essay, Field investigates the relationship between substantivalism, relationism, and the philosophy of mathematics in greater depth, providing a number of arguments that qualify and refine his earlier claims. Besides granting a number of concessions to the more sophisticated brands of relationism, he nonetheless insists that relationism can only succeed as an alternative to substantivalism if it takes "the relation between physical things and numbers to be a brute fact, not explainable in other terms" (Field 1989, 186). This view—which he dubs, "heavy duty platonism" (hereafter HDP)—would seem to denote the truth-based nominalist conception defended above, since a "brute fact", if interpreted minimally, has (or is) a truth-value. Furthermore, he states that HDP "is a much more radical position than mere platonism, i.e., the mere acceptance of mathematical objects" (185), an admission that would seem to suggest that HDP's brute facts are a separate issue from the question of abstract objects. Whether or not abstract objects are a necessary ingredient of Field's HDP remains unclear, it should be noted, but henceforth HDP will be taken as synonymous with truth-based nominalism alone (and not abstract objects). Overall, Field does conclude that, if one accepts HDP, then a relationist can successfully account for, say, acceleration in Newtonian physics: "one could simply take co-ordinate functors defined on points of matter as primitive, and define an acceleration functor from them in the usual way, without ever having to appeal to unoccupied regions" (191). Hence, Field reasons that sophisticated modal relationist proposals, such as Mundy (1983), tacitly endorse HDP (199), even though HDP "violates the whole spirit of relationalism" (192). This line of argument is quite telling, moreover, since it links together Field's HDP, our truth-based nominalism, and sophisticated relationism, with the common element among these three being mathematical modality and truth-values, and not abstract objects.

Yet, from a philosophy of mathematics perspective, Field's (1989) case against the utilization of HDP by relationists simply begs the question, for there have been numerous nominalist formulations that forsake the existence of abstract objects but defend a conception of mathematical truth and modality that goes beyond fictionalism: besides those mentioned above, such as Azzouni (2004), one can add Salmon (1998), Priest (2005), and Bueno (2005), to name only a few. Accordingly, why must a relationist forsake HDP (= truth-based nominalism) if many philosophers of mathematics embrace hypotheses that are identical to, or closely aligned with, truth-based nominalism?

Does truth-based nominalism (i.e., Field's HDP, sans abstract objects) specifically flout relationist doctrine? While it obviously depends on how you define relationism, the evidence of past and present relationist hypotheses does not support Field's definition. Besides the numerous modern relationist proposals that uphold the less strict, non-eliminative versions of sophisticated relationism explored above and in Chap. 5 (such as by Sklar, Teller, Huggett, and many more), an important precedent for admitting the truth-values associated with spatiotemporal structures, but not their existence as an entity or object, can be found in Leibniz (see §7.3.2 above). Invoking Leibniz as a counter-example to Field's conception of nominalism is particularly warranted, moreover, since one of Field's arguments mirrors Clarke's case, in the correspondence with Leibniz, that geometric ratios or proportions are not quantities. We have already examined much of this historical material, but it is worth drawing attention once again to Leibniz' response to Clarke's contention that relationism is limited to geometric ratios or proportions, and not quantities, since Field takes Clarke's side in this dispute. Calling it "the problem of quantities for the relationalist" (1989, 196), Field brings up the well-known fact that spatial ratios or proportions, which he regards as the relationist's only acceptable form of spatial relationship, underdetermines the quantitative (distance, metric) relationships among things unless one accepts the allegedly anti-relationist HDP (185). Leibniz' "order of coexistences" account of space (L.III.4) provokes the same argument from Clarke: "the distance, interval, or quantity of time or space…is entirely a distinct thing from the situation or order and does not constitute any quantity of situation of order; the situation or order may be the same when the quantity of time or space intervening is very different" (C.V.54). Yet, Leibniz insists that his conception of spatial relationships ("order of coexistences") includes quantity, thereby sanctioning the type of "brute fact" reading of quantity (cf. Field's HDP) that constitutes truth-based nominalism:

> As for the objection that space and time are quantities, or rather things endowed with quantity, and that situation or order are not so, I answer that order also has its quantity: there is in it that which goes before and that which follows; there is distance or interval. Relative things have their quantity as well as absolute ones. For instance, ratios or proportions in mathematics have their quantity and are measured by logarithms, and yet they are relations. And therefore though time and space consist in relations, still they have their quantity. (L.V.54)[15]

To recall our earlier discussion, Leibniz' repeatedly attempts "to confute the fancy of those who take space to be a substance or at least an absolute being" (L.III.5, or "absolute reality", L.V.62), even though he concedes that space still expresses "real truth" (L.V.47).[16] Contra Field, the modality associated with truth-based nominalism has long been favored by spatial relationists, as well as by Leibniz, although the latter's views differ from modern relationism in significant ways (see Chap. 3).

---

[15] While the origins of Leibniz' view remain uncertain, it is highly likely that this particular dispute between Leibniz and Clarke stems, at least in part, from a more basic dichotomy concerning the definition of ratios in (Euclid's) geometry, a dichotomy that found many important advocates in the seventeenth century. One school of thought, represented by Wallis, accepts that ratios are quantities that can be explained purely algebraically using an intricate exponent method; whereas others, such as Barrow, argue that ratios are not quantities, hence algebraic relationships fail in this regard (i.e., a geometric conception of quantity is primary). What is intriguing is that, like Wallis, Leibniz defends the primacy of algebra over geometry (e.g., AG 251–252), while Clarke and Newton side with Barrow in defending the primacy of geometry over algebra. In short, the foundations debate in the philosophy of mathematics, i.e., whether algebra or geometry is more fundamental, or at least superior, may be the origins of this particular aspect of the Leibniz-Clarke correspondence—and, in turn, it may ultimately reflect a more basic difference in their respective visions of the deep metaphysics of spatial ontology: Newton and Clarke's spatially extended God conception versus Leibniz' non-spatial God/monad hypothesis. More historical work is required on the origin of Leibniz' view, but see Jesseph (2016) for an analysis of the ratio dispute between Wallis and Barrow.

[16] In the L.III.5 quote above, the contrast between "space as absolute being/reality" and "space as a real truth" could be taken to correspond to, respectively, a commitment to abstract geometric objects versus the truths of geometry. This inference gains support given Leibniz' rejection of abstract objects in general (see footnote 13). Yet, leaving aside abstract objects, Newton's conception of space is actually similar to Leibniz' theory: both accept nominalism, with the main difference being the ontological foundation of that nominalist spatial structure, God (Newton/Clarke) versus matter and God (Leibniz). Since God is central to the ontology of space for both Newton and Leibniz, the L.III.5 quote can also be viewed as the difference between positing a conception of space that is either entirely independent of matter (Newton/Clarke) or dependent on matter for instantiating the God-grounded truths of spatial geometry (Leibniz).

7.4.2   Nominalist Counter-Arguments

The reasons that Field offers for linking substantivalism to nominalism reveal his unique, and controversial, conception of modern field theories in physics. As he explains, "a field is usually described as an assignment of some property, or some number or vector or tensor, to each point of space-time; obviously this assumes that there are space-time points", but, on Field's construction of Newtonian gravitation theory, "it does without the properties or the numbers or vectors or tensors, but it does not do without the space-time points" (1980, 35). A host of objections have been raised against this conception of spacetime points as physical entities: unlike ordinary objects, spacetime points do not endure through time, have no mass, cannot move or be empirically detected, and have no location (since they comprise location; see Resnik 1985). Furthermore, Field's ontology is committed to space-time regions as much as spacetime points (1980, 36–37), and this helps to explain his later admission that both the substantivalist and relationalist can construct space-time points from a mereological postulate, so that points become minimal parts of a spacetime region (1989, 172–174). Field concedes, in effect, that the conception of the metric field as a physical entity, namely, metric field relationism, is identical to his substantivalist rendering of the metric: it is for these reasons that Field claims that relationism must admit the possibility of unoccupied spatial regions (174), for only on this scenario can a distinction be drawn between a nominalism based on substantivalism as opposed to relationism. Yet, as we have seen, since the metric field of GR is defined over the entire manifold, there are no, as it were, "metrical voids" (i.e., a region of the manifold where the metric is absent); and, since the metric alone, absent the stress-energy of matter and other fields, has the potential to induce physical effects via its dual role as the gravitational field, a matter-less, energy-carrying metric field in the context of GR fits the definition of a nominalist relationism as much as it does a nominalist substantivalism (see, once again, Earman and Norton 1987, 519, who detail some of the physical consequences of the energy carried by gravity waves). Consequently, not only is Field's case for a substantivalist nominalism predicated on an outdated theory, Newtonian gravity, but our best theory of the large scale structure of spacetime, GR, does not admit the type of energy-less void space that he requires to distinguish nominalist substantivalism from nominalist relationism.

The difference between an ontological commitment to points versus a commitment to extended regions, so that points become parts of a larger entity, is also evident in Arntzenius (2012), a work that strives to extend Field's project in order to encompass GR and its mathematics built upon differential geometry. Arntzenius defines substantivalism as the view that "space and time (relativistically, spacetime) are objects that exist in addition to ordinary material objects such as tables and chairs", whereas relationism "rejects the existence of space and time (spacetime) and maintains that all that exists is material objects" (2012, 153; and, it is unclear if Arntzenius thinks nominalism can be paired with relationism at all). Unlike Field, who confines his fictionalist nominalism to Newtonian gravitation theory and its flat space, Arntzenius' goal is to extend the fictionalist-based nominalization strategy to curved spacetime and GR, thus Field's nominalist ontology of manifold points is insufficient (e.g., one can no longer treat vectors as pairs of points, as does Field). However, like Field's substantival conception of spacetime points, Arntzenius denies the existence of mathematical entities, which is how he defines platonism (215), while putting forward an approach to nominalism that simply "substantival-izes" those mathematical entities, i.e., he treats mathematical entities as substances and thereby secures by fiat a nominalist classification. For instance, Arntzenius argues that, given "the need to quantify over functions from spacetime points to real numbers, the obvious strategy to consider is to enrich the ontology in such a way as to provide nominalistic surrogates for such functions....[W]e will do this by posit-ing, for each spacetime point, a miniature one-dimensional space—a 'scalar value line'—endowed with a rich structure making it in effect a copy of the real line" (243–244). Whereas truth-based nominalism can secure the real line structure via, say, the possible divisions of an extended body (a la Aristotle), Arntzenius' fictional-ist nominalism demands an underlying physical entity that is actually identical (iso-morphic) to the real number line: this strategy is identical to Field's approach save for the choice of nominalist basis, scalar value line versus point manifold. He continues:

> This fundamental ontology does not merely contain many collections of entities isomorphic to the real numbers. It also contains some entities we might think of as uniquely natural candidates to *be* the real numbers; namely, the constant scalar fields.... [O]ne might doubt whether this fundamental ontology properly deserves to be thought of as 'nominalistic'. But let us not get hung up on the label....We have entities that behave just like the real numbers; indeed, we also have entities—namely, the *fusions* of constant scalar fields—that behave like sets of real numbers. But we are still far from having to accept the full set-theoretic hierarchy. This seems to us like a genuine theoretical gain. (246)

Ultimately, besides suggesting "fibre-bundle substantivalism" (185), and even "sophisticated fibre-bundle substantivalism" (194), the application of Arntzenius' version of nominalism to quantum theory leads to "Hilbert space substantivalism" (269). With regard to the latter, Arntzenius repeats his concern that treating a math-ematical structure (in this case, a Hilbert space) as a substance would seem quite removed from traditional nominalism (269). Nevertheless, his strategy does fit the minimal definition of nominalism advanced in this investigation, namely, the rejec-tion of non-spatiotemporal abstract objects. If one posits entities that stand for the real numbers and sets of real numbers, with all higher set-theoretical structures reduced to these entities, then nominalism is satisfied. Hence, one of the advantages of our analysis is that Arntzenius' nominalist project is indeed upheld as consistent with nominalism (since there is no appeal to abstract objects).

On the other hand, a major problem with Arntzenius' scheme, as well as Field's, is that it fits the definition of relationism as equally as it does substantivalism. A component of his nominalist strategy is to show "how all the 'mixed' vocabulary of some platonistic physical theory can be eliminated in favour of 'pure' predicates all of whose arguments are concrete physical entities" (215). But, the elimination of spacetime structure in favor of "concrete physical entities" describes relationism as well as substantivalism. Therefore, as with Field, Arntzenius' project would seem equally suited to those sophisticated relationists who conceive mathematical fields as physical entities, such as metric field relationism. Moreover, as discussed previously, the absolutist/substantivalist standpoint traditionally regards space as lacking the ability to interact causally with bodies or other fields (e.g., Newton), but the type of fibre-bundle structures that Arntzenius develops includes the metric of GR (213–270), which can partake in these mutual interactions; and the same is true of Field, whose approach to field theories is motivated by the idea that spacetime points have causal powers ("space-time points or regions are full-fledged causal agents"; Field 1980, 114). Ironically, Arntzenius criticizes some sophisticated versions of relationism, specifically, Huggett (2006b), for usurping the substantivalist's position (or, more accurately, for "piggybacking" on substantivalism; 2012, 169–170)—but, as just noted, his own approach to nominalism is open to the same charge, namely, as appropriating a strategy of "elimination of abstract spatiotemporal structure in favor of physical entities with causal powers" that one would normally associate with relationism.

Transferring substantivalism to the realm of quantum mechanics, via Arntzenius' Hilbert space substantivalism, also opens the door for a powerful nominalist critique. If, as Field insists, relationism must admit the possibility of a void space, then, to be fair to the relationist, a closely related scenario should also apply to the substantivalist—namely, the existence of spatial structure in the complete absence of all matter or physical fields. Yet, would the Hilbert space substantivalist that Arntzenius posits (2012, 269) tolerate the existence of QM's Hilbert space, an abstract vector space, given the non-existence of the quantum mechanical wavefunction that is an element of that vector space? If an affirmative answer is given, as would seem necessary to render Hilbert space substantivalism truly substantivalist in the customary sense that we associate with the possibility of a universal vacuum state in Newtonian theory, then Arntzenius would be placed in the unenviable position of having to defend the existence of a sort of platonic or Pythagorean bare mathematical entity; i.e., an abstract vector space without the wavefunction. As a counter-reply, the committed substantivalist may invoke an analogy between GR's metric field and QM's wavefunction, insisting that, just like metric field substantivalism, Hilbert space substantivalism also posits a congruence of the physical component (gravitational field in GR, wavefunction in QM) and the spatial component (metric in GR, Hilbert space in QM). Put differently, a Hilbert space substantivalist can claim that there are no "wavefunction voids" in QM just as there are no metrical voids in GR. So, the relationist's claim that substantivalism must allow for the possibility of an empty space (i.e., entirely void of matter/fields) simply fails to fit the details of these contemporary theories, GR and QM.[17] However, this response, while

---

[17] The analogy developed here is a bit loose, as a commentator has pointed out in an earlier version of this material. Since a wavefunction is an element of a Hilbert space, and is thus not defined on a Hilbert space in the same way that a metric is defined on a manifold, a more correct analogy would be between a wavefunction as an element of a Hilbert space and a manifold point as an element of a manifold. However, substituting an analogy between the wavefunction/Hilbert space and a point/manifold, rather than employing an analogy between the wavefunction/Hilbert space and a metric/manifold, works just as well for our purposes. On a related note, Dieks correctly points out that, as regard non-relativistic QM, "the Hilbert space formalism does not start from a space-time

perfectly correct, only demonstrates the weakness of Field's contention that relationism must also admit a void—i.e., given the congruence of the physical and geometric/mathematical components in each of these theories, GR's metric field and QM's wavefunction, the problematic potential for a completely energy-less void space is ruled out on *both* the substantivalist and the relationist interpretations.

There are other difficulties linked to Arntzenius' Hilbert space substantivalism. Since a Hilbert space is often categorized as the analogue of the six-dimensional phase-space employed in classical physics (with 3 position and 3 momentum variables for each particle in the classical system), the "reality", if any, that can be attributed to the state-space in classical mechanics should be identical to the reality assigned to QM's version of a state-space, i.e., the infinite dimensional Hilbert space. Yet, the six-dimensional state-space used in classical physics has traditionally been given an instrumentalist interpretation: it is a mere tool for extracting information on the behavior of the system. Accordingly, QM's Hilbert space should be given an equally instrumentalist interpretation, and not accorded the status of a substantival entity. As Friedman notes:

> Scientists themselves distinguish between aspects of theoretical structure that are intended to be taken literally and aspects that serve a purely representational function. No one believes, for example, that the so-called 'state space' of mechanics—phase space in classical mechanics and Hilbert space in quantum mechanics—are part of the furniture of the physical world. Quantum theorists do not believe that the observable world is literally embedded in some huge Hilbert space.... Rather, the function of Hilbert space is to represent the properties of physical quantities like position and momentum by means of a correlation with Hermitian operators. Similar remarks apply to other auxiliary 'spaces' such as 'color space', 'sound space', etc. (Friedman 1981, 4)

Many philosophers, such as the contributors to the current debate on the interpretation of the wavefunction in QM (see Ney and Albert 2013), might reject Friedman's purely instrumentalist interpretation of the various spaces (Hilbert, configuration, etc.) that can be employed in QM. Nevertheless, wavefunction realism, or antirealism, does not imply substantivalism or relationism. Indeed, despite Wallace and Timpson's defense of the view they call "spacetime state realism" (as a rival to wavefunction realism), they concede that both substantivalist and relationist interpretations are admissible (2010, 700, n.2). Rather, the debate on the reality and/or fundamental nature of a configuration space versus a Hilbert space would seem to pertain to the non-eliminative character of one of these structure in QM: in other words, either configuration space or Hilbert space, or possibly another space, is a feature of QM that cannot be reduced to other structures, nor eliminated in favor of these other structures. Yet, whether a *mathematical* structure is reducible or irreducible no more supports substantivalism than it does relationism, but it does impact the disputes among the different factions that make up nominalism, truth-based versus fictionalist, as we have seen.

---

manifold in which particles are located. The quantum state is given by a vector in Hilbert space, and has in general no special relation to specific space-time points. Rather, 'position' is treated in the same way as 'spin' or other quantities that are direct particle properties: all the quantities are 'observables', represented by Hermetian operators in Hilbert space" (Dieks 2001a, 16).

7.4.3 Scientific Realism and Platonism/Nominalism

To conclude this portion of our investigation, it would be worthwhile to briefly examine the confluence of platonism/nominalism and contemporary scientific realism, in particular, the approach to spatial or geometric structures in Psillos (2010, 2011, 2012), for our more fine-grained three-part platonist/nominalist scheme can provide useful insights on the presuppositions of this program.

In several works, Psillos defends a platonist conception of abstract objects in opposition to the fictionalist position dubbed "nominalistic scientific realism" (NSR), an approach that strives to rid theories of mathematical objects by drawing on Field's scheme described above. Because Psillos equates NSR with anti-realism and its commitment to the empirical adequacy of theories, he concludes that non-causal abstract objects should be embraced by the scientific realist in much the same way that they accept unobservable theoretical entities. Among these abstract objects, Psillos lists several items familiar to spacetime ontologists, "like phase spaces, vector spaces, and groups" (2010, 951), as well as inertial frames (2011, 4). Leaving aside Psillos' justifiable critique of a fictionalist interpretation of scientific theories, his embrace of platonism should raise a number of concerns for scientific realists. First of all, it is unclear what Psillos means by platonism: Does he endorse Plato's notion, where abstract objects exist outside of space and time, or immanent realism, where abstract objects exist in material objects? Since it is difficult to envision a scientific realist embracing an hypothesis whose domain is akin to Plato's heaven of non-spatiotemporal Forms, immanent realism would seem the best option, although Psillos does leave open the possibility that his "physical abstract entities", as he calls them, lack spatiotemporal location (2010, 950). Regardless of the domain quandary concerning abstract objects, the main obstacle confronting a platonist scientific realism is that it would seem committed to an immense number of abstract objects due to the well-known fact that modern theories in foundational physics can be given many different mathematical formulations involving different abstract objects. This topic will be taken up at length in Chap. 8, but a commonly cited example from the history of classical gravitation theories is the choice between the standard spacetime formulation of GR, with its non-flat spacetime structure representing one type of abstract object, as opposed to the Newton-Cartan formulation, with its flat spacetime structure representing a second brand of abstract object.

While platonists may happily tolerate this profusion of abstract objects, a better option for a scientific realist would seem to lie in a commitment to the truth-values associated with abstract objects (truth-based nominalism), as opposed to their existence as entities (immanent realism). That is, admitting a plethora of truth-values linked to the different formulations of a given scientific theory and its *single* ontology seems much more congenial to scientific realism than admitting an ontology populated by a plethora of different, undetectable abstract objects for each of its many possible formulations. Besides, Psillos' case for platonism is largely centered on the perceived need to uphold the truth of mixed physical-mathematical scientific statements—and, since he concedes that "we are ignorant" of the manner by which "the concrete [physical] and the abstract co-operate to render mixed statements true" (2012, 80), it follows that the realist stance favored by truth-based nominalism as regards the truth-values of possible structures should constitute a solution to Psillos' quest. In all fairness, it should be further noted that some of the truth-based nominalist positions described in §7.1.2, in particular, Lewis and Hellman, are occasionally categorized as platonist due to their realism concerning truth-values (see Colyvan 2001, 142; and footnote 3), so there is a good reason to conclude that truth-based nominalism would meet Psillos' anti-NSR requirements.

7.5   Conclusion

At this point, it is worth briefly summarizing some of the main conclusions of this chapter. As argued above, the non-causal role that spacetime structures play in scientific theories closely resembles the non-causal relevance of mathematics in general, hence the application of platonism and nominalism gains support in the assessment of spatial ontologies. Once these similarities between mathematics and spacetime structures are established, the attempts to link sophisticated relationism to instrumentalism (i.e., Earman 1989) can be seen as inadequate since they stem from an erroneous conception that spacetime structures are akin to the scientific realist's theoretical entities, rather than as another manifestation of the deeper philosophical issues associated with how mathematics and geometry relate to experience. In Chap. 8, the relevance of platonism/nominalism for scientific realism will be explored further within a more comprehensive examination of structural realism, and OSR and ESR. As will be demonstrated, some supporters of OSR also traffic in the ambiguities that lie at the intersection of mathematics and physics, but proponents of ESR can benefit from the resources provided by truth-based nominalism to overcome the underdetermination problems that plagues OSR.

Turning to our three-part division among nominalist concepts, not only can it provide useful insights that elude other methods of evaluating spatial ontology, in particular, substantivalism and relationism, but it can also assist the scientific realist's approach to spatial structures. As regards Newton and Leibniz, who are often presumed to be the most important representatives of substantivalism and relationism, we have argued that different versions of truth-based nominalism, Newton's incorporeal versus Leibniz' unique incorporeal-corporeal hybrid, reveal salient features concerning their respective God-based, P(O-dep) spatial ontologies—and these insights, moreover, are apparently not obtainable using other conceptual resources. Truth-based nominalism can also aid Psillos' case against a fictionalist-based scientific realism, but without requiring a realism about abstract objects or a commitment to substantivalism. On the other hand, Field's effort to block a relationist from espousing truth-based nominalism begs the question and runs counter to the historical record, while his and Arntzenius' fictionalist strategy is no more or less amenable to substantivalism than to eliminativist versions of relationism in the context of GR and QM.

In surveying the results of our investigation, truth-based nominalism clearly stands out as the most attractive and versatile options for understanding the relationship between the underlying physical ontology of spatial theories and their mathematical/geometrical structure—and, importantly, truth-based nominalism would seem to be a position that both the sophisticated substantivalists and sophisticated relationists can equally accept. Part of the reason for truth-based nominalism's success may stem from the fact that it represents a sort of intermediate approach between the extremes of immanent realism and fictionalism, a middle way that is analogous to the position that the sophisticated forms of both substantivalism and relationism take with respect to manifold substantivalism and a strict eliminative relationism. And, of course, the property theory of space, especially the ontological dependence version promulgated throughout our investigation, P(O-dep), is also compatible with truth-based nominalism.

Chapter 8
The Multiple Paths Towards an Epistemic Structural Realist Spatial Ontology

The structural realist approach to spacetime theories, explored in Chap. 5, represents one of the more influential third-way spatial ontologies to have recently appeared on the scene, but there are various difficulties associated with this strategy that have the potential to undercut its viability. This chapter will examine what most likely constitutes the primary obstacle, namely, the numerous underdetermination problems that plague structuralist conceptions of spacetime theories, a significant and contentious development in the structuralist literature that, as will be argued, also demonstrates the advantages that the spacetime version of epistemic structural realism (ESR) holds over its ontic competitor (OSR). Recent non-realist structuralist accounts, by Friedman and van Fraassen, have touted the fact that different structures can accommodate the same evidence as a virtue vis-à-vis their realist counterparts; but, while problematic for OSR, and possibly ESR, these claims gain little traction against a properly constructed liberal version of epistemic structural realism (see Chap. 5 for an introduction to these structuralist concepts). Overall, a broad construal of spacetime theories along epistemic structural realist lines will be defended which draws upon both Friedman's earlier work and the convergence of approximate structure over theory change, but which also challenges various claims of the ontic structural realists.

The contemporary revival of structuralism in the philosophy of science can be traced to a host of early twentieth century structuralist philosophies, principally, the epistemology and ontology of mathematical physics put forward by Poincaré, Eddington, Weyl, Russell, and Cassirer, to name just a few. Given the structuralist's fixation on the mathematical and/or empirical structure of theories, as opposed to, say, the underlying theoretical entities alleged to bring about and sustain those structures, it is therefore not surprising that the underdetermination of structure became an obstacle for the new program; e.g., Newman's critique of Russell (see footnote 13). That is, since more than one structure is likely consistent with the same evidence, a theory's structure is as underdetermined as its ontology (for, as is well known, several different sets of unobservable entities can be reconciled with the same observational evidence). This chapter will examine the current crop of

© Springer International Publishing Switzerland 2016
E. Slowik, *The Deep Metaphysics of Space*, European Studies in Philosophy of Science, DOI 10.1007/978-3-319-44868-8_8

structuralist conceptions of science as regards the underdetermination problem, focusing specifically on spacetime theories. Although much of the literature on structural realism has focused on quantum mechanics (QM), it is arguably the case that the spacetime structures used in classical gravitation theories more readily disclose the tensions between the competing ontic and epistemic orientations, hence this chapter will devote as much attention to general relativity (GR) as to QM and quantum gravity (QG).

In §8.1, the rival strains of structural realism, ontological and epistemological, will be evaluated in the light of the underdetermination problem regarding spacetime theories, for the difficulty is manifest by these approaches in diverse ways. The platonism/nominalism categories, introduced in Chap. 7, will be discussed in §8.2, for it can help to disclose the differences between ESR and OSR. In §8.3, theories that posit spacetime as an emergent effect of a non-spatiotemporal foundation will be briefly assessed against the backdrop of structural realism and the issue of the locality of theoretical entities. After demonstrating, in §8.4, the similarities between epistemic structural realism, early twentieth century structuralism, and Michael Friedman's idea of the relativized a priori, the role that underdetermination plays in his assessment, and, to a lesser extent, in van Fraassen's, will be the main subject of §8.5. Developing an epistemic brand of structural realism that can account for the various underdetermination problems surveyed in earlier sections will also be the goal of §8.5, along with a critique of a sophisticated form of ontic structural realism.

Building on the discussions in Chaps. 5, 6, and 7, the goal of this chapter is to demonstrate that a properly construed broad or liberal conception of epistemic spacetime structural realism is best equipped to handle the underdetermination problem; or, put differently, that all other structuralist approaches must fall back on conceptual resources that are strikingly similar to the liberal brand of epistemic structuralism. Consequently, the "multiple paths" that a scientific theory can take as regards distinct mathematical formulations or different ontological interpretations (i.e., underdeterminism) is not inconsistent with epistemic structural realism, but it does draw our attention to the convergence and constraints manifest in scientific theorizing, as well as Friedman's own early defense of realism. These last ingredients in the structural realist account have often received scant consideration, but are central to our goal of establishing the viability of epistemic structural realism for spacetime theories. Finally, although it is not directly addressed in this chapter, the property theory of space, of either the P(O-dep) or P(TL-dep) type, is compatible with the liberal construal of epistemic structural realism developed below. But, as argued in Chap. 5, this due to the simple fact that this liberal brand of ESR would envision all of the different sophisticated spatial ontologies (sophisticated substantivalism, sophisticated relationism, property theory) as the same ontology (that is, just as long as they make the same predictions on possible states of motion; see, §5.2.2).

8.1 Structural Realism, Underdetermination, and Spacetime Theories

The underdetermination problem was briefly discussed in Chap. 5, but we will now explore its ramification in more detail. Since the fortunes of both ESR and OSR depend in no small measure on how they handle the underdetermination of theoretical and mathematical structure, it is imperative that the relevant differences between OSR and ESR, and variant formulations of these positions, be examined as they pertain to underdeterminism.

8.1.1 OSR and ESR-L

As explained in previous chapters, structural realism (SR) holds that what is preserved in successive theory change is the abstract mathematical or structural content of a scientific theory, hence SR departs from a straightforward realism involving theoretical entities. There are two chief benefits associated with SR. First, SR can explain the progressive empirical success of scientific theorizing, and thus accommodates the "no miracles argument" (NMA), since the structures present in our best scientific theories connect with, or in some manner capture, the "real" world (i.e., NMA holds that the success of science would be a miracle without some realist commitment). Yet, secondly, SR evades the "pessimistic meta-induction" (that plagues standard scientific realism) because it does not make an ontological commitment to the theoretical entities used in specific theories: in brief, the pessimistic meta-induction contends that, since the theoretical entities that appeared in past successful scientific theories were ultimately discarded for different theoretical entities in a successor theory, the likely continuation of this process should dissuade any realist commitment to theoretical entities. Worrall's well-known example that strives to counter the pessimistic meta-induction involves the evolution in nineteenth century optics from Fresnel's elastic solid ether to Maxwell's electromagnetic field: "Fresnel's equations are taken over completely intact into the superseding theory [Maxwell's]—reappearing there newly interpreted but, as mathematical equations, entirely unchanged" (Worrall 1989, 120). Hence, while entities may come and go, an invariant structure underlying these theories, past, present, and future, provides a basis for scientific realism. What is less well-known, though, is Worrall's comment that this example is "unrepresentative", and that "the more common pattern is that the old equations reappear as *limiting cases* of the new—that is, the old and new equations are strictly inconsistent, but the new tend to the old as some quantity tends to some limit" (120, original emphasis). The case that is offered to establish this point is rather telling for the would-be spacetime structural realist, as will be examined more carefully in §8.2: "Einstein's equations undeniably go over to Newton's in certain limiting special cases. In this sense, there is 'approximate continuity' of *structure* in this case" (121). Inspired by Poincaré, Worrall's original formulation of SR laid the groundwork for ESR, for it regards structure from a purely epistemological perspective, such that the mathematical structures that turn up in our best scientific theories do not provide any information on the actual ontology of entities and processes that underlie the observed structural relationships. In short, ESR makes a realist commitment to the invariance of structure across scientific change.

OSR, in contrast, rejects an epistemological interpretation of structure; rather, structures *do* reveal facts or truths about the underlying ontology—and may, in fact, *be* the underlying ontology. Since OSR incorporates mathematical structures and relations within its ontological assessment, ESR might seem closer to traditional scientific realism, for an ESR theorist *could* regard the mathematical structures in our best theories as merely the epistemologically-assessable relations between the relata, i.e., the entities, so that only the entities are real (and not the relations themselves). But, a more liberal form of ESR seems preferable, namely, a form of ESR that leaves open the possibility that the underlying ontology may include the relations alongside the relata in the same manner as OSR, even if the precise ontological details are epistemologically inaccessible. There are no grounds for denying this formulation of ESR, furthermore, since it is in keeping with the skeptical orientation of ESR concerning our knowledge of nature's deep ontology. Henceforth, following the convention introduced in Chap. 5, all references to ESR will take this more liberal or broad form, dubbed "ESR-L"; i.e., the underlying ontology can include only relata, only relations, or both relations and relata (and, unless otherwise noted, ESR-L will refer to ESR in what follows). The NMA would seem to constitute the essential realist content of ESR-L, for it holds that the approximate continuity of mathematical structure manifest over the history of any one science is not an accident, but is due to the "constraints" that reality imposes on our scientific endeavors (see Brading and Landry 2006 for a similar construal of ESR, and §8.3).[1] Of course, all of the other forms of structuralism explored in this chapter, with the likely exception of van Fraassen's, would also agree with the claim that nature's constraints are non-accidental features of our best scientific theories.

As recounted in Rickles and French (2006), Cassirer and Eddington's philosophical reflections on the group structure of GR and QM motivates contemporary OSR, since the structural role that, say, quantum particles play within the larger mathematical structure can admit a straightforward ontological reading. That is, QM particles no longer retain the autonomous, individual status usually associated with scientific realism; instead, it is the group structure itself which directly represents the ontology for OSR advocates. Transferred to the modern spacetime ontology

---

[1] Saatsi (2010) has argued that a scientific realism requires more than mere approximate continuity, but should also explain the success of the earlier theory from the vantage point of the succeeding theory. Yet, while correct as a final goal or heuristic of scientific realism, explanations of this sort would seem to require access to the ontologies underlying these theories (or, at least, the succeeding theory's ontology). ESR-L denies that this requirement is necessary to establish the success of ESR-L *relative* to its non-realist rivals—the reason being that the non-realist alternatives have no grounds for claiming a future directed convergence of approximate continuity over the course of science (see, §8.4 and §8.5).

debate, OSR seeks to handle the points of the spacetime manifold in an analogous fashion: the points are either a derived, secondary aspect of the primary ontological unit, in this case, the metric; or, the points and metric, as relata and relation, are construed as being ontologically "on a par", so that neither is more basic. Until recently, these interpretations have been understood as a form of sophisticated substantivalism (e.g., Hoefer 1996), or even sophisticated relationism (as discussed in Dorato 2000), but Rickles and French (2006, 24) claim that they are closer in spirit to SR.

Before proceeding to examine the underdetermination problem associated with SR, especially OSR, it would be useful to contrast the more traditional reading of ESR with ESR-L. In the context of spacetime theories, Esfeld and Lam (2008) interpret ESR as a commitment to autonomous spacetime points (the relata): "applied to the framework of the standard tensor representation of space-time, epistemic structural realism implies that the identity of the space-time points is constituted by their fundamental intrinsic properties, independently of the space-time structure—that is, independently of the metric" (2008, 35; see, also, Dorato 2008, 24, for a similar reading of ESR). In other words, Esfeld and Lam regard ESR as somewhat like Hartry Field's (1980) mathematical structure-eliminating version of manifold substantivalism, since the manifold points retain an intrinsic identity apart from the higher mathematical structures in the spacetime, in particular, the metric. But, while this interpretation may possibly be true for older, positivist-leaning conceptions of ESR, it is not applicable to ESR-L. Since, as described above, ESR-L remains agnostic regarding the underlying ontology, the manifold points may have an ontological status that is either intrinsic or, like OSR, is derived from, or on a par with, the metric. In short, Esfeld and Lam's reading of ESR foists a commitment to a *particular* ontology (i.e., independent manifold points)—ESR-L, in contrast, can choose from among a number of underlying ontologies, just like OSR.

8.1.2   Entity Underdetermination

While Rickles and French' assertion has much merit, a major drawback with their attempt to link sophisticated forms of both substantivalism and relationism to OSR is that the main goal of Worrall's initial plan for SR—namely, to uphold the NMA and defeat the pessimistic meta-induction—is thereby neglected or forsaken altogether. That is, sophisticated substantivalism never attempted to counter anti-realist worries, since it is a metaphysical interpretation of spacetime theories with different goals in mind, but defeating anti-realism was the motivation behind Worrall's conception of SR. The contribution of Esfeld and Lam (2008) is, once again, a case in point, for they strives to apply OSR exclusively to spacetime theories, and they are quite candid in their assessment of ESR and the pessimistic meta-induction. After rehashing SR's desire to defeat the pessimistic meta-induction a la Worrall, and correctly noting that the cumulative progress employed by SR pertains to approximate structure (2008, 28), they reckon that the arguments that support the pessimistic meta-induction apply to "views about structure as well", and thus "structural realism as such does not rescue scientific realism" (2008, 29). Discarding Worrall's original intentions for SR, Esfeld and Lam thus conclude: "we pre-suppose scientific realism, but we do not intend to use structural realism in support of scientific realism" (29).

Yet, Esfeld and Lam's negative appraisal of the prospects for countering the pessimistic meta-induction by way of SR is unwarranted, and is not endorsed by all advocates of OSR (see, French 2014, 40–41). ESR, in particular, *does* try to locate an invariant within the historical progress of science: hypothetical entities will come and go, but the retention of mathematical structure, both as an approximation and a limiting case, has been upheld over the course of past scientific theory change, and is predicted to be upheld over the future course of theory change. What the OSR theorists have provided, in contrast, is yet another form of scientific realism motivated by the presumed existence of an unobservable entity, but which this time incorporates various aspects of mathematical/conceptual structure alongside all of the usual drawbacks attendant on a belief in unobservable entities and their intrinsic properties (more on this below). Returning to Esfeld and Lam, they defend a moderate form of OSR within the context of GR, such that the points and metric/fibres are on a par ontologically (employing both the standard tensor and fibre bundle formalisms), but then later concede that "in the framework of certain candidates for QG [quantum gravity], such as loop quantum gravity or the algebraic generalization of GR, there may be no reference anymore to space-time points" (44). Accordingly, by folding OSR into sophisticated substantivalism, the advocates of OSR for spacetime theories have ceded the field, albeit unintentionally, to the anti-realist's pessimistic meta-induction argument against all forms of SR, and hence against their own spacetime version of OSR. The anti-realist will happily point out that, once again, the "entities" (in this case, spacetime points) that appear in our best current theory will be replaced eventually by a different set of entities in the successor theory; so, one should draw an anti-realist conclusion about the existence of all theoretical entities, whether those entities are viewed as distinct individuals (e.g., spacetime points, electrons) or those entities subsume the individuals into their larger structure (e.g., metric, fibre bundles, group-theoretic structure in QM).

One should not misconstrue the meaning or goal of this last criticism, however. It is not simply the argument that spacetime points will (or may) be discarded by a successful QG hypothesis: even if one adopts an eliminativist form of OSR, which denies the reality of spacetime points for the metric/fibre structures of GR, it will still be the case that those mathematical structures will almost certainly be replaced by the *different* mathematical structures used in QG; e.g., replacing tensors or fibre bundles in GR for the more fundamental Hilbert space, or replacing the standard formalisms in *both* GR and QM for some deeper mathematical structure employed by an underlying successor theory (so that GR and QM are now the limiting cases of this deeper theory). In brief, underdetermination is a much greater threat to OSR, which *reifies* structure to some degree, as opposed to ESR-L, which does not. Although the platonist/nominalist question will be addressed later, the use of the term "reifies" at this point in our analysis is not meant to denote the immanent realist position discussed in Chap. 7, where mathematical entities or abstract objects are posited as existing alongside the physical ones—rather, it is meant to signify that specific aspects or general characteristics of the mathematical structure of our best current theories, say, GR or QM, are taken to directly represent, or actually correspond to, the underlying ontology. For instance, since the group structure in QM engenders a sort of holistic picture of particles, wherein the individual particles cannot be separated from the field encoded in the group structure representation (i.e., permutation invariance), most QM-inspired OSR theorists have drawn the lesson that the world's underlying ontology must be identically non-individualistic or holistic: e.g., "world-structure" (or "world-as-structure").[2] The underdetermination that affects the ontic structural realist, consequently, is but another instance of the same underdetermination problem common to all brands of scientific realism that reify theoretical entities, a problem we will dub "entity underdetermination".[3] In short, while our best contemporary spacetime theories confirm GR, as opposed to, e.g., a Machian alternative, it will likely be superseded by a more fundamental theory, such as one of the many competing QG hypotheses. Given this likely outcome (since GR is a classical gravitational field theory that is strictly inconsistent with quantum mechanics), GR will thus be seen as *approximately* true at large scales of space and time in the sense that its equations are mere limiting cases of the more fundamental QG theory and its *different* class of entities and mathematical structures—and this, of course, is exactly what ESR-L predicts and is designed to handle, hence the rationale for ESR-L over OSR.

The same reasoning applies to QM. While group-theoretic structure is an entrenched component of modern particle physics, it is possible that a successor theory may come along that replaces group structure with a non-holistic mathematical or conceptual alternative that retains a robust, non-eliminativist role for individual particles or other entities. This individuals-based QM successor theory would thus overturn the motivation for the holistic world-structure ontology: i.e., group structure would then be seen as a higher level aspect of the phenomena that is reducible to the more fundamental, non-holistic, individuals-preserving structure of the hypothetical successor theory. As a result, the anti-realist's claim that "all entities are eventually replaced" will be vindicated, since the world-structure ontology will have been called into question by the ascension of the non-holistic successor theory. There are a host of speculative QG hypotheses that would seem to raise this very

---

[2] See, e.g., French and Ladyman (2003), and Ladyman and Ross (2007) for more on the ontology of OSR. The characterization of this ontology as holistic, along with other descriptions, are outlined in Ladyman (2009). "World-Structure" is the term mentioned in Ladyman and Ross (2007, 158). French (2014, 183) mentions the "blobject" (see, Horgan and Potrc 2008) as a potential candidate for a monistic reading of the world's structure.

[3] This is different from the "metaphysical underdetermination" mentioned in Ladyman (1998), which is generated by alternative realist interpretations of a single theory. Entity underdetermination, in contrast, is the thesis that different theories, with different ontologies, will likely replace our currently successful theories. In essence, entity underdetermination is the ontological consequence of the pessimistic meta-induction. However, metaphysical underdetermination is a major problem for a fictional nominalist construal of OSR, as will be discussed in §8.2.

challenge for QM and its standard formalism, prompted in no small part by the enduring obstacles faced by the more prominent QG theories, e.g., string theory and loop quantum gravity. According to Butterfield and Isham, "[t]he idea here is that both classical general relativity *and* standard quantum theory emerge from a theory that looks very different from both....presumably by not being a quantum theory, even in a broad sense—for example, in the sense of states giving amplitudes to the values of quantities, whose norms squared give probabilities" (Butterfield and Isham 2001, 60). Some of these alternatives involve the use of quantum computational techniques modeled on quantum cellular automata, the latter being discrete, cell-like computational devices with local neighborhoods patterned on electronic circuits: "The probably best-known of these emergent quantum approaches goes back to 't Hooft, [who] proposes a deterministic, pregeometric, non-quantum substrate, which should possibly be modeled by something like cellular automata" (Hedrich 2009, 25). Other examples, investigated by Butterfield and Isham, describe QG strategies that reject the real and complex number structures central to the group-theoretic formalism of QM:

> So, according to this line of thought, the use of real numbers (and similarly, complex numbers) in quantum theory in effect involves a prior assumption that space should be modeled as a continuum. If so, then the suggestion that standard spacetime concepts break down at the Planck length and time, and must be replaced by some discrete structure that only 'looks like' a differentiable manifold at large scales, means that we cannot expect to construct a theory of this discrete structure using standard quantum theory—with its real and complex numbers. (2001, 85)

Of course, the OSR theorist might claim that all of these potential successor theories still exemplify their world-structure ontology, but this rejoinder would only draw attention to the indeterminate and vague nature of that ontology, so that the very meaning of "world-structure" becomes strained beyond credibility. For example, even granting that it is information that connects the cellular automata in the quantum computational QG theory mentioned above, a series of local, discrete quantum computers seems more akin to an individuals-based conception of reality—and hence the particles-as-individuals interpretation of QM—than the holistic world-structure ontology based on group structure.[4] Likewise, Wheeler's lawless, chaotic non-QM substrate proposal, often categorized as an early instance of a QG-type theory, would seem to be the very antithesis of a world-structure ontology (see, once again, Hedrich 2009, 25).

Of course, it night be the case that the OSR theorist is correct, and that all successor theories will uphold the group-theoretic structure in contemporary QM (or an equivalent holistic conception), so that the world-structure ontology is vindicated, contra anti-realism. The ESR-L theorist, on the other hand, takes the anti-realist

---

[4] That is, given a QG ontology that posits a collection of discrete objects (quantum cellular automata) connected by information, the best QM analogue would seem to be the more traditional conception that relies on individual objects (electrons, etc.) and their interconnections (relations). Prospective QG theories of this sort may fail, of course, but the point is that they cannot be ruled out, hence OSR remains subject to the pessimistic meta-induction.

challenge of the pessimistic meta-induction as a motivating principle, a rationale that prompts their skeptical approach to ontology. On this issue, ESR-L may reflect the more "anti-realist friendly" environment that prevails in spacetime theories, where different structural approaches abound, unlike QM and its seemingly ubiquitous, at least thus far, mathematics of group structure.

8.1.3 Formalism Underdetermination

Returning to the analysis of spatial theories and SR, there is another form of underdetermination implicit in Esfeld and Lam's comments on the possible demise of spacetime points. The reference to algebraic forms of GR raises the specter of a "formalism underdetermination", a problem that can be seen as a variant of the entity underdetermination just divulged for OSR, although it is not necessarily tied to the inevitable succession of theories over time (e.g., GR replacing Newtonian theory). Put briefly, there are many competing mathematical formulations of any given spacetime theory: as mentioned in Chap. 5, Cartan showed that Newtonian gravitation theory could employ a mathematical treatment similar to that advanced in GR, such as the use of a non-flat connection and a geometrized gravitational potential, a formulation that is strictly inconsistent with the flat inertial structure of standard Newtonian Gravitation Theory (see, e.g., Pooley 2006).[5] In the case of GR, there are a number of competing formalisms from which to choose—besides the typical tensors on manifold method, there are the twistor, Einstein algebra, and Dirac algebra formulations surveyed in Bain (2006).[6] This formalism underdetermi-

---

[5] Pooley (2006, 88) raises the formalism underdetermination issue, along with a number of metaphysical underdetermination objections (see also footnote 3). However, if OSR theorists fall back upon their world-structure ontology, it is not clear that these metaphysical underdetermination cases can gain much traction against OSR. For instance, Pooley raises the specter that different interpretations of the measurement problem (e.g., de Broglie-Bohm versus other interpretations) lead to the underdetermination of the exact nature of the realist ontology, despite the use of the same mathematical formalism by these different interpretations (of the collapse of the wave function). Yet, since the same mathematical formalism is utilized, both collapse interpretations uphold the same world-structure ontology, with the difference lying in the mere details of how that ontology functions in the collapse case—and this is a much less troubling underdetermination than the entity or formalism underdetermination problems raised above, which do call into question the world-structure ontology itself.

[6] Bain (2009) strives to address the formalism underdetermination discussed in Pooley (2006) (dubbed, "Jones underdetermination" by both authors). Bain's very promising analysis employs a category-theoretic approach in order to model a theory's dynamical structure (say, solutions of the field equations in GR) via the symplectic manifold that encodes the phase space of dynamically possible states. This conception may blunt the worries associated with formalism underdetermination, but it depends on a category theory framework which some may find problematic (Bain 2009, 17). More importantly, the many different constructions of GR occupy different symplectic manifolds (e.g., twistor models are different from tensor models with/without boundary conditions; 19), and so neither the entity nor the metaphysical underdetermination problems, as we have called them, have been diminished (see footnote 3). The OSR theorist can, of course, always claim that

nation also afflicts ESR and ESR-L, although we will postpone that discussion until §8.3.

Furthermore, in the context of QM, one cannot rule out the possibility that an alternative to the group structure formalism of QM may be devised, an alternative formalism that challenges the world-structure hypothesis by admitting, in some fashion, a non-holistic role for individual particles (i.e., a new approach that is without the drawbacks that are common for individuals-based interpretations of the standard QM formalism; see, e.g., French 1989, for individuals versus non-individuals interpretations of QM). Now, the OSR advocate of the world-structure ontology could reply that these potential individuals-based formalisms are simply limiting cases of the holistic group structure formalism (see French 2014, 41), but then the question remains open as to which formalism correctly represents the underlying ontology. Maybe the new individuals-preserving formalism corresponds to the actual ontology, with the group structure formalism constituting a mere limiting case of a non-holistic realm of individual quantum particles?

Returning to the general topic of formalism underdetermination in the context of spacetime theories, another response that the defenders of OSR might offer is that these alternative mathematical formulations, if they are truly grounded in an identical physical theory, are much more akin to different hierarchical arrangements of the same geometric component structures (manifold, metric, affine, conformal, etc.), rather than different geometrical structures altogether: e.g., conformal structure is basic for the twistor theorist, with the manifold and metric as derived structures, whereas manifold and metric structure is basic for the traditional tensor theorist, and conformal structure is derivative. As Bain comments in his explanation of these basic/higher level orderings, "what is real, the spacetime structuralist will claim, is the structure itself, and not the manner in which the alternative formalisms instantiate it" (2006, 64). Although this response would appear to be neutral as regards the OSR/ESR dispute, it would seem more appropriate coming from the ESR camp, since they take an epistemological stance on mathematical structure. In addition, how does an OSR theorist accept an ontology of spacetime structures *per se*, or *in general*, as opposed to a particular mathematical treatment of spacetime structures? That is, a spacetime theorist who sides with an ontology of "the geometry of shapes and angles" (conformal structure) or any of the other general mathematical structure, rather than an ontology tied to a specific mathematical theory and its formalism, say, twistor theory, would seem to be advocating a metaphysics of universals, and not particulars (i.e., twistor theory would be a particular instantiation of those universals). Would the OSR theorist gladly accept this outcome, and sanction general geometric structures over particular instantiations of those structures? The world-structure ontology in QM would also appear susceptible to the same problem, since a general non-individuals metaphysics is the motivation and grounds of any specific mathematical formulation of OSR's ontology, such as group

---

future empirical data may favor some of these competing models, given their different dynamical structure—but the ESR-L theorist will interpret this state of affairs as supporting their more cautious brand of SR, since the evidence may forever fail to decide among these alternatives.

structure. Put differently, "a non-specific, general *ontological* structure" might be a contradiction in terms, and would appear to be much more appropriate if translated into "a non-specific, general epistemological structure that is grounded in a specific, non-general ontology", i.e., ESR-L (see, also, French 2014, 209–210, which discusses these specific worries).

Summing up, ESR-L has a more palatable account to offer, since it maintains a strict neutrality on the underlying ontology. The ontology that undergirds our best spacetime theory may coincide with either traditional scientific realism, so that the mathematical structures used in that theory are entirely phenomenal and are mere relations among old-fashioned physical entities (which may explain why Psillos 2006 finds less to fault in ESR); or the underlying ontology may ultimately coincide with a *particular* spacetime formulation of OSR, such as twistor theory—but it need not endorse an ontology of spacetime structures "in general", or world-structure. In other words, both the entity and formalism underdetermination problems are, for ESR-L, simply a result of our inability to empirically and theoretically determine the nature of the underlying ontology, but they do not need to counter those under-determination worries by saying that the underlying ontology *really is* just a loosely-defined collection of generalized geometric properties, or world-structure. Consequently, the metaphysical implications of ESR-L seem much less radical than for OSR.

As will be discussed below, the OSR theorist might reply that the underdetermination of structure in our best spacetime theories is merely descriptive or representational, for there are many ways to characterize the world-structure ontology via different mathematical formalisms. The use of correspondence principles that can correlate different mathematical formulations of the phenomenon, say, classical physics and QM, would be an instance of this approach (e.g., Saunders 1993). Of course, the OSR theorist is free to endorse these maneuvers, and they are certainly not contradictory. Yet, it would seem that the lesson to be gleaned from the competing hierarchical arrangements of component geometric structures, as well as from the use of correspondence principles in QM, is that one should refrain from conceiving mathematical formalisms as directly representing the underlying physical ontology—but, this good advice apparently contradicts what OSR counsels, since its world-structure ontology is directly based on an important feature of the group structure formalism in QM, namely, the non-individuality of particles via permutation invariance, on both the moderate and eliminativist construals of OSR. So, it could be argued, OSR is trying to have it both ways: on the one hand, the appeal to correspondence principles and/or general geometric structures signals a caution or wariness in reading the ontology directly off the formalism, but, on the other hand, they do take aspects of the formalism directly into their ontology, i.e., group structure as the basis for their non-individuals world-structure ontology. As a further counter-reply, the OSR theorist could claim that the non-individuals, holistic conception that informs their underlying ontology is, in fact, the best interpretation of QM, and possibly, QG as well, thus their choice of which element of the formalism to "reify" is both motivated and defensible. The ESR-L theorist will respond, need-less to say, by recalling the long history of allegedly fundamental theories—and their ontological *interpretations*—that have been overturned, and caution that the world-structure hypothesis may yet undergo the same fate.

Consequently, despite the fact that a specific metaphysical interpretation of a mathematical formalism evokes the dreaded pessimistic meta-induction, the OSR theorist is not without a viable defense against the allegations put forward above. The main point of these criticisms, to put it somewhat differently, is that by making structure ontic, the underdetermination of structure (of either the entity or formalism variety) should equate with an underdetermination of ontology, and thereby undercut the world-structure ontology actually offered by the OSR theorists. In contrast, the implications for the underlying ontology connected with ESR-L has the advantage that it is, first, less metaphysically controversial than its OSR rival, and, second, that it is also a much more defensible *scientific realist* position (i.e., in opposition to scientific anti-realism). Since these structures need not reveal the specific ontology that underlies our best theories, and the knowledge that they do provide is both fallible and revisable just as long as approximate structural continuity is preserved, ESR-L would seem to be the better scientific realist option. These considerations, finally, would seem to be in line with many of the recent defenses of ESR that raise analogous, if more general and less spacetime specific, themes (Morganti 2004, Saatsi 2010).

8.2   ESR, OSR, and Platonism/Nominalism

Can the platonist/nominalist categories, first developed in Chap. 7, assist the OSR theorist in defusing the problems raised above? Overall, while our platonist/nominalist categories are relevant for assessing the OSR/ESR dichotomy, it will be argued that the benefits largely accrue to the advantage of ESR-L, not OSR. Since ESR and ESR-L are epistemologically based, and truth is an epistemological notion, it naturally follows that truth-based nominalism aligns with ESR/ESR-L. As regards OSR, however, the situation is more difficult. In what follows, we will confine our attention to the work of Steven French, for he has specifically commented upon the relationship between his interpretation of OSR and structuralism in the philosophy of mathematics, i.e., *ante rem* (platonist) and *in re* (nominalist) structuralism (see Shapiro 2000). In short, French is skeptical of the applicability of mathematical structuralism, claiming that "the classification of '*ante rem*' vs. '*in re*' structuralism may not be appropriate in the context of OSR, where the world-as-structure is neither abstract, in the sense of being non-causal nor a system that is structured, in the sense that there is a system that is ontologically prior to the structure" (2012, 25; also, French 2006). "World-as-structure", once again, seems to signify a non-individuals, holistic conception of quantum particles. Rejecting the view that the group structure of QM represents the relations among autonomous individual quantum particles, OSR offers, instead, an "entirely structural" account, whereby physical objects, such as electrons or quarks, are "reduced to mere 'nodes' of the structure, or 'intersections' of the relevant relations, in some sense" (French 2006, 171, 183).

Turning to the platonism/nominalism distinction, French's comment that OSR rejects abstract structure thereby rules out platonism and immanent realism, but perhaps truth-based nominalism as well, since the latter accepts a realism concerning the truth-values related to these non-causal abstract objects (although not to their existence as entities; see Chap. 7, once again). On the other hand, since fictionalist nominalism takes an eliminativist stance on mathematical structure, it fits nicely with French's claim that world-structure is neither an abstract object nor "a system [of individual particles] that is ontologically prior to the [group-theoretic] structure". The support for a fictionalist-based interpretation of French's philosophy is, indeed, quite compelling; e.g., "the idea is that instead of conceiving our ontology in terms of objects, and then having to face the dilemma of whether to regard them as individuals or not, we focus on the relevant group-theoretical structures underpinning quantum statistics and effectively re-conceptualize (or eliminate) our putative objects in terms of these structures" (French 2011, 217). Group-theoretic structure is not the underlying ontology—rather, the underlying ontology is a physical structure that is precisely described using group-theoretic structures. This response is, moreover, exactly what fictionalists would counsel, for they take mathematical structure to be a replaceable means of describing an isomorphic physical ontology. Additionally, unlike platonism, immanent realism, and truth-based nominalism, causality is incorporated into French's version of OSR (2012, 25), a feature that he shares with Field and Arntzenius' fictionalism (as discussed in Chap. 7).

Nonetheless, French (2011) suggests that his world-structure ontology is "multifeatured", a development that would appear to undermine a fictionalist classification for his brand of OSR. Since the fictionalist holds that the physical ontology and its mathematical representation are structurally isomorphic, it follows that different mathematical formulations of a theory correspond to a different structured ontologies, thus inducing the same kind of underdetermination problem that inspires the anti-realist's case against the existence of theoretical entities (i.e., the pessimistic meta-induction described in §8.1). Consequently, if truly fictionalist in the manner of Field and Arntzenius' project, French's OSR would be equally susceptible to an ontological underdeterminism resulting from an underdeterminism of structure. Bueno has demonstrated this point with respect to the different structures required for QM's overall mathematical representation: while both group-theoretic and Hilbert space structures are essential, he concludes that "when we combine both mathematical frameworks, no clear picture of what is going on at the quantum level emerges" (Bueno 2012, 94).[7] French (2011) responds to the general underdetermination challenge by seeking a "common structure" that lies beneath the different mathematical formulations, a common structure that, besides his preference for a group-theoretic account, may require the addition of dynamical (symplectic) structure for its full characterization. Motivated in part by the work of Bain (e.g., 2009) and Belot (2006), French concludes:

---

[7] The type of underdeterminism raised by Bueno would seem to be of the "metaphysical underdeterminism" variety (see footnote 3), since the same physical theory is open to many different ontological interpretations.

Hence, the structuralist still has some work to do in supplementing the 'object structure' [particles vs. non-particles] with the relevant dynamical structure and fleshing out the 'world structure' as multi-featured. In effect what we have is an appropriately complex ontology that includes both the group-theoretically characterized structure underlying the particles-as-individuals and particles-as-non-individuals packages, and the common symplectic structure underlying the Hamiltonian and Lagrangian formulations. (2011, 219; see, also, French 2014, 156)

Therefore, since there is more than one structure that tracks his "complex ontology" from different mathematical perspectives, the fictionalist prescribed isomorphism between mathematical and physical structures is no longer assured; i.e., the same physical ontology may not be capable of accommodating these different mathematical structures in a coherent fashion, as Bueno's case makes clear. French's "why worry" strategy as regards underdeterminism would thus not be acceptable for a fictionalist, for fictionalism requires an isomorphism between mathematical structure and ontology, whereas French seems to hold that a simple commitment to structure, no matter how complex or multi-featured, is ontologically sufficient.

A host of related difficulties are embodied in French's new multi-featured world-structure ontology, although some of these problems have been mentioned previously. First, despite French's causation-based misgivings about the relevance of mathematical structuralism for OSR, philosophers of mathematics will maintain, quite correctly, that all theories that posit a relationship between mathematics and physics fall within the scope of traditional platonist/nominalist categories, hence it is worthwhile to evaluate OSR for clues pertaining to the most accurate classification. Among the categories in our tripartite scheme introduced in Chap. 7, truth-based nominalism would appear to be the best, and maybe only, option: that is, leaving aside his rejection of abstract objects, the central theme of French's "multi-featured" OSR is his hypothesis that different mathematical structures can be used to characterize the underlying physical ontology, and this overall conception is best captured by truth-based nominalism, and not fictionalism. Second, the revelation that the world-structure ontology is multi-featured places OSR fairly close to ESR-L, if not identical to ESR-L, save for OSR's allegiance to a particles-as-non-individuals QM holism. In brief, there is tension between French's claim that world-structure is not "a system that is ontologically prior to the [group-theoretic] structure", and the fact that (i) a particles-as-individuals interpretation is compatible with the group-theoretic account of QM, and (ii) the recognition that QM's "complex ontology" necessitates supplementary structures beyond the group-theoretic for a more accurate characterization (in other words, group structure alone cannot disclose QM's complex ontology). Provided (i) and (ii), a more accurate category for French's theory would thus seem to lie in the liberal form of ESR, ESR-L, designated "ESR_2" by French and Ladyman (2012), where the invariance of structure (perhaps approximate structure) over theory change fails to disclose the full nature of the underlying ontology; i.e., whether the ontology is only structure (relations), only particles, or both particles and structure on a par. Yet, French and Ladyman (2012, 27) reject ESR_2 (ESR-L) and posit instead a particles-as-non-individuals interpretation, a specific metaphysical reading of an acknowledged complex, multi-faceted ontology. As a direct result, this OSR strategy must inevitably prompt the underdetermination anxieties described above (such as the anti-realist's pessimistic meta-induction), especially in an environment where a projected theory of quantum gravity may upend the prevailing mathematical and ontological conceptions of the micro-realm (see §8.1).

Moreover, since French's multi-featured OSR starts from a fixed metaphysical assumption about the underlying ontology, in contrast to the skeptical position of ESR-L, there is a corresponding difference in their attitudes towards the invariance of structure: OSR projects an invariance of structure with the aim of upholding a non-individuals conception of QM particles, whereas ESR and ESR-L project an invariance of structure to defeat anti-realism (i.e. the pessimistic meta-induction). For French, one of the consequences of this diversity in the goal or purpose of invariant structure is "that it is not at all clear that the properties that the structuralist should focus on in theory change will be the same as those involved in [OSR's] structural reconceptualization of objects" (French 2006, 171). These different standpoints on invariant structure may also be reflected in Esfeld (2013), where ESR, and hence ESR-L, counts as "partial-realism" since it intends to defeat anti-realism without specifying an underlying ontology, but OSR seeks to obtain "complete realist" status since it does posit a specific underlying entity (i.e., the particles-as-non-individuals package). Framed in this manner, Esfeld's insight pinpoints the main difference between OSR and ESR-L via their respective realist aspirations, as also argued above.

8.3 First Reflections on Non-spacetime Hypotheses

Before proceeding to examine other aspects of the underdetermination issue, it would be useful at this juncture to examine some of the arguments that have been raised against QG theories. Although string theory, loop quantum gravity, and a host of other QG hypotheses that advance an alternative foundation for theoretical physics are often criticized on the (somewhat prosaic) grounds that there is little or no supporting evidence, several commentators have put forward the more ambitious claim that a certain class of these hypotheses are problematic, and potentially incoherent, both conceptually and empirically. In particular, the skeptical assessment is aimed at those QG strategies, and interpretations of QM, that posit an underlying ontology that does not possess the 4-dimensional spacetime properties, chiefly metrical and topological, associated with twentieth century field theories, such as GR. Rather, these QG theories, some previously discussed in §8.1, claim that the spacetime structures employed by GR and other higher level field theories emerge from the non-spacetime QG entities and processes posited at a more fundamental ontological level. Hereafter, a "non-spacetime" theory denotes a QG proposal (or QM interpretation) that does not take the continuous 4-dimensional metrical and topological structure of spacetime as fundamental, although the term "pregeometric" will also be used to signify these theories (especially in Chap. 10). Lam and Esfeld (2013) confine their study to canonical QG theories, specifically, geometrodynamics and loop quantum gravity, but we will include a host of other QG and non-QG proposals in our discussion as well. The analysis in this section will also help to set the stage for a more detailed investigation of QG hypotheses in Chap. 10.

8.3.1   *Quantum Mechanics, Quantum Gravity, and Spacetime*

Put in simplest terms, several commentators would appear to reject these non-spacetime QG proposals, insisting that "it is unclear how to make sense of concrete physical entities that are not in spacetime and of the notion of ontological emergence that is involved" (Lam and Esfeld 2013, 287). In league with similar criticisms put forward earlier by Maudlin (2007b) against configuration space interpretations of QM (but applicable to QG as well, see below), Bell's notion of a "local beable" is adopted by Lam and Esfeld to counter these non-spacetime QG proposals (see, Bell 1987, 234). A "beable" refers to, in this case, the fundamental objects of a theory's ontology, whereas a "local beable" is that object's association—locality—within a definite spacetime region. The problem, put roughly, is that non-spatiotemporal entities are not localizable, or their localization has yet to be determined, and this predicament has lead Maudlin, as well as Lam and Esfeld, to render a negative verdict on these non-spacetime hypotheses. On Maudlin's estimation, "local beables do not merely exist: they exist somewhere" (Maudlin 2007b, 3157), so it follows that any theory which admits beables that cannot be localized does not achieve "physical salience" (3167). Likewise, Lam and Esfeld declare that "there are no beables without local beables" (Lam and Esfeld 2013, 290).

Viewed in historical context, the local beables quandary can be seen as a skirmish in the larger battle over the relationship between spacetime physics and matter/field theory, i.e., whether spacetime relationships are derived from, or are independent of, the dynamical processes at the microphysical level. This debate need not involve the age old absolute/substantival versus relationism question (as will be explained in Chap. 10), and it is not restricted to QG, but it has been a major issue in the development of twentieth century physics, and in the relationship between QM and GR in particular. According to Friedman, "spacetime physics, on Einstein's view, must precede, and then constrain, the development of microphysics" (Friedman 2013a, 195), although many have drawn the opposite conclusion, e.g., Brown regards the "geometrical structures of Minkowski space-time as parasitic on the relativistic properties of the dynamical matter fields" (Brown 2005, 100).

Returning to the specific case at hand, a major part of Lam and Esfeld's "no local beables" argument against non-spacetime QG theories is based on their interpretation of the role of spacetime in quantum mechanics. Put briefly, they argue that non-seperability and/or entanglement in QM, in both the non-relativistic and quantum field theory settings, depends crucially on spacetime, thus QG theories (which are QM-based) face potentially insurmountable difficulties. These allegations raise legitimate concerns, it must be admitted, but they also involve questionable assumptions, as the work of Dieks (2001b) helps to make clear. First, only some interpretations of non-relativistic QM posit a spatiotemporal position (trajectory) for quantum systems, e.g., Bohmian mechanics, whereas the more traditional interpretation first developed by Bohr does not. Under the Copenhagen interpretation, and its Hilbert space formalism, the complementarity of position and momentum entails that a quantum system can lack a spacetime position under some experimental arrangements. As Dieks notes, "[Q]uantum mechanics is not a spacetime theory. The Hilbert space formalism is self-sufficient, and does not need a spacetime manifold as a background. The quantum-mechanical states are defined directly as elements of a Hilbert space. Furthermore, it is possible to interpret these abstract mathematical states in terms of systems which do not always possess positions" (2001b, 232).[8] Turning to quantum field theory (QFT), there have been several interpretations of algebraic QFT that forsake the point manifold of Minkowski spacetime for a reconstruction based on overlapping sets of subalgebras that represent physical subsystems (see, e.g., Bannier 1994, Schroer and Wiesbrock 2000, and Dieks 2000). These strategies can be seen as favouring a sophisticated relationist or property interpretation of QFT since the point manifold is replaced (or recaptured) by a particular ordering of these physical subsystems—but, more generally, these algebraic QFT constructions are akin to non-spacetime QG hypotheses in the sense that the relevant structures of the physical subsystem are encoded in a Hilbert space. That is, a Hilbert space structure, which possesses neither manifold nor metric, gives rise to QFT's Minkowski spacetime in a supervenience or emergence fashion. Accordingly, *if* the details can be worked out, these interpretations of QM and QFT do not necessitate spacetime at the foundational level. Likewise, the advocates of "wave-function realism", such as Albert (1996) and Ney (2012), claim that the complex-valued $3N$-dimensional configuration space (for an N-particle quantum theory) is the fundamental space, with the 3-dimensional space of macrolevel processes (and common experience) either emergent or, in Albert's words, "illusory" (Albert 1996, 277). To sum up, the idea that macrolevel spacetime emerges in some manner from a deeper and quite different level of reality has gained many advocates, hence the claim that QM and the related QM-based theories (QFT, QG) require the standard spacetime backdrop common to classical and relativistic physics is itself a contentious claim.

---

[8] On Diek's analysis, which rejects substantivalism for a property theory or a sophisticated relationist proposal, "one should take the Hilbert space formalism as basic. All features which are traditionally associated with attributes of space should be distilled from this Hilbert space description. Obviously, Hilbert space is here not seen as something substantial, replacing absolute space, but rather as a mathematical device with the aid of which we give a systematical account of physical properties and their evolution" (2001b, 235). See, also, Chaps. 5 and 7.

8.3.2 Non-spacetime Theories and Inference to the Best Explanation

While both Lam and Esfeld (2013, 291) and Maudlin (2007b, 3160) admit the possibility of constructing successful non-spacetime QG theories, Maudlin insists that these proposals lack "physical salience" relative to those approaches that retain the standard background spacetime structure. Maudlin concludes that non-spacetime interpretations of QM depend crucially on what "a derivation of something isomorphic to local structure would look like, where the derived structure deserves to be regarded as physically salient (rather then merely mathematically definable). Until we know how to identify physically serious derivative structure, it is not clear how to implement [a non-spacetime] strategy" (2007b, 3161). Huggett and Wüthrich (2013, 277) interpret this passage as invoking a form of "empirical incoherence" argument, presumably, in the sense that the evidence for such a theory would be local, and thus inconsistent with non-local beables. On Huggett and Wüthrich's estimate, this kind of reasoning simply begs the question, and they offer a blueprint for how to understand the scheme underlying spacetime emergence:

> [S]uppose we have a theory, $T(\tau_1, \tau_2, \ldots, \tau_n)$, of some non-spatiotemporal entities, $\tau_1, \tau_2, \ldots, \tau_n$, and a demonstration that, given suitable idealizations, some formal structure can be derived in which certain variables are functionally related just as phenomenal—'old'—spacetime quantities....[T]he τs are defined to be the unique collection of things satisfying the theory, such that *the structure in question veridically represents the spatiotemporal quantities*. So, by definition, if the τs exist, there is no further question of whether spacetime emerges from them, since they just are (in part) the things from which spacetime emerges. (Huggett and Wüthrich 2013, 284; original italics)

In short, an ontology gains physical salience if it successfully "saves the phenomena", in this case, by deriving macrolevel spacetime from a more fundamental microlevel ontology that is not spatiotemporal.

From a slightly different angle, perhaps the same point can be expressed as an "inference to the best explanation": if a non-spacetime QG theory can be formulated that integrates QM and GR—a goal that remains one of the elusive holy grails of theoretical physics—then the success of that venture provides the basis for embracing the theory's fundamental ontology, regardless of its spatiotemporal or non-spatiotemporal status. Maudlin himself interprets Newton's arguments for absolute space along the same lines, i.e., as an inference to unobservable entities that effectively explains the phenomena, just as the atomic hypothesis successfully explains the macrolevel behaviour of bodies:

> [P]hysics is evidently in the business of postulating unobservable entities in service of explaining observable behaviour. The postulation is always risky, but, as the atomic hypothesis illustrates, the risk can sometimes pay off handsomely. Newton knew that absolute space and time are not, in themselves, observable, but he also explained how postulating them could help explain the observable facts. Why is this any worse than postulating atoms? (Maudlin 2012, 46)

Well, following the same logic, why is postulating a non-spatiotemporal fundamental ontology (from which 4-dimensional spacetime emerges) any worse than postulating absolute space? Like absolute space, a non-spatiotemporal QG ontology is unobservable, but, if successful, it also "explains the observable facts" previously captured, separately, by QM and GR. Yet, by integrating QM and GR, it would gain a level of empirical support that both QM and GR lack individually.

Another analogy between a well-known scientific theory or project and the local beables argument against non-spacetime QG lies in the neurobiological account of mental phenomena, an analogy that Lam and Esfeld examine as well. Berkeley's idealism, like Descartes' dualism, stems from a deep-seated scepticism of the mind's material origins: "What connexion is there between a motion in the nerves, and the sensations of sound or colour in the mind? Or how is it possible these should be the effects of that?" (WGB I 184). In other words, there is a close parallel between the claim that spacetime cannot emerge from a non-spatiotemporal ontology and the belief that a mind cannot emerge from a non-mental ontology. Nonetheless, there are few idealists or dualists among contemporary philosophers of mind, despite the fact that the mind-brain relationship remains fairly opaque. The available evidence, such as brain damage impairing mental function, does offer indirect support for a mind-brain connection; but, presumably, a successful and complete correlation of brain events and mental events is the only way to counter the scepticism embodied in the "How can a mind come from a non-mind?" credo. Similarly, a non-spacetime QG theory that successfully combines QM and GR, following the guidelines set down by Huggett and Wüthrich above, is likely the only defence against the local beables argument. Lam and Esfeld likewise mention the mind-body problem while entertaining a possible supervenience interpretation of non-spacetime QG theories:

> [I]f properties of type B (e.g. mental properties) supervene on properties of type A (e.g. neurobiological properties), one may in a loose and somewhat misleading sense say that the properties of type B emerge from properties of type A....Supervenience implies covariation in the following sense: any variation in type B-properties necessarily involves a variation in type A-properties. However, there is no account available how a variation in spatio-temporal properties could involve a variation in the properties of a more fundamental entity that is not spatiotemporal;...(Lam and Esfeld 2013, 292)

But there is no "account available" how a variation in neurobiological properties could involve a variation in mental properties either. So, assuming that Lam and Esfeld accept a mind-body link, the failure to provide a successful supervenience account (or an emergence, causal, or temporal explanation, as they also discuss) must not undermine the overall plausibility of the neurobiological hypothesis. Once again, it is the indirect evidence (from, e.g., brain injuries) in conjunction with the implausibility of the remaining options (such as idealism and dualism) that is largely responsible for why the neurobiological hypothesis is the best inference (see, e.g., Churchland 1988, who makes this exact case against folk psychology). If a particular non-spacetime QG theory were to actually incorporate QM and GR successfully along the lines set down by Huggett and Wüthrich above, the lack of a supervenience, or emergence, etc., explanation of the manner by which a non-spatiotemporal ontology brings about spacetime would similarly fail to undermine that QG theory's status as the best inference. In short, the manner by which a non-spacetime theory gains physical salience seems commensurate with the way other theoretical entities in different physical theories have achieved that same status, whether it is the atomic hypothesis or the neurobiological basis of the mind. That is, just as a material body seems to have properties (color, solidity, etc.) that are not possessed by its subatomic constituent elements (protons, electrons, neutrons, which are neither colored nor solid), so it is possible that the macrolevel spatiotemporal structure of the world is an emergent feature of a hidden realm of entities that possess radically different spatiotemporal properties, or maybe none at all.

8.3.3   Structural Realism and Beables

Finally, returning to the main topic of this chapter, SR and its ontic and epistemic variants, Lam and Esfeld also associate non-spatiotemporal entities with ESR, but not OSR: "the structures that OSR admits are concrete physical structures through their being embedded, implemented or instantiated in spacetime. Without the commitment to spacetime, it would simply be unknown as in ESR what the entities are that implement or instantiate the mathematical structure of the theory in question" (Lam and Esfeld 2013, 289). Yet, besides possibly illuminating their opposition to these non-spacetime QG theories, this assessment prompts the question previously discussed in §8.1: What are the entities in spacetime that instantiate OSR? Leaving aside QM's measurement problem, wave-particle duality, and a host of other mysteries, the evidence is compatible with both the individuals and non-individuals (world-structure) interpretations of QM, as even French (2011, 219) concedes, hence the ontology of QM seems as problematic and indeterminate as QG.[9] In brief, if Lam and Esfeld hold that ESR is the proper categorization for a theory with an "unknown" ontology, then QM should also qualify as ESR (or ESR-L), as opposed to OSR, regardless of its spacetime or non-spacetime setting. And, in fact, Esfeld (2013) would seem to be in line with this type of criticism, for he concludes that OSR *does* face major hurdles as regards the interpretation of QM, i.e., between the many-worlds, collapse, and hidden-variables approaches, and hence OSR is not, at least at present, "sufficient to answer the question of what the world is like if quantum mechanics is correct" (Esfeld 2013, 19).

Furthermore, since OSR's ontological interpretation seems equivalent to treating QM's standard mathematical formalism, group structure, as isomorphic to, or a description of, a physical entity (see, once again, §8.1), an argument is thus required to explain why the mathematical structures employed by other theories, including non-spacetime QG theories, cannot qualify as the hypothesized OSR entity as well. A non-spacetime ontology is still an ontology, and it possesses a mathematical

---

[9] Lam and Esfeld, in fact, cite the Everett interpretation of QM in their analysis (2013, 289–290). But, one might reasonably ask: Is a branching universe (or spacetime) really more ontologically palatable than a fundamental ontology that lacks the metrical and topological properties of spacetime? Many would, I suspect, demur.

structure. Put differently: How exactly does spacetime, an elaborate 4-dimensional structure, convert an "unknown" ontology into a "known" ontology? Until this question is answered, the view that spacetime renders the mathematics of QM amenable to a scientific realist interpretation would seem to beg the question against those QG theories that posit non-spatiotemporal entities (from which spacetime emerges) which also seek a scientific realist construal.

Nevertheless, there is another investigation that apparently sides with the conclusion that the strategy underlying QG theories is closer to ESR (or, better yet, ESR-L), although not for the reasons offered by Lam and Esfeld. In Wüthrich (2012), an interpretation of structural realism in the context of causal set theory, a non-spacetime QG proposal, prompts the following assessment: "for the wholesale structural realist to meet the antirealist challenge, there must be isomorphisms between substructures of the models of succeeding theories in the relevant sense in order to underwrite the necessary structural continuity across scientific revolutions", where the structural continuity "must manifest itself in the form of partial isomorphisms between their models, i.e., of isomorphisms between [the QM-based] causal sets and substructures of the general-relativistic spacetimes" (Wüthrich 2012, 239). As argued in §8.1, many of advocates of OSR are not particularly motivated by the anti-realist challenge of the pessimistic meta-induction, nor do they deem structural continuity across scientific revolutions to be their primary goal; instead, they often provide a relata-less ontology that denies the individuality of, say, quantum particles or spacetime points. Accordingly, despite the fact that Wüthrich raises concerns for both OSR and Worrall's original formulation of ESR (2012, 226–229), the quotation above—alongside his espousal of *in re* structuralism as the most plausible structural realist strategy (238–239)—seems better suited to the brand of ESR advanced in this chapter, i.e., ESR-L.

8.4 Non-realist Structuralisms and ESR-L

This section will compare and contrast various "non-realist" (i.e., neither realist nor anti-realist) structuralisms with the ESR-L hypothesis. As will be demonstrated, issues pertaining to the convergence of structure raise problems for these non-realist interpretations, but support ESR-L.

8.4.1 Early Structuralism and ESR/OSR

Given its obvious epistemic credentials, ESR-L bears a rough resemblance with the various conceptions of objectivity advanced by the early structuralists, as well as by Friedman and van Fraassen.[10] Since the objective/subjective dichotomy is more

---

[10] There is, of course, a great deal of subjective license involved in attempting to translate early twentieth century neo-Kantian theories into the modern OSR/ESR dichotomy, but their collective approach would seem to fall clearly on the epistemological side of that dichotomy, and not the

naturally grouped among epistemological notions, it may have appealed to the early structuralists as a means of overcoming the thorny ontological issues that emerged in the new physics, with objectivity serving as a neutral position between the rival realist and anti-realist camps. As such, the transcendental idealism that characterizes the theories of space by Cassirer (via Marburg neo-Kantianism), Weyl (via Husserlian phenomenology), and Eddington can be characterized as "non-realist", i.e., neither realist nor anti-realist, but it is probably best judged as simply opposed to subjective idealism or relativism: see, Ryckman 2005, who refers to Eddington's approach as "transcendental idealism shorn of 'noumenalism'" (2005, 234), where "transcendental idealism" is "a *metaphilosophical* standpoint, beyond realism, idealism (or anti-realism)" (287). As is well-known, the structuralists employed the geometrical concept of an invariant, as a geometrical object preserved within a group of coordinate transformations, to capture their newfound understanding of the objectivity of knowledge. This approach drew inspiration from Klein's Erlangen program and Helmholtz, but likely received its most important impetus in the tensor formalism of GR. On the whole, this branch of mathematics can be loosely described as the study of "what remains the same" (invariant) under different spatial perspectives, thereby demonstrating how the objective features of geometry emerge from the subjective, with a structure (the transformation group) providing this objective knowledge (see, e.g., Weyl 1949, 116, Eddington 1939, 85–87). This group theoretic notion of an invariant structure, consequently, is of a far different kind than the so-called "invariant" of ESR-L, where the approximate continuity of structure over theory change is the intended meaning. Moreover, the various Kantian, Husserlian, etc., motivations that prompted the early structuralist theories are absent from contemporary ESR-L.

Cassirer entitled his system, a "universal invariant theory of experience", contending that the "procedure of the 'transcendental philosophy' can be directly compared" to geometry: "Just as the geometrician selects for investigation those relations of a definite figure, which remain unchanged by certain transformations, so here the attempt is made to discover those universal elements of form, that persist through all change in the particular material content of experience" (1952, 268–269). In *Substance and Function*, Cassirer argues that the usual conception of science/philosophy, as disclosing the world's underlying ontology (substances), must now be replaced by the more accurate appraisal that regards science as obtaining a knowledge of the invariant mathematical interrelationships, framed by physical laws (functions), manifest over the course of the empirical sciences. In the context of

---

ontological. That is, given the predominant emphasis on conceptual categories and their like in shaping our experience of the world, drawing purely ontological lessons, apart from these epistemic components, seems quite problematic. Friedman's discussion of the goals of Cassirer's *Substance and Function* makes this point clear: "[I]n accordance with the 'critical' theory of knowledge,...convergence, on this view, does not take place towards a mind- or theory-independent 'reality' of ultimate substantial 'things'....'Reality', on this view, is simply the purely ideal limit or endpoint towards which the sequence of [theoretical] structures is mathematically converging—or, to put it another way, it is simply the series itself, taken as a whole" (Friedman 2005, 75).

spatial theories, Cassirer identified the metric tensor, g (formally, $ds^2 = g_{ab}dx^a dx^b$, the line-element), as the geometric invariant corresponding to an a priori spatial component of knowledge (for the then current state of scientific theorizing, although a deeper invariant may be revealed in the future; 1952, 433). Because the coefficient g_{ab} of the metric tensor can encode any number of different geometries, the flat geometry of Euclidean space in standard Newtonian theory can be thus seen as a particular instance of the larger metric group. Yet, it is not an "approximation" of g in some limiting process sense: it is an instance of g. It is only when the metric is coupled to the field equations, for both the Newtonian theory and GR, that you get an approximation to GR in special cases (slow speeds and far away from massive bodies) via a limiting process. This distinction between regarding g as either alone or in the context of the larger theory is essential, and it remains unclear to what extent the early structuralists addressed this issue. (More research is required here, although it is beyond the bounds of this work.) The case of Cassirer is instructive: besides the retrospective search for functional invariants over the course of scientific theorizing, which could support either ESR or OSR, he also furnishes examples of spatial invariants that draw upon the notion of a transformation group so as to model the subjective/objective dichotomy in the manner of Weyl and Eddington (e.g., Cassirer 1979)—and this latter facet of invariance might reflect the OSR approach more closely since objectivity is attached to the whole group of transformations encoded by g. On the other hand, the Kantian element in all of these earlier structuralist schemes is much more in league with ESR (ESR-L), so finding clear-cut analogues among modern SR variants and early structuralism is quite problematic (see footnote 10).

8.4.2   Freidman's Relativized a Priori Structuralism

In several essays, Michael Friedman has strived to resuscitate a conception of scientific knowledge that borrows elements from both Cassirer's neo-Kantianism and the notion of a revisable a priori endorsed by some of the logical positivists (e.g., the early Reichenbach, and Carnap). As Friedman notes (2000, 116–118), Cassirer's approach to scientific knowledge fits the Kantian classification of a purely "regulative" ideal, since the intellectual faculties alone determine the invariants of experience in a sort of abstract or reflective manner (so that this invariant content can only be established retrospectively, as the ESR-L theorist would also counsel). Friedman's theory of scientific knowledge, conversely, strives to preserve some facet of the "constitutive" process in cognition, which for Kant involved both the understanding and sensibility in a synthesis that renders comprehensible human sense experience. Rejecting Kant's fixed (Euclidean) synthetic a priori framework, Friedman's "relativized a priori" scheme seeks to counter both Quine's holism, which purportedly undermines all applications of an analytic/synthetic distinction, and the relativism implicit in Kuhn's theory of scientific revolutions.

In GR, Friedman's three-part division of theoretical structures can be delineated as follows: the mathematical structure is the semi-Riemannian manifold, which constitutes the spatiotemporal framework of the theory; the equivalence and light principles of GR constitute the coordinating principles that link the mathematical structures to the theory's empirical laws (respectively, the constancy of the velocity of light coordinates physical phenomena with the manifold's infinitesimally Minkowskian metric, and the equivalence principle coordinates the paths of test particles with the metric's geodesics); finally, there are the empirical laws, such as Einstein's field equations, which are only rendered empirically meaningful by the constitutive function of the mathematical structures and coordinating principles (and thereby serve as a relativized a priori framework, contra Quine). Friedman maintains that, through an analysis of the progressive historical development of these constitutive structures, "we can thus view the evolution of succeeding paradigms or frameworks as a convergent series, as it were, in which we successively refine our constitutive principles in the direction of ever greater generality and adequacy" (2001, 63), which thereby undermines Kuhnian relativism. While Friedman tends to steer clear of the issue, he does briefly comment on the relationship between his approach to rationality and the realism/anti-realism quandary as it pertains to scientific truth: toward the end of his Stanford lectures, he adds that his conception of scientific rationality "does not proceed on the basis of 'scientific realism'" (2001, 118), and, in fact, "is *consistent* with…'anti-realist' conceptions of truth" (68n, original emphasis).

8.4.3   Convergence, the Relativized a Priori, and Realism

Yet, even if the examination of science is transferred to the rarefied heights of rationality, there still remains an important "realist" consequence associated with Friedman's theory. In particular, Friedman simply assumes that the historical evolution of scientific theories will converge to a, presumably, finite set of relativized a priori structures (or, for Cassirer, mathematical/physical invariants, since Cassirer appears to make the same assumption). The relativized a priori hypothesis requires, first, "that earlier constitutive frameworks are exhibited as *limiting* cases, holding *approximately* in certain precisely defined special conditions, of later ones", and, second, that "the concepts and principles of later paradigms…evolve continuously, by a series of natural transformations, from those of earlier ones" (63, emphasis added). Friedman's reference to earlier frameworks as limiting cases that approximate later frameworks is especially intriguing, since it is identical to Worrall's claims for SR, a view that would be eventually become ESR. Friedman must hold, accordingly, that non-realism (neither realist nor anti-realist) is compatible with requiring a limiting approximation of successive frameworks, whereas Worrall reads this same demand as supporting structural realism. What are we to make of this predicament? Is one person's realism another's non-realism?

While Friedman is certainly justified in regarding the relativized a priori as non-realist, there would seem to be no justification for his assumption that the series of structures will continue to converge over the course of future scientific theorizing. For instance, it is conceivable that, within a particular science, an earlier theory (relativized a priori framework) may not be subsumed within a later theory (framework) as a special case of that later theory; i.e., the unity of science may be violated (see Kitcher 1981 on unification). Therefore, as the defender of ESR-L will insist, the presumption of a converging series has an unmistakable, if slender, realist implication; where "converge" is defined as the retention of earlier theories/frameworks as limiting cases that approximate the latter theories/frameworks. In other words, stipulating that the structures of theories converge is tantamount to endorsing the "no miracles" argument, NMA, favored by ESR-L advocates, since the NMA argument, construed in general terms, predicts this convergence (while simultaneously rejecting the anti-realist assumption that it is merely an historical accident). Consequently, given that Friedman's neo-Kantianism and ESR-L agree on this convergence prediction, Friedman appears to be in need of an explanation for his convergence prediction, whereas ESR-L has a built-in explanation grounded on the NMA (we will address an objection to this critique of Friedman in §8.5).

An anti-realist, moreover, would likely refrain from making any such convergence prediction, thus Friedman is not correct in maintaining that his theory is *neutral* between the realist and anti-realist positions. Unless the anti-realists would welcome a blurring of their thesis with ESR, invoking a future-directed convergence of scientific structures falls clearly on the realist side of the realism/anti-realism debate, since "empirical adequacy" would seem to be satisfied by any and all future patterns of scientific theory construction as regards convergence or non-convergence. To give an example, the anti-realist philosophy seems consistent with the situation, described above, where future successful theories in a particular science fragment into a series of incommensurable "local" theories, so that each is confined to a restricted set of non-overlapping phenomena, and where the combined empirical data of all of these local theories are not subsumable within a larger theory: e.g., if no successor quantum gravity theory is ever found to incorporate GR and quantum theory; or, if Kepler's planetary laws and Galileo's free-fall law were never subsumed within Newton's gravitational theory.[11] In fact, van Fraassen would seem to endorse this very possibility, for he has insisted that "[t]here cannot be in principle,

---

[11] More problematically, some of these fragmented local theories may violate the laws and data of other local theories: e.g., a theory confined to just biology could undermine the conservation of mass-energy, the latter constituting just another local theory. Now, a number of objections might be raised in response. First, the neo-Kantian might claim that a mathematical invariant can always be found that will link the totality of empirical data, thus rejecting the possibility just noted (although, as a counter-reply, this would be difficult to prove). Second, the anti-realist may insist that empirical adequacy dictates that all empirical data must be subsumed under more general theories (since empirical adequacy would be sacrificed if they failed to be subsumed in that manner). However, as argued previously, this demand would render anti-realism practically indistinguishable from ESR-L, since it basically admits that scientific theories *must* converge, and it hard to imagine an anti-realist embracing that position.

but only as a historical accident, convergence to a single story about our world" (1991, 482), and, more recently, he has leveled this form of underdetermination argument specifically against structural realism (see van Fraassen 2008). Admissions of this sort therefore undermine Friedman's attempt to declare the neutrality of his version of approximate continuity in relation to the realism/anti-realism dichotomy. Indeed, van Fraassen's interest in "empiricist structuralism" seems rooted in the controversy surrounding the complex relationship between theory, data models, and the phenomena (on the semantic view of theories), with all that it entails for the venerable observation/theory puzzle for his brand of positivist-influenced empiricism, but which seems irrelevant for the convergence debate (see van Fraassen 2006, 2008: and, e.g., Bueno 1997, for a sophisticated treatment of structural empiricism that explores the relationship among these models).

8.5 Multiple Routes and ESR-L

This section will further explore the role of convergence in scientific theories, as well as the related issues involving constraints and compensatory adjustments, with the ultimate goal of developing ESR-L.

8.5.1 *Convergence and the Multiple Routes Hypothesis*

In all fairness to Friedman and his relativized a priori hypothesis, he does attempt to secure a rationale for his non-realism (neither realist nor anti-realist), and his arguments call to mind van Fraassen's allegations of the historical contingency of a "single story". Friedman, in effect, calls into question the view that science is evolving towards a final, *single* conception of entities and/or structures, a view that he associates with scientific realism. There is "an essential element of convergence in the historical evolution of successive constitutive frameworks", he concludes, but asserts that "this is explicitly not convergence to an entirely independent "reality" (however conceived) but rather convergence *within* the evolving sequence of constitutive frameworks itself" (Friedman 2001, 118). Furthermore, his "conception of scientific rationality does not even require that there be a *uniquely* correct sequence of convergent successor theories—something that would certainly be required by any version of 'scientific realism'" (2001, 118). Friedman does acknowledge "an element of 'internal' realism or what Kant called 'empirical realism'", since "once a given constitutive framework is already in place, there is a perfectly precise sense in which we can then speak of a 'matching' or 'correspondence' between a theory formulated within that framework and the empirical or phenomenal world" (2001, 118). To be precise, Friedman only demands "that *any* reasonable route through [constitutive principles] be convergent", which implies there can be more than one route (2001, 68).

This is a persuasive line of argument, it should be noted: convergence need not be conceived as evolving towards a final structure or entity, but can be interpreted as simply aiming at greater generality and empirical adequacy. Yet, the ESR-L supporter will point out that merely allowing for the possibility of multiple constitutive frameworks does not free Friedman's theory from the necessity of acknowledging the constraints imposed on theoretical constructions, i.e., that there are "external" (real world) factors that constrain theory construction but which cannot be accurately tracked from within theories. These constraints are manifest in both (i) the fact that each framework in an evolving sequence appears as limiting approximation of the later framework, and (ii), that not all frameworks are constructible. We have discussed (i) above, but (ii) requires more elaboration. Put simply, Friedman's "multiple routes" hypothesis, as we will call it, still faces the threat of subjective idealism (or relativism, solipsism) unless, of course, he admits that there are constraints in place to guarantee that *not all* constitutive frameworks, or all routes within a single framework, are candidates for his inter-theoretical version of empirical realism. If any and all frameworks, and/or routes through a framework, are equally constructible and successful, then scientific theorizing would support a virulent form of subjectivism or relativism, because frameworks that directly contradict one another on some specific prediction or theoretical commitment, say, T and $\sim T$, would be equally compatible with the same evidence. To borrow Laudan's distinction (Laudan 1996, 42), Friedman's appeal to multiple routes clearly constitutes a case of nonuniqueness, where there are more than one successful frameworks, but it does not follow that it upholds egalitarianism, where all frameworks are equally successful. Yet, even if the successful frameworks comprise a finite set, i.e., nonuniqueness, as opposed to just one, that still means that the majority of constructible frameworks are not successful—and thus an explanation is required to account for why some are successful and other are not. The realist will explain this failure of egalitarianism in terms of the constraints imposed by the world's ontology (as will be further discussed below). Of course, Friedman could reply that the needed constraints are supplied by the demand for greater communicative rationality and, especially, empirical adequacy at each step in a converging series of constitutive frameworks. So, Friedman is not without an explanation of constraints, although the manner by which these constraints operate to limit the egalitarian option would need to be explained in more detail than he has hitherto provided.[12] The realist, in contrast, has the benefit of a built-in explanation of the source of these constraints, especially ESR-L, as we will explain shortly.

---

[12] Can Friedman's neo-Kantian claim, as does the realist, that the world provides the needed constraints to rule out the egalitarian option? While the world's ontology obviously plays a major role in scientific theorizing for Friedman (i.e., given his admission of empirical realism), to claim that it *explains* convergence would clash with his insistence that his view is consistent with anti-realism. As explained previously, van Fraassen's anti-realism holds that convergence is contingent; so, if Friedman really accepts that his view is consistent with anti-realism, then he would need to add that convergence may also be a mere accident. Given Friedman's past history of supporting realism, along with his demand for convergence (explored above), it appears unlikely that he would regard convergence as merely a contingent accident.

8.5.2   Constraints, Compensatory Adjustments, and the Two Requirements of Multiple Routes ESR-L

The multiple routes objection runs counter to traditional scientific realism, needless to say, since that thesis must defend the view that science is ultimately grounded upon a single, final ontological class of theoretical entities. But structural realism, and ESR-L in particular, need not endorse a corresponding final structure—in fact, given the complexity of theoretical constructions of spacetime theories (such as formalism underdetermination), there may be good reasons for endorsing the possibility of multiple routes. First, ESR-L does not rule out the possibility that there can be more than one approximate limiting structure or formalism preserved over the course of scientific change; rather, it only holds that these structures are grounded upon the world's ontology. The construction of different spacetime theories from the same empirical basis is well-known in philosophy of physics circles, of course, as famously initiated by Poincaré's example of the alternative spatiotemporal interpretations of his disc-world inhabitants (Poincaré 1905, Chap. 4). Two theoretical constructions are consistent with the evidence obtained on Poincaré's disc-world: (1) that the geometry is Euclidean but "universal forces" distort the measuring apparatus, or (2), the geometry is non-Euclidean and there are no universal forces (see, e.g., Kosso 1997). Besides comprising an early instance of the underdetermination of theories, the lesson for the structuralist is that it is the "geometry + physics", $G + P$, that faces the tribunal of experience, and not simply the geometrical structure, G, alone. Spacetime structures, as with all mathematical apparatus applied to empirical results (see Chap. 5), must be coordinated to physical processes—and this leaves open the possibility that, in the limit of scientific theorizing, more than one $G + P$ combination may be consistent with the empirical evidence.[13]

In light of these discussions, the multiple routes version of ESR-L has two requirements. First, the $G + P$ combinations must be nonunique rather than egalitarian, since, as argued above, admitting any and all such combinations would threaten realism/objectivism. In Poincaré's example, not all aspects of the choice between (1) and (2) are conventional. If one chooses to retain a flat space, then one *must* postulate universal forces that distort the measuring instruments. Alternatively, if one decides to straightforwardly accept these measurements, then one *must* conclude that the space is non-Euclidean. Accordingly, once a framework is conventionally chosen (respectively, that the space is flat or that the meter sticks provide accurate

---

[13] A famous counter-argument against structural realism, originally introduced by M. Newman against Russell, should be briefly addressed at this point, although we will only explore a version of the argument adapted for the semantic view of theories: if a mathematical structure represents the world by standing in an isomorphic relationship with the world's structure, then one can undermine the uniqueness of this representation by introducing another world domain of the same cardinality and carving out a structure that is also isomorphic to the mathematical structure in this new domain. As argued by French and Saatsi (2006), however, this problem can be surmounted by including interpretations of the mathematical structure's theoretical variables, so that they refer to a particular group of properties and relations. This reply to the Newman problem would seem to parallel Poincaré's insight, namely, that structures are always linked to the world via coordinating principles—and hence our $G + P$ approach to structure naturally includes French and Saatsi's defense. Friedman himself regards the Newman argument as quite problematic for structuralism; see, Demopoulos and Friedman (1985).

measurements), the conclusions that can be drawn about the world are, in large part, determined. On a relativist/subjectivist construal of Poincaré's disc-world, however, any geometry, and all assumptions about measuring instruments, should equally apply: for instance, option (3), that the geometry is Euclidean but no universal forces distort the apparatus. That all past observational evidence fails to confirm option (3) thus discloses the presence of constraints imposed on theory construction, constraints that seemingly come from "outside" the particular theoretical framework chosen by the scientist (since the framework itself cannot determine the nature and precise occasion of these constraints in advance). Put differently, by appealing to the underlying physical ontology, the scientific realist can thus provide a natural explanation for the manifest experience of theoretical constraints, whereas Friedman's rationality hypothesis must simply accept these constraints as merely brute facts of the conjunction of empirical adequacy and communicative rationality.

In Friedman's *Foundations of Space-Time Theories* (1983), the nonuniqueness of spacetime constructions is, in fact, openly acknowledged, for it comprises an important part of Friedman's case against metric conventionalism. In this earlier work, Friedman seems content to defend a general commitment to spacetime realism or absolutism, but the ontological implications of his analysis are largely absent. That is, while relationism and conventionalism are subjected to a lengthy critique, Friedman's reluctance to engage the deeper ontology issues relating to (what we now call) "substantivalism", as well as his contention that various spacetime structures are preserved over the course spacetime theorizing (so that these structures receive repeated boosts in confirmation due to their unifying power), bear an uncanny resemblance with the ESR-L approach developed above. For example, Friedman reckons that we should endorse "a realistic attitude toward the space-time structure $\langle M, D \rangle$ [the topological manifold M, and the affine connection, D, with the latter covering both the absolute and dynamical cases]…since the unifying power of this structure [i.e., $\langle M, D \rangle$] has steadily increased [from Newtonian physics through GR]" (1983, 261). That is, Friedman (1983) endorses a realist commitment to spacetime structures, such as $\langle M, D \rangle$, but he (apparently) never sanctions the existence of a distinct *physical* entity or substance that corresponds to $\langle M, D \rangle$. ESR-L takes a similar stance on the ontology question, although it is more amenable to sophisticated forms of relationism than Friedman allows in his (1983) investigation.[14] As argued in Chap. 5, sophisticated relationist proposals are consistent with

---

[14] To be precise, Friedman tends to identify relationism with our eliminativist and super-eliminativist brands of relationism first defined in Chap. 1 (see, e.g., Friedman 1983, 217). He does entertain the proposal, a la Sklar, that absolute acceleration is a "primitive quantity", and which he classifies as relationist, but he rejects it on the grounds that it lacks the unifying power of absolute space: since the domain of absolutist spacetime is the entire manifold, M, whereas the domain of the "primitive quantity" relationist is just the spacetime points occupied by physical events, p, Friedman argues that this Sklar-type relationism still suffers from a lack of unifying power in comparison with the absolutist due to the limitations of p's domain (257). Yet, if one embraces the type of modality implicit in Teller (1991), where a single object can give rise to the inertial structure of the entire manifold, then Friedman's argument is no longer applicable. For example, Teller concludes that his "liberalized relationist" is "committed to the same inertial structure to which the substantivalists feel committed", but that "coordinate systems are now taken to describe actual and *possible* relations to actually existing objects, instead of objectively existing space-time points" (1991, 381; emphasis added). In all fairness, Friedman does seem to acknowledge that the relationist could appeal to a "primitive law" (1983, 257) to capture the unifying power enjoyed by the

a realism and objectivity concerning spacetime structures, but reject the notion that space is a special independently existing entity of some sort (rather, it is just another physical field). For the advocates of ESR-L, the substantival/relational dispute in ontology should be viewed as a separate issue from the reality (objectivity) of the spacetime structures themselves (in conjunction with the coordinating principles). In other words, both substantivalist and relationist hypotheses are compatible with the minimalist ontological implications of ESR-L (i.e., the convergence of space-time theories and the manifest constraints/nonuniqueness of spacetime construc-tions), although both will obviously differ on the deeper ontological foundations.

Given the basic similarity of Friedman (1983) and ESR-L, it is therefore not surprising that one can locate a similar appeal to the nonuniqueness of spacetime constructions (as opposed to egalitarianism) in Friedman's discussion of the "compensatory adjustments" that are required when changing among empirically equivalent spacetime theories, i.e., our Poincaré-style geometric underdetermination. He demonstrates that the aspects of theories that actually contribute to empirical success cannot be merely changed or discharged without making "compensatory adjustments" to the rest of the theory (1983, 292)—hence, our cases (1) and (2) above furnish a simple instance of compensatory adjustments as applied to Poincaré's hypothetical disc world. Friedman's own example that involves these same compensatory adjustments centers upon the decision to graft a non-Euclidean metric h' to a Newtonian spacetime with a flat connection $°D$. Overall, this new theory, $°D+h'$, will no longer be empirically equivalent to the earlier version of the theory, $°D+h$, that retains the Euclidean metric h (since the particles that comprise an expanding or shrinking body, due to h', will fail to follow straight paths as deter-mined by $°D$, consequently there will be no adequate Newtonian explanation for this fact in $°D+h'$). To preserve the empirical equivalency of these theories, one must introduce a non-flat affine connection D compatible with h': explicitly, one must change the laws of motion (i.e., the components of the affine connection) from $(d^2x^i/dt^2)+°\Gamma^i_{jk}(dx_j/dt)(dx_k/dt)=0$, in the earlier Newtonian theory, to $(d^2x^i/dt^2)+\Gamma^i_{jk}(dx_j/dt)(dx_k/dt)=F^i$, in the revised theory. The introduction of a universal force, F^i, that distorts the inertial paths, as in case (1), thereby secures for this new theory, $D+h'$, empirically equivalent predictions in keeping with Newtonianism.[15] By vividly portraying the constraints imposed within various Newtonian spacetime constructions (whether $°D+h$, $°D+h'$, or $D+h'$), Friedman's argument thus invokes our nonuniqueness clause on the construction of spacetime theories, and is thus in keeping with the multiple routes interpretation of ESR-L.

---

absolutist (and thus bring M and p into congruence, albeit indirectly through the primitive law), but he elsewhere deems this general strategy (of appealing to primitive quantities) as *ad hoc* (248), a description that foreshadows Earman's later "instrumentalist rip-off" allegation (see Chap. 5).

[15] Friedman (1983, 297–299). That is, the force F^i now explains why the particles that comprise an expanding or shrinking body, due to h', deviate from straight-line trajectories as the body moves, apparently inertially, through various regions of space. Here, the $°\Gamma^i_{jk}$ are the flat components of $°D$, and the Γ^i_{jk} are the non-flat components of D, such that, $F^i=(\Gamma^i_{jk}-°\Gamma^i_{jk})\,(dx_j/dt)(dx_k/dt)$.

The second requirement for a multiple routes formulation of ESR-L is that each $G+P$ combination must, from its perspective, be an invariant feature across all of the other combinations, $G'+P'$, that makeup the nonunique class of successful spacetime theories (so as to defeat the pessimistic meta-induction). In other words, each $G+P$ combination must approximately contain all of the empirical data and predictions of all other $G'+P'$ combinations; and, returning to Poincaré's example, the "flat space plus forces" theory does indeed approximate empirically the "curved space without forces" theory.[16] (A similar restriction must also be in place for Friedman's multiple routes hypothesis of scientific rationality.)

To summarize our conclusions, the multiple routes form of ESR-L supports the idea that there may remain a select handful of geometry plus physics combinations (and not just one) that save the phenomena. Friedman's analysis is thus informative for ESR-L since it discloses the importance of all the constitutive elements in scientific theories, both mathematical structures *and* coordinating principles, when searching for invariant structure in spacetime theories. The multiple routes formulation of ESR-L does, moreover, allow for the nonuniqueness of multiple routes in a much more satisfactory manner than Friedman's (2001) rationality method. Finally, it should be noted that, in the limit of scientific theorizing, it may turn out that only one geometric structure, G, from within one $G+P$ pair, is consistent with the evidence and/or the scientific method's list of theoretical virtues, so ESR-L is not committed to a multiple routes outcome.

8.5.3 OSR and Multiple Routes

Before concluding, there is one more topic of interest to the structural realist that concerns underdetermination and spacetime theories, namely, the possibility that OSR could adopt the multiple routes notion examined above. This possibility has been suggested by Brading and Skiles (2012) under the rubric of a "law-constitutive" account of objects: "what it is to *be* a physical object *at all* is to satisfy a certain system of physical laws", even though "[t]his is *not* to say that objects ontologically depend upon our *theories* about what those laws are, or even upon the laws

---

[16] An anti-realist or OSR realist may object, at this point, that the multiple routes form of ESR-L is tantamount to an equivalence of structure at the empirical level only, and thus it is simply insufficient to qualify as a realist theory. This objection, nevertheless, fails to take into account the fact that ESR-L is, indeed, a very liberal brand of realism. Given the two requirements of the ESR-L theory advocated above—nonuniqueness, and that each nonunique $G+P$ combination must be an invariant feature across all of the other nonunique combinations, $G'+P'$—the "realism" in this theory is manifest in two important ways: (1) via predictions on future theoretical constructions (which runs counter to anti-realism, or at least is quite difficult for the anti-realist to explain), namely, that theories *will* converge; and (2) that there *will* exist constraints on the theoretical constructions (thus eliminating the egalitarian option along with its relativist implications for epistemology and ontology). Finally, it should be noted that a famous species of underdetermination concerning spacetime theories is consistent with this Poincaré-inspired exegesis, as well as with ESR-L; namely, the topological underdetermination manifest by observationally indistinguishable spacetime theories, as explored by Malament (1977) and Glymour (1977).

themselves" (2012, 104; original emphasis). However, they caution that "there is no guarantee that this strategy will generate one unified kind of physical body: perhaps the bodies that serve as the subject matter of the laws when gravitation is included will turn out not to be identical to those that serve as the subject matter of the laws when electrical phenomena are at issue" (104, n.9).

While an intriguing idea, a good case can be made that this law-constitutive view more properly falls within the ESR-L conception, rather than OSR. The example they provide, of a gravitational law-constitutive versus an electrical law-constitutive conception of ontology, ultimately involves other laws that are common to both, and which binds or constrains the application and scope of the two former laws, gravitational and electrical; e.g., the conservation of mass-energy. While individually the application of these separate laws may not directly refer to the deeper conservation law (and GR, in fact, does not have a conservation law), it is widely believed that their application will satisfy the conservation principle (as will our grand theories that unify these more specialized laws). This strongly suggests, consequently, that the separate constitutive laws are only capturing a mere facet of reality, and that they are constrained at a deeper level of reality by these other factors (e.g., the conservation law). Hence, like French's "complex ontology" interpretation of OSR (see §8.2), it is difficult to interpret the scenario offered by Brading and Skiles as upholding the ontic form of structural realism, since a deeper level of ontology constrains the application and scope of these separate laws (gravity and electrical). Given these circumstances, which are arguably endemic to scientific theorizing in general, it seems more plausible to say that these separate constitutive laws are different *epistemological* perspectives of the world's *single* ontology (ESR-L), and not different ontologies altogether. The statement by Brading and Skiles that objects do not depend upon our theories would seem consistent with this last point.

On the other hand, a more radical version of this law-constitutive approach might actually support the view that all aspects of ontology (objects) do depend upon these laws/theories—but this would be tantamount to the radical subjectivism worries previously examined, since there would be no explanation for, nor any basis for presupposing, the existence of constraints in theory construction unless one assumes the mere brute fact of these constraints. Indeed, this last option would place the law-constitutive form of OSR in the same camp as Friedman's non-realist and van Fraassen's anti-realist hypotheses, as well as violate Saatsi's reasonable demand that any variant of scientific realism procure an explanation of past theoretical success (see footnote 1). All told, what remains that is worthy of the SR label in this version of OSR? ESR-L, in comparison, at least offers the goal of a final explanation of past theoretical success, predicated, once again, on our experience of the approximate continuity of mathematical structure over the succession of past theories, through either a single route or multiple routes.

To recap, the best option for the law-constitutive OSR theorist is to declare that ontology is itself hierarchical, with laws revealing perspectival aspects of the one ontology by way of various law-constitutive frameworks, but which thus posits (via the one, underlying and interconnected ontology) a vehicle for the constraints imposed on these higher-level frameworks. Yet, the viewpoint just described *is* ESR-L, since it accepts that our theories have not, and possibly may never, reveal the world's ontology in complete detail. Interpreted in this manner, the law-constitutive form of OSR could thus be viewed as a different, ontology-oriented "path" to the same conclusions reached via ESR-L. This last concession, in addition, helps to support the main theme of this chapter: namely, that the underdetermination dilemma forces the various structural realist interpretations into adopting something that is suspiciously like ESR-L, or, as in the case of Friedman's nonrealist neo-Kantian theory, mandates a similar convergence of approximate structure over theory change (which, as argued above, is the chief realist commitment of ESR-L).

8.6   Conclusion: OSR, ESR, and Historical Analogies

Finally, since our overall investigation has accorded a special place to analogies between seventeenth century natural philosophies of space and contemporary physical theories, one might wonder if the material from the last chapter concerning the ESR/OSR categories and platonism/nominalism can be applied to, say, Newton and Leibniz as well.

Interestingly, if one brings together many of the results of our investigation from Part I into one final historical analogy, then it would appear that ESR-L (or ESR) is much closer to Leibniz' spatial theory than OSR, since truth-based nominalism fits naturally with both Leibniz' theory and ESR-L (see ch. 7 and §8.2). Newton, on the other hand, is more difficult to assess, for while his incorporeal nominalism appears closer to truth-based nominalism, an immanent realist interpretation could also apply (as noted in Chap. 7), and similar qualms can be raised about the relevance of the OSR/ESR dichotomy in Newton's case. Are there historical precedents among seventeenth century natural philosophies for a spatial ontology that is analogous to our multiple routes version of ESR-L (or multiple routes OSR)? On the whole, the conception of Leibniz' theory of motion advanced in Chap. 4 would seem amenable to a multiple routes interpretation, for it posits an ontology wherein different physical interpretations of the same world are simultaneously upheld (e.g., Ptolemaic and Copernican).

Nevertheless, a word of caution should be introduced at this point regarding the viability of these attempted analogies. In short, while applying platonism and nominalism to seventeenth century spatial ontologies is justifiable, since these concepts were part of the seventeenth century's metaphysical landscape, the use of modern structural realist concepts in these analogies seems forced and anachronistic. The motivation for modern structural realist approaches stems from the realism/antirealism debate in late twentieth century philosophy of science (with a legacy that can be traced to early positivism), so it is probably not surprising that structural realism does not naturally align with our seventeenth century spatial ontologies. On the other hand, Newton and Leibniz' spatial theories do reflect, as just noted, the influence of platonist and nominalist concepts, and this may help to explain the more abundant evidence for platonist and nominalist ideas in their respective spatial ontologies.

Part III
The Deep Metaphysics of Space from the
Seventeenth Century to Quantum Gravity

Chapter 9
A New Taxonomy Beyond Substantivalism and Relationism I: Early Modern Spatial Ontologies

This chapter will forge a new third-way taxonomy or classificational system for spatial ontologies based on the many themes, both historical and conceptual, that have been featured in our investigation. After an introduction to the limitations of the standard dichotomy in a quantum gravity setting (§9.1), the new classificational scheme will be developed (§9.2) and then applied to Newton, Leibniz, and a host of other natural philosophers from the sixteenth through eighteenth centuries (§9.3). Next, the lessons gathered from our analysis will be juxtaposed with rival assessments of spatial ontologies and employed to resolve various perplexities relating to seventeenth century theories (§9.4 and §9.5). The new taxonomy will then be applied, in Chap. 10, to the strategies proposed among competing quantum gravity hypotheses and to examine various issues, such as background independence and nominalism. All told, the new taxonomy will be demonstrated to be in harmony with the property theory of spatial ontology, whether of the P(O-dep) or P(TL-dep) type.

9.1 Introduction: Quantum Gravity and The Substantivalist/ Relationist Dichotomy

As first remarked at the beginning of our investigation, even granted the most basic and clear-cut conception of substantivalism and relationism—where the substantivalist holds, and the relationist rejects, that space/spacetime is an independently existing entity—it has become quite evident that the attempts to ascribe either a substantivalist or relationist interpretation to classical gravitation theories is an exercise fraught with perils, since the sophisticated forms of both ontologies are seemingly identical as regards their content in the modern setting of general

© Springer International Publishing Switzerland 2016
E. Slowik, *The Deep Metaphysics of Space*, European Studies in Philosophy of Science, DOI 10.1007/978-3-319-44868-8_9

relativity (GR).[1] What is less well-known, however, is that this metaphysical quagmire has likewise ensnarled philosophers concerned with the ontology of quantum gravity (QG), an assortment of strategies whose goal is to connect the physics at the micro-realm of quantum mechanics (QM) with the large-scale structure of space and time in GR. In short, the general consensus would seem to be that sophisticated versions of both substantivalism and relationism are equally consistent, or equally problematic, interpretations of QG (e.g., Rickles 2005, Earman 2006), a conclusion that is apparently reflected in the rival appropriations of an important QG hypothesis, loop quantum gravity (LQG), for either Leibnizian relationism or Newtonian substantivalism. For example, a Leibnizian lineage for LQG has been put forward by Smolin (2000, 119–120; 2006, 200–203), among many others. Yet, in Dainton (2010), which defends the relevance of the substantival/relational dichotomy in GR (2010, 380–381), it is argued that the ontology of LQG "seems as substantival as any conception", prompting Dainton to ask, "What could be less Leibnizian?", despite the fact that LQG is "very different from Newton's absolute space" (405–406). Since the substantival/relational dichotomy *is* the reigning ontological classification or scheme in the philosophy of space and time, it is imperative to investigate why it leads to such conflicting assessments, and to examine if there are better alternatives.

This chapter, and the next, will begin to meet this challenge by offering an alternative range of conceptual distinctions that, in conjunction with the platonism/nominalism categories first presented in Chap. 7, lie below the rather imprecise and obscure dichotomy imposed by contemporary substantivalism and relationism. In particular, an examination of a range of seventeenth century metaphysical speculation on the deep ontology of space, by Gassendi, Newton, Leibniz, and others, will reveal a host of uncanny similarities with the modern QG program: these similarities concern (i) the spatial geometry at both the foundational level of ontology and at the derived or resulting levels of ontology, and (ii) forms of platonism and nominalism as regards the spatial geometry at these two levels. As will be demonstrated, (i) and (ii) more directly concern both the seventeenth century and the contemporary QG approaches to spatial ontology than the manifest equivocalities of the substantivalist/relationist division—and, quite importantly, (i) and (ii) also obviate the dubious ontological distinctions between the sophisticated substantivalist and sophisticated relationist interpretations of LQG and other QG hypotheses, as well as for the metric field in GR. Indeed, since these new sets of issues and distinctions are not congruent with the modern dichotomy, the conclusions of this chapter and Chap. 10, which build on our previous efforts in earlier chapters, can be interpreted as challenging the utility of substantivalism and relationism (although a fairly trivial and imprecise version of the dichotomy, parasitic on the platonism/nominalism distinction, can always be invoked, needless to say). The results of our investigation will also question whether the contemporary division between a background inde-

---

[1] A nice overview is Belot (2000). As for examples, see, once again, Hoefer (1996) for metric field substantivalism; Rovelli (1997) for metric field relationism; and Dorato for a structural realist (2000) reading that is similar, in an ontological sense, to metric field relationism. The "out-moded" nature of the dichotomy, as regards classical gravitational theories, is defended in Rynasiewicz (1996).

pendent or dependent formulation of a theory's spacetime geometry, which is a central theme in the search for QG, is the modern equivalent of the substantival/relational divide. The conclusions reached might strike the reader as controversial, but, on both historical and philosophical grounds, it will be argued that our new taxonomy that incorporates (i) and (ii) is far superior to the substantival/relational distinction as regards the analysis of the deep ontology of space.

9.2 A New Taxonomy for Spatial Ontologies

If anything is disclosed in the competing ontological interpretations of Smolin and Dainton, as recounted above, it has little to do with the adequacy of the substantival/relational dichotomy. Rather, it exposes the enduring aspiration among latter-day thinkers to appropriate either Newton (a presumed substantivalist) or Leibniz (a presumed relationist) as the proper historical ancestor of a particular modern theory of spacetime substantivalism or relationism, so that a clear line of descent can be established that buttresses the substantivalist or relationist credentials of that modern ontology. Yet, the actual details of the spatial hypotheses of seventeenth (and sixteenth and eighteenth) century natural philosophers undermine such attempts, as we will now begin to explore. Besides Newton and Leibniz, who are the alleged historical representatives of, respectively, substantivalism and relationism, other important natural philosophers will also be examined, in particular, Descartes, More, Patrizi, Gassendi, and the pre-critical Kant. Furthermore, since the respective spatial ontologies of both Newton and Leibniz have been explored at length in previous chapters, the ensuing discussion will largely summarize our earlier findings, yet the new classificational system will disclose new details as well as provide a more comprehensive basis for comparison with other spatial ontologies.

In order to more accurately pinpoint the differences between Newton, Leibniz and other natural philosophers of the Early Modern period concerning the deep ontology of space, the new taxonomy will focus on the two main issues that, as revealed in our investigation, are in contention: first, in §9.2.1, whether the geometric properties at the various levels of spatial ontology are the same or different; and second, in §9.2.2, whether those geometric properties support platonism or nominalism. As will become quite evident once the details are provided, these two main issues naturally align with a property theory of spatial ontology as opposed to substantivalism and relationism, hence justifying an eventual third-way classification for this taxonomy.

9.2.1 Equivalent Geometric Structure and Ontological Levels

The first item in the new taxonomy concerns the geometric structures posited at (a) the "foundational level" of ontology (entity/entities associated with that theory) that are *identical* to (b) the geometric structures posited at the "secondary level" of ontology (entity/entities associated with that theory), where the secondary level ontology (or theory) is grounded on, emerges, or results from, the foundational level of ontology (or theory). We will dub this distinction, FGS, for "foundational geometric structure", with the foundational level usually, but not always, linked to the microphysical realm ("microlevel"), and the resultant, secondary level entities often, but not always, associated with the observable macroscopic level ("macrolevel"). In the seventeenth century, the geometry of space, i.e., the geometry at the macrolevel of material bodies, is three-dimensional and (usually) Euclidean, but the nature of the geometric features at the foundational level, i.e., God and Leibniz' monads, could take different forms.[2] As first disclosed in Chap. 4, Leibniz' *New Essays* puts forth three ways that a being can be related to place/space at the secondary macrolevel, a three-part division that will be of much service in our subsequent analysis:

> The Scholastics have three sorts of *ubeity,* or ways of being somewhere. The first is called *circumscriptive.* It is attributed to bodies in space which are in it point for point, so that measuring them depends on being able to specify points in the located thing corresponding to points in space. The second is the *definitive.* In this case, one can "define"—i.e. determine—that the located thing lies within a given space without being able to specify exact points or places which it occupies exclusively. That is how some people have thought that the soul is in the body, because they have not thought it possible to specify an exact point such that the soul or something pertaining to it is there and at no other point....The third kind of ubeity is *repletive.* God is said to have it, because he fills the entire universe in a more perfect way than minds fill bodies, for he operates immediately on all created things, continually producing them, whereas finite minds cannot immediately influence or operate upon them. (NE II.xxiii.21)

In what follows, we will explore how these three types of ubeity relate to the spatial geometry at the material macrolevel.

As for "circumscriptive ubeity", Leibniz mentions only bodies, but the idea is that the entity is mapped to three-dimensional Euclidean space in a point by point manner, much like the modern notion of an isomorphism. Leibniz' analysis also assumes that the entity fully shares in the geometric properties intrinsic to macrolevel space, the most important of these properties being the metric (distance), as Leibniz specifically mentions: "measuring them depends on being able to specify points in the located thing corresponding to points in space". The second way that a being can be related to space is "definitive ubeity", wherein "the located thing lies within a given space without being able to specify exact points or places which it occupies exclusively". Unlike the metrical structure implicit in circumscriptive ubeity, which

---

[2] One of the reasons that the foundational/secondary levels dichotomy is introduced is that it is more general than a microlevel/macrolevel distinction; in particular, the latter distinction is never employed by Leibniz as regards the relationship between monads and matter, nor is it applicable in any way concerning Newton or Gassendi's theory. Furthermore, the Euclidean structure of Leibnizian space seems certain. In the context of assessing Euclid's axiomatic geometry, and the fact that our senses seem to conflict with its geometric conclusions, Leibniz comments that "what I value most in geometry, considered as a contemplative study, [is] its letting us glimpse the true source of eternal truths and of the way in which we can come to grasp their necessity, which is something that the confused ideas of sensory images can never distinctly reveal" (NE IV.xii.6).

also incorporates the topology of space, definitive ubeity is a topological conception alone, for the length or extension of the entity is indeterminate, i.e., it is not "possible to specify an exact point such that the soul or something pertaining to it is there and at no other point". More carefully, if an entity obtains a Euclidean metrical determination, then an exact set of continuously structured spatial points needs to be specified for that entity, but, since Leibniz states that the exact points cannot be determined, the continuously extended regions needed for Euclidean metrical space cannot be applied to that entity. Hence, because the being's spatial properties are limited to individual parts/points within a certain region of space, which in the limit of the division of a part converts to non-dimensional points, all that definitive ubeity can furnish is something akin to the topological notion of a neighborhood (which is non-metrical).[3] As discussed in Chap. 4, a notion that is closely aligned with definitive ubeity, especially as regards its topological emphasis, is More's "holenmerism", the thesis that a being is whole in every part of space, namely, the points of space.

Finally, there is "repletive ubeity", which Leibniz assigns to God, and who "operates immediately on all created things, continually producing them, whereas finite minds cannot immediately influence or operate upon them". Traditionally, repletive ubeity was synonymous with God's omnipresence alone, unlike circumscriptive and definitive ubeity, which only pertain to finite beings (see, e.g., Grant 1981, 368, n.125). Leibniz captures this denotation of the term in his comment that God "fills the entire universe", but he mainly emphasizes God's immediate operation on all created beings, which he equates with God's continual production of all being. Although it is not mentioned in this passage, Leibniz' further associates God's immediate operation (continual production) with the "extension of power" doctrine (EP), the latter concept predicated on the assumption that God's being is not situated in space (see §9.3.4 and §4.2), rather, only God's act of continually producing the world can be situated in space. So, leaving aside the effects of God's actions, Leibniz' unique application of repletive ubeity involves the absence of all macrolevel geometric properties as regards God's being itself. A similar non-spatial conception also seems to hold true for Leibniz' monads, as will be discussed in §9.3.4. Throughout the remainder of our investigation, we will employ the term "repletive ubeity" to designate non-spatiality alone, whether for God's omnipresent being or finite monads. More's term, "nullibism", first investigated in Chap. 2, is

---

[3] As Grant recounts, the use of circumscriptive and definitive ubeity to signify the manner by which, on the one hand, bodies, and on the other, souls and angels, relate to space was common in the Scholastic period (Grant 1981, 130). Likewise, in exploring Peter Lombard's use of these distinctions, Grant describes circumscriptive ubeity as the view that the entity "fully occupies and fills its place" and is "fully delimited and circumscribed by the termini of that place", whereas definitive ubeity is the view that the entity "is delimited and defined locally only by the terminus of a place…but does not occupy its place as a dimensional [i.e., extended] entity" (342, n.66). Although slightly different in emphasis, Grant's analysis upholds the fact that circumscriptive and definitive ubeity correlate, respectively, with the metrical and topological components of space, since an entity's non-dimensional "occupation" of place/space entails that it is only the (non-dimensional) points of place/space that can be ascribed to that entity (and not a dimensional length). Grant also concludes that definitive ubeity came to be associated with the "whole in every part" doctrine, i.e., holenmerism (343, n.67).

thus equivalent to our notion of repletive ubeity, as is, of course, EP (see, also, footnote 13 in §4.2.2). Furthermore, we will not restrict circumscriptive and definitive ubeity to just finite beings, but likewise use these categories as applicable to God's omnipresence. In brief, since Leibniz' use of the three-part ubeity distinction in the NE II.xxiii.21 quotation centers more upon the way that a being is manifest in space, as opposed to the being's finite/infinite extent, our investigation will follow Leibniz' lead and interpret the ubeity distinction as pertaining to the form of spatial presence alone, regardless of whether the being is finite or infinite.

Returning to the issue of spatial geometry at the foundational and secondary levels, the three types of ubeity presented in Leibniz' discussion—circumscriptive, definitive, and repletive—therefore correlate with, respectively, three types of geometrical properties that are shared between the foundational entity and the entity/entities at the secondary level; metrical, topological, and pregeometric—where "pregeometric" (or "pregeometry") signifies that there are no identical bi-level geometric structures, i.e., the foundational entity's metrical and topological properties differ significantly from the metrical and topological properties manifest at the secondary level, or, that the foundational entity lacks geometric properties altogether. Hence, pregeometry captures the non-spatiality implicit in the seventeenth century's repletive ubeity, nullibism, and the extension of power doctrines (and, although pregeometry is a contemporary physics-based concept, it will often be used interchangeably when describing the three seventeenth century doctrines just mentioned). The resultant secondary level of spatial geometry, as noted previously, is the three-dimensional Euclidean geometry and continuous topology assumed by most (but not all) seventeenth century theorists, and, for modern QG theories, it is often the geometry assumed in GR or Quantum Field Theory (QFT, the field version of QM), namely, the standard four-dimensional Lorentzian metric and its topological manifold. In what follows, we will dub these three positions, in their order of presentation; FGS(met), FGS(top), FGS(prg). Accordingly, in the seventeenth century: circumscriptive ubeity, FGS(met), holds that the spatial properties of the foundational entity are identical with the metric of a Euclidean or similar three-dimensional space (and which includes the continuous topology of Euclidean space, unless otherwise noted); definitive ubeity, FGS(top), contends that the foundational entity only possesses the topological properties of a Euclidean or similar three-dimensional space; and repletive ubeity, FGS(prg), is the thesis that the foundational entity is either non-spatial (non-spatiotemporal, non-spacetime) or manifests unique metrical and topological spatial properties not found at the secondary three-dimensional macrolevel Euclidean (or a similar three-dimensional) space. In Chap. 10, when we turn to modern QG theories, the emphasis will be the same, with FGS denoting the geometric structures of the foundational level theory that are identical to the geometric structures utilized by the secondary level theory, with the latter grounded upon, and resulting from, the foundational theory (although there may be no such identical structures shared among these levels). There are further differences that will need to be addressed, in particular, the type of property theory of space that is relevant to the context, P(O-dep) or P(TL-dep), although that topic will be taken up in the following chapter.

Finally, and foreshadowing some of the analysis in the remainder of Part III, one of the advantages of the interpretation of ubeity presented above is that it provides a convenient means of classifying God's presence for past spatial ontologies: circumscriptive ubeity is supported by Bruno, Campanella, (possibly) Patrizi, and many Cambridge Neoplatonists (the later More, Cudworth, Raphson, Newton); definitive ubeity, the most popular view in the Middle Ages, by Aquinas, Oresme, the Coimbrans, Suárez, Gassendi, the early More, and many others; and repletive ubeity, which stems from Scotus' writings, by Thomas of Strassburg and other late Medieval thinkers (Funkenstein 1986, 59–62), Descartes, Leibniz, a few Cartesians, some Leibniz-Wolffians, such as the pre-critical Kant, and Euler.

9.2.2 Platonism/Nominalism and Ontological Levels

On both historical and conceptual grounds, the second component of our new taxonomy, platonism and nominalism, provides a more accurate substitute for the ill-defined standard dichotomy, substantivalism and relationism. As explained in Chap. 7, the underlying spatial ontology for both Newton and Leibniz, i.e., God (as well as Leibniz' monads), support a nominalist classification. For Newton, God is present in space, and God's being instantiates spatial geometry, hence this theory fits what we have dubbed, "incorporeal nominalism". For Leibniz, God grounds the truths of space, but only matter (by way of monads) instantiates bodily extension, and thus space, and so this theory represents a sort of mixture of incorporeal and corporeal nominalism (where "corporeal nominalism" signifies that matter alone instantiates space). Nevertheless, as also discussed in Chap. 7, Newton's spatial ontology is very similar to, and functions like, platonism, since space is neither instantiated nor affected by material bodies: this view we dubbed, "virtual-platonism". Since, as regards matter, incorporeal nominalism implies virtual-platonism,"virtual-platonism" will be used henceforth in place of "incorporeal nominalism" in the context of Early Modern spatial ontologies, and simply "nominalism" in place of "corporeal nominalism". The rationale behind this terminological switch is that, first, this new dichotomy, virtual-platonism versus nominalism, better conveys the difference between a theory that regards space as, respectively, independent of, or in some way dependent on, corporeal entities (such as macro-level material bodies/fields). And the second (and more important) reason is that the virtual-platonism/nominalism distinction is better suited for application to the foundational and secondary level distinction, as will be explained below. Like the corporeal/incorporeal division in nominalist ontologies, the virtual-platonism/nominalism distinction is also more historically accurate than the substantivalism/relationism distinction since Newton and Leibniz were influenced by the platonist and nominalist traditions prevalent in their day, e.g., the platonism that underlies the type of Neoplatonism practiced by More and Newton (and, earlier, Patrizi, as will be explained), as well as the nominalist line of argument that Leibniz uses to discredit absolute space (see, Chap. 3). Finally, and rather obviously, incorporeal nominalism only applies to pre-nineteenth century theories, hence virtual-platonism in the context of modern theories of physics only signifies that the spatial geometry at a given spatial level is independent of matter or physical fields at that level.

Therefore, our second general issue concerns whether spatial geometry at the secondary level is either independent of, or dependent on, the entities that arise at that level, a distinction that we will dub, SL(v-plt), for secondary level geometric virtual-platonism, and SL(nom), for secondary level geometric nominalism. A further rationale behind the virtual-platonism designation stems from its limited range: virtual-platonism does not support nor deny the existence of abstract objects, whether in the world (immanent realism) or outside the world (platonism), but simply claims that the spatial geometry at the secondary level can exist in the absence of the entities, usually matter or fields, that arise or emerge at the secondary level (from the foundational level), i.e., secondary level entities do not instantiate (bring into existence) space at the secondary level. SL(nom), in contrast, holds that space (spatial geometry) at the secondary level only exists when the secondary level entities exist. How virtual-platonism and nominalism apply to the geometry, if any, at the foundational level of ontology, i.e., FL(v-plt) and FL(nom), will be postponed until §10.5.3. Employing these definitions, we can now present the real issues that separate Newton, Leibniz, and other seventeenth century natural philosophers on the ontological foundations of space. In contrast to the substantivalist/relationist dichotomy, these new conceptual distinctions will also allow a more accurate assessment of the true lineage of these seventeenth century theories among the competing contemporary QG theories, although this analysis will be deferred to Chap. 10.[4]

9.3   Applying the New Taxonomy: The Early Modern Period

Given the preceding outline and justification for the two elements that comprise the new classificational system for spatial ontologies, we are now in a position to apply these categories to several major theories from, roughly, the late sixteenth century up to the end of the eighteenth century, a selection of spatial hypotheses that represent important trends, both historically and in terms of their essential structural features.

---

[4]Throughout the following analysis, the reference to an ontological level is taken quite generally, and is intended to include a scale difference as well, although a difference in scale is not applicable in the case of the Early Modern spatial ontologies (or even some QG theories, if geometrodynamics is included; see, also, footnote 2). Hence, while arguments are put forward that pertain to the FGS and SL categories, there are no further conclusions offered on the difficult philosophical problems associated with how level or scale differences should be understood generally, i.e., as regards the implicit metaphysics of level/scale differences *per se*. These issues lie outside the scope of our investigation.

9.3.1 Newton and the (Later) More on Spatial Ontology

Using the terminology from the previous section, Newton's spatial hypotheses favor circumscriptive ubeity for God, for he assigns the same geometric structure at both the bodily (secondary) level and for God (foundational level), with the latter directly providing the foundation of space, and he also sides with a virtual-platonism as regards spatial geometry at the secondary level: FGS(met), and SL(v-plt). With respect to the latter, SL(v-plt), an oft cited passage from *De grav* discloses that spatial geometric structures would exist even in the absence of all bodies (secondary level entities), thus supporting virtual-platonism: "[F]or the delineation of any material figure is not a new production of that figure with respect to space, but only a corporeal representation of it, so that what was formerly insensible in space now appears before the senses" (N 22). There are many passages that we have previously examined that confirm FGS(met), e.g., God contains "all other substances in Him as their underlying principle and place" (TeL 132), but the best evidence is, once again, the "determined quantities of extension" (DQE) hypothesis (see §2.3). To recap, Newton's DQE hypothesis accepts a conception of material bodies that denies the existence of corporeal substance, and where God directly grounds bodily properties, including extension, rather than corporeal substance: "extension [i.e. God's extension] takes the place of the substantial subject in which the form of the body [i.e., the determined quantities] is conserved by the divine will" (N 29). He rejects earlier views of substance, especially Descartes' dualism, by reasoning that "if the distinction of substances between thinking and extended is legitimate and complete, God does not eminently contain extension within himself", and, "hence it is not surprising that atheists arise ascribing to corporeal substance that [i.e., extension] which solely belongs to the divine" (31–32). Accordingly, if there is no difference between corporeal and incorporeal substance, since God is the only true substance, then there is only one attribute of extension that all beings share, namely, God's extension; therefore, FGS(met).[5]

Finally, it should be noted that Henry More is the main advocate of FGS(met) and SL(v-plt) in seventeenth century England, and his conception has close similarities with Newton's approach. After embracing definitive ubeity in his early work (see the next section), More changed course and adopted the circumscriptive outlook, ultimately becoming its most forthright defender. The metrical structure and virtual-platonism of his God-based spatial ontology is nicely encapsulated in the following:

> For if after the removal of corporeal matter out of the world, there will be still Space and distance in which this very matter, while it was there, was also conceived to lye, and this distant Space cannot but be something, and yet not corporeal, because neither impenetrable nor tangible, it must of necessity be a substance Incorporeal, necessarily and eternally existent of it self: which the clearer *Idea* of a *Being absolutely perfect* will more fully and punctually inform us to be the *Self-Subsisting God.* (AAA 338; see also, EM 56–57).

---

[5] Furthermore, God's infinite "quantity of existence" with respect to the attribute space is a "substantial", or actual, presence, and not merely a figurative or "virtual" presence. In a passage from the 1713 *Principia,* he contends that "He is omnipresent not only *virtually* but also *substantially;* for action requires substance" (N 91). As used in this context, "virtual" could stand for either a definitive presence in space (holenmerism), or Leibniz' brand of repletive ubeity, or both.

9.3.2   Gassendi and (Early) More on Spatial Ontology

Unlike many of his contemporaries and predecessors, Newton and More deny holenmerism (definitive ubeity), a doctrine that does admit a difference in geometric properties with regard to various incorporeal and corporeal substances. On holenmerism, incorporeal beings are "whole in every part", which thereby allays any concerns about their actual or metaphysical divisibility. Gassendi accepts this holenmerist view of God, which is equivalent to FGS(top), by declaring:

> [W]e conceive an infinity as if of extension, which we call [God's] immensity, by which we hold that he is everywhere. But, I say *as if* of extension, lest we imagine that the divine substance were extended through space like bodies are. Indeed, although the divine substance is supremely indivisible and whole at any time and any place, yet doubtless as corporeal substance is said to be extended—that is not at one point only but is spread out through many parts of space—so there is a kind of divine extension, which does not exist in one place only, but in many, indeed, in all places. (RIV 94)

On Gassendi's estimation, bodies occupy space by being extended ("spread out") across many points, but his qualification, "as if of extension", with respect to God implies that God only shares with bodies the property of occupying the points of space, i.e., that "divine extension", as he also calls it, lacks the dimensional extension of body across the points of space ("lest we imagine that the divine substance were extended through space like bodies are"), even though "[God] is not at one point only". Since space is continuous ("space…remains continuous, the same, and motionless", SWG 395), and since God only occupies the points of this continuous space (e.g., he elsewhere refers to God's "immensity according to which he is *present* in every place", 396; emphasis added), it thus follows that only the topological properties of space are applicable to God, but not the metrical properties of space.[6] Put differently, while God and matter share topological structure, the holenmerist doctrine (definitive ubeity) that God is "whole in every part" undermines the ascription of Euclidean metrical structure to this being—hence, FGS(top).

Furthermore, since Gassendi accepts that space is three-dimensional and Euclidean in structure at the ontological level of bodies (see, also, Grant 1981, 210), and is independent of matter (secondary level entities), he sides with Newton in accepting our form of geometric virtual-platonism, SL(v-plt). Specifically, while God is not really extended, God grounds a form of incorporeal extension that is congruent to the corporeal dimensions of body at that ontological level, and this fact accounts for the dimensionality of any vacuum: space's "dimensions are incorporeal; so place is an interval, or incorporeal space, or incorporeal quantity" (SWG 386); and, space is "an incorporeal and immobile extension in which it is possible to designate length, width, and depth so that every object might have its place"

---

[6] On the continuity of Gassendian space, see LoLordo (2007, 119–124). Holden (2004, 105) contends that the Gassendi-influenced Walter Charleton departs from Gassendi on this issue, for Charleton argues for a discrete minimal structure for both bodies and space. Gassendi, on the other hand, accepts only an atomic minima, or least part, for matter. Patrizi, as argued in De Risi (2015, 13), accepts that space is discrete as well.

(391). Hence, the congruence of incorporeal and corporeal dimensionality and the possibility of a vacuum justifies our virtual-platonism designation as regards his spatial geometry at the secondary macrolevel, since the dimensionality of space and the place of every body are independent of material existents.[7] In conclusion: FGS(top), SL(v-plt).

In addition, God's ontological role is unique in Gassendi's natural philosophy, given his further contention that space and time, which are infinite, are neither substances nor accidents but a more general category of being: "space and time must be added as two members of the classification [of being], as if to say that all being is either substance or accident or place, in which all substances and all accidents exist, or time, in which all substances and all accidents endure"; and, "space and time must be considered real things, or actual entities" (384). He reasons that "since it follows from the perfection of the divine essence that it be eternal and immense, all time and space are therefore connoted", and insists that God "both exists supremely in Himself...[and] also necessarily exists in all time and in every place" (RIV 94). In short, there is a form of co-dependence between God and space: "That God be in space is thought to be a characteristic external to His essence, but not with respect to His immensity, the conception of which necessarily involves the conception of space" (94; cf. Newton above).[8] Therefore, although God is not the cause of space and time for Gassendi (or for Newton), the fact that God's immensity entails infinite space, and God exists necessarily, hence justifies our FGS(top) classification.

---

[7] There is a further possibility, however, that three-dimensional incorporeal extension or space may be an emergent secondary level entity for Gassendi—in that case, SL(nom) would apply to his spatial ontology, since spatial geometry would be instantiated by that 3-dimensional incorporeal entity, space. Overall, Gassendi seems conflicted on space's ontological status. He occasionally claims that space is a "real thing" or "actual entity" (SWG 384), which would favor space as an emergent secondary level entity. But he also claims, more insistently, that space is nothing positive, a mere "negative quality" associated with an entity's *absence* (389). This view was held by some Scholastics who endorsed the imaginary space idea (see, Grant 1981, chap. 6), and it would seem to fit the SL(v-plt) conception, since space as negative quality would not be an emergent entity, but at most an emergent secondary level fact that can exist in the absence of the actual secondary level entity, matter. Since Gassendi also accepts that space does not classify as a substance or property, a possible explanation is that space as "real thing" pertains to God's holenmerism, FGS(top), and his "negative quality" comments pertain to 3-dimensional incorporeal extension, and thus SL(v-plt) would be upheld once again.

[8] Grant claims that, for Gassendi, space is "coeternal with and independent of God" (1981, 212), which seems to imply that space could exist in the absence of God. But, this reading is refuted in the passage quoted above, which employs the seventeenth century's internal/external attribute dichotomy: "That God be in space is thought to be a characteristic external to His essence, but not with respect to His immensity, the conception of which necessarily involves the conception of space" (RIV 94). So, given this internal/external dichotomy, by denying that space is external to God's immensity, it follows that space is "internal" to God's immensity. A philosophical tract that employs this distinction is Magirus 1642. Accordingly, it is implausible to infer that space retains any real independence from God, especially given Gassendi's further claim that God "necessarily exists in all time and in every place" (94). Additionally, Gassendi holds that both space and God are infinite and immobile (91–95), which is also indicative of their close interdependence (once again, cf. Newton's *De grav*).

Henry More also accepted definitive ubeity (holenmerism) in his early work, most notably in his correspondence with Descartes, where he argues that God is "whole everywhere, and His whole essence is present in all places or spaces, and in all points of space", such that "it does not follow that He has parts outside parts, or, consequently, that He is divisible" (More 1662, 76–77, trans. in Reid 2007, 93).

9.3.3   Patrizi and the Origins of Seventeenth Century Virtual-Platonism

The precursor who greatly influenced the type of virtual-platonism about space evident in Gassendi, More, Newton, and many others, is Francesco Patrizi (1529–1597), especially through his *De Spacio Physico*. Patrizi posited an indivisible, immovable space that underlies all entities, whether incorporeal or corporeal; and, not only is a vacuum possible within the world, but God could also move the entire world through the vacuum that lies outside the world. Nevertheless, "the vacuum itself is nothing else than three-dimensional space" (PS 231). As with the later Neoplatonists, Patrizi maintains that space is prior to all entities, except, of course, God: "[S]pace is by its nature prior to the world, and the first of all things in the world, before which there was nothing and after which, everything" (240); yet, "Space was brought forth by the First One before all other things, as if breathed out by the breath of His mouth and diffused into the finite and the infinite" (227). In more detail, space is not independent of God, a view that he holds in common with every other seventeenth century thinker examined in our investigation, and, like More and Newton, he postulates a spatially extended God:

> But if Divinity as a whole is indivisible, as it is, it will exist in indivisible Space and be enveloped all around by divisible Space [i.e., the space of corporeal entities and, possibly, lesser incorporeal entities]. Even if it exists nowhere, it cannot be conceived without Space. If it exists anywhere, either at the summit of the sky or above it, it is sure to be in Space. If it exists everywhere, it cannot but be in Space. All things (*entia*), therefore, and whatever stands above things, are in Space and cannot but be in Space. Hence before things could go forth from the Divine Deep they needed Space, with which the Deep itself is surrounded, in order to go forth and exist—if they were to be, or have the power to be, things (226–227).

While other works by Patrizi may be consistent with either repletive or definitive ubeity (see De Risi 2015, 15), *De Spacio Physico*, his most influential tract on space, seems to defend circumscriptive ubeity for God. Specifically, given his claim that God and God's space are indivisible (cf. N 26), that God "will *exist* in indivisible Space", and that space, place, and void are three-dimensional (PS 229–238), it would seem to follow that FGS(met) is a better candidate to capture Patrizi's spatial ontology than definitive ubeity, FGS(top). However, it must be admitted that Patrizi's conception of the God-space relationship is rather difficult to pin down.

Moreover, Patrizi rejects the latent nominalist view that space is the mere "capacity" to receive bodies, which is the interpretation of the corporeal void typically paired with definitive ubeity by the Scholastic imaginary space theorists, e.g., Oresme, but also Barrow (as will be explained below): "Is [space] merely an aptitude for holding bodies, and nothing more than that? To be sure, this Space holds all bodies, but it does not seem to be nothing but an aptitude" (240). Patrizi ultimately concludes "that Space is above all a substance, but not the 'substance' of the [Aristotelian] category" (241), a view that would inform the nearly identical taxonomies put forward by Gassendi, Newton, and many other Neoplatonists in the seventeenth century. Patrizi's own unique way of describing this peculiar "substance" is that space "is an incorporeal body and a corporeal non-body" (241). Interestingly, Patrizi reckons that geometry is prior to arithmetic:

> It likewise becomes clear that the continuum is older than the discrete, since no division can be made by any power without an antecedent continuum. Hence there is solved that difficult question..., namely, which is prior in nature, the science of the continuum, which is called 'geometry'..., or the science of the discrete, which is called arithmetic. The latter I of course say stems from the former, and hence is *secondary*. And since Space is the first of all natural things, it is clear that the science of space both of the continuous and the discrete, is prior to matter. (244)

Like Newton (as discussed in §9.3.1), mathematics is actually *in* space, and it is by contemplation of space, with the assistance of matter, that we comprehend mathematics:

> [I]n reality itself there are lengths breadths, and depths, without number and without end. And our mind selects those finite spaces that can be accommodated to the spaces of earthly bodies. The mind does not separate these spaces from bodies by abstractions, as some contend, since these spaces are not primarily and *per se* in earthly bodies, but are prior to bodies, and are actualized in primary Space....But the mind by its own power cuts off from primary Space those parts that will be of future use to it either in contemplation or operation (243–244).

Hence, in true incorporeal fashion, geometry is not actualized in matter, but is "actualized in primary Space". That is, like Newton's spatially extended God, mathematics/geometry is brought about by a unique "entity", in Patrizi's case, space, which is closer to the incorporeal category of being (since, as above, Patrizi describes space as "an incorporeal body"). Therefore, like Newton, Patrizi accepts both incorporeal nominalism, the view that space is grounded in an incorporeal entity, as well as virtual-platonism, SL(v-plt), the view that space is independent of the entities (bodies) at the secondary level.[9] However, if the divisible space of body is an emergent secondary level entity, then SL(nom)—but there is no textual evidence to support this view (or in the case of Gassendi, for that matter, where a similar inference could be drawn; see footnote 7).

---

[9] Unlike Patrizi, who holds that our knowledge of mathematics is a nominalist abstraction from space, another virtual-platonist concerning space, Gassendi, believes that our knowledge of mathematics is a nominalist abstraction from matter (SWG 246). Hence, it is important to note that a nominalist account of mathematics can differ from a nominalist account of space; see, also, Lennon (1993, 117–135), which explores Bernier's nominalist critique of the many non-nominalist elements of Gassendian space. Furthermore, Patrizi and Newton favor the view that geometry is prior, or at least superior, to algebra, and both are traditionally categorized as absolutists/substantivalists; whereas Descartes (Pr I 48) and Leibniz (AG 251–252) side with algebra/numbers over geometry/extension, and both are traditionally classified as relationists (i.e., Descartes lists numbers as more general than extension). This prompts an intriguing question: Is the foundations debate in mathematics—i.e., which is more basic, algebra or geometry?—an unacknowledged source of the standard dichotomy? See, Jesseph (1993, chap. 3), for a succinct overview of the foundations debate in the Early Modern period. Finally, the platonic orientation of Patrizi's spatial hypotheses is discussed in Henry (1979).

9.3.4   Leibniz on Spatial Ontology

As explained in Chap. 3, God plays a foundational role in Leibniz' deep ontology of space: he contends that space's "truth and reality are grounded in God, like all eternal truths", and that "space is an order [of situations] but that God is the source" (NE II.xiii.17). Yet, in contrast to both Newton's hypothesis of a spatially extended God (circumscriptive ubeity) and Gassendi's holenmerist idea of an unextended but spatially situated God (definitive ubeity), Leibniz rebuffs the notion that "God discerns what passes in the world by being present to the things", rather, God discerns things "by the dependence on him of the continuation of their existence, which may be said to involve a continual production of them" (L.V.85). In other words, Leibniz sides with repletive ubeity, a view that, at least on Leibniz' estimation, denies that God is situated in space, for he insists that "God is not present to things by situation but by essence; his presence is manifested by his immediate operation" (L.III.12). All of these themes are nicely encapsulated in a passage that we have frequently quoted:

> Where space is in question, we must attribute immensity to God, and this also gives parts and order to his immediate operations. He is the source of possibilities and of existents alike, the one by his essence and the other by his will. So that space like time derives its reality only from him, and he can fill up the void whenever he pleases. It is in this way that he is omnipresent. (NE II.xv.2)

Therefore, while Leibniz' accepts that God's immensity grounds space, his acceptance of a non-spatial God (repletive ubeity) clearly rules out a similarity of geometric structure at the foundational and secondary (material) levels of reality—and the same holds for Leibniz' basic ontological unit (other than God), namely, monads. As a simple substance, a monad, like God, has no spatial parts, "[b]ut where there are no parts, neither extension, nor shape, nor divisibility is possible" (AG 213). The non-spatiality of the monads is, in fact, a common theme in Leibniz' late work, and this includes both the metrical and topological aspects of space: e.g., "there is no spatial or absolute nearness or distance among monads. And to say that they are crowded together in a point or disseminated in space is to use certain fictions of our mind when we seek to visualize freely what can only be understood" (Lm 604); and, "monads, in and of themselves, have no position with respect to one another" (AG 201).

Leibniz, moreover, rejects the FGS(top) conception of incorporeal beings, i.e., where God and souls are situated in the points of space, for he explicitly rejects holenmerism (definitive ubeity) in the correspondence with Clarke: "[t]o say [a soul] is, the whole of it, in every part of the body is to make it divisible of itself. To fix it to a point, to diffuse it all over many points, are only abusive expressions, *idola tribus*" (L.III.12). Accordingly, for both God and monads, Leibniz advocates FGS(prg). Furthermore, in Leibniz' claim that "God is not present to things by situation but by essence; his presence is manifested by his immediate operation" (L. III.12), God's "essence", on Leibniz' unique interpretation, is associated with the possibility of existing things (see the NE II.xv.2 passage above), and "immediate operation" is correlated with the continual conservation or reproduction of the world via God's will. As discussed in §4.2, immediate operation seems equivalent to the extension of power doctrine in Leibniz' late output, a conception that situates God's actions in space but not God's being. A process that is somewhat analogous to immediate operation (and thus analogous to repletive ubeity, extension of power, and nullibism as well) may also explain how monads relate to the secondary macro-level of bodies: "I do not think it appropriate to regard souls [monads] as though in points. Perhaps someone might say that souls are not in place but through operation, speaking here according to the old system of influx; or rather, according to the new system of preestablished harmony, that they are in place through correspondence" (LDB 123–125). As also noted in Chap. 4, whether or not a monad's operation is similar to God's operation is a difficult question, but it does not affect the main point above, namely, that Leibniz adopts a monadic version of the repletive ubeity concept that he also applies to God, so that monads are similarly not situated in space although the "results of" monads, i.e., bodies, are situated.[10]

---

[10] Given passages like L.III.12, Pasnau (2011, 338, n.21) entertains the possibility that a definitive ubeity (holenmerist) interpretation is applicable to Leibniz' God, even if it does not apply to finite souls/monads. Presumably, Pasnau's inference is based on the standard Scholastic theology of presence , which holds that a being's presence by essence entails a being's actual presence in space (see also §4.2.2). Yet, the textual evidence refutes this holenmerist/definitve ubeity interpretation. Besides the quotation above, where Leibniz rejects the claim that "God discerns what passes in the world by being present to the things" (L.V.85), he argues against Clarke that "if God must be in space, if being in space is a property of God, he will in some measure depend on time and space and stand in need of them" (L.V.50). This passage explicitly refers to (and rejects) the possibility that God is "in space"; moreover, Clarke understands Leibniz' view as straightforwardly denying God's presence/situation in space, as his reply to L.V.85 makes clear: "That God (sec. 85) perceives and knows all things not by being present to them, but by continually producing them anew, is a mere fiction of the schoolmen, without any proof" (C.V.83-88). As discussed earlier (§4.2.2), the only explanation that is consistent with the evidence is that Leibniz has reinterpreted the metaphysics of presence by essence to entail, or at least allow, a presence by power alone (see, Adams 1994, 357, an exhaustive study which draws a similar conclusion). Indeed, a precedent for Leibniz' approach, that accepts a non-spatially situated God that operates on bodies by conserving them, can be found in the fourteenth century thinker, Thomas of Strassburg (see, Funkenstein 1986, 61). In an effort to counter the abundant textual evidence of God's non-spatiality, Pasnau suggests that the references to God lacking situation are due to a classificational distinction: "he likely has in mind the scholastic category of *situs* or Position, which implies the sort of spatial arrangement typified in the canonical example of *sitting*. God is not extended in this way because God lacks corpuscular structure" (2011, 338, n.21). First, "*situs*" was often used to describe the situation of immaterial beings in the world in the late scholastic and Renaissance periods, as a perusal of the literature quickly reveals. Second, and more importantly, there is no evidence to support the contention that Leibniz is using a scholastic, technical sense of "situation" or "presence" in the correspondence (i.e., that only applies to corporeal things). Leibniz, moreover, explains his use of "presence" in the continuation of the L.V.85 discussion: "A mere presence or proximity of coexistence [where space is "an order of coexistences", L.III.4] is not sufficient to make us understand how that which passes in one being should answer to what passes in another". Leibniz references "presence" in a quite general way, here, such that it applies to any "being" (and not just corporeal being), a reading that also applies to his definition of "situation": "*Situs* is a certain relationship of coexistence between a plurality of entities" (Lm 671, c. 1714).

Since "there is no space where there is no matter" (L.V.62), and space is not "an absolute being" (L.III.5), Leibniz's spatial ontology sides with geometric nominalism at the secondary level, SL(nom). Nevertheless, space represents "real truths" (L.V.47), and these truths are independent of matter: "[t]ime and space are of the nature of eternal truths, which equally concern the possible and the actual" (NE II.xiv.26). Given that these truths are independent of existing bodies, this form of explanation, in effect, betrays a strong penchant for absolutism—but, it is an absolutism about the truths of geometry conceived in a nominalist fashion, secured via God's immensity, and not an absolutism that posits space as an entity that can exist in the absence of matter.[11] The similarities between an absolutist conception of space and Leibniz' conception were first addressed in Chap. 3, where the "universal place" hypothesis was explored (NE II.xiii.8).

9.3.5   The Pre-critical Kant on Spatial Ontology

Leibniz' monadic project became an essential element of the natural philosophy of space of the eighteenth century "Leibniz-Wolff" school, although many of the members of this tradition apparently situated their monads, or soul-like simple substances, in the points of space, unlike Leibniz (see, De Risi 2007, 309 n. 8; and Watkins 2005, 58). On the other hand, these philosophers followed Leibniz by associating their simple substances with force, and by conceiving the material world as an emergent or supervenient effect of monadic force. Given the new taxonomy developed above, the Wolffian approach to space fits the FGS(top) and SL(nom) categories, the latter classification due to their rejection of absolute space; e.g., Wolff declares that "*space* is the order of things that co-exist. And therefore *space* cannot exist if things are not present to occupy it, although it is still different from these things" (Wolff 2009, 15). Roger Boscovich may be the most prominent natural philosopher to espouse this line of thought, although he conceived his point forces as discontinuous, i.e., the point forces are not contiguous but are separated by a repulsive force (which is counterbalanced by an attractive force). For Boscovich, accordingly, the relevant topology in his version of FGS(top) is not the continuous structure accepted by Gassendi, although SL(nom) remains (see, Boscovich 1966, 19–67).

In a series of works, starting with his very first, *Thoughts on the True Estimation of Living Forces* (1749), and including the *Physical Monadology* (1756), the precritical Kant puts forward his own brand of monadic hypothesis that upholds SL(nom), but he ultimately endorses a decidedly Leibnizian form of FGS(prg)

---

[11] To summarize some of the points from Part I, the seventeenth century theories of motion that allegedly support relationism (Descartes, Leibniz) are largely irrelevant as regards the deep metaphysics that underwrites their respective spatial hypotheses, namely, God (and monads). Furthermore, it is the analysis of this neglected metaphysical component, in correlation with the higher level structures, that will allow comparisons between seventeenth century and QG theories (although the deep metaphysics underlying QG theories is natural, and not supernatural).

rather than the FGS(top) favored by the Wolffians. On Kant's reckoning in the *True Estimation*, the forces that comprise his simple substances/monads, which are prior to bodily extension (NS 1:17), can interact with each other, and by these interactions they give rise to bodily extension and space, hence SL(nom): "there would be no space and no extension if substances had no force to act external to themselves. For without this force there is no connection, without connection, no order, and, finally, without order, no space" (1:23). Furthermore, since Kant's monads, which are associated with souls in the standard Leibniz-Wolff fashion (1:21), are not in space *per se*, it follows that a substance that does not partake in their mutual interconnections would not be in space.

> A substance is either connected with and related to other substances external to it, or it is not. Because every independent entity contains within itself the complete source of all its determinations, it is not necessary for its existence that it should stand in any connection with other things. That is why substances can exist and nonetheless have no external relation to other substances, or have no real connection with them. Now since there can be no location without external connections, positions, and relations, it is quite possible that a thing actually exists, yet is not present anywhere in the entire world (1:22–23).

Given this hypothesis, Kant further contends that "God may have created many millions of worlds" (1:22), presumably, by interconnecting different sets of substances to form different worlds along with their different spaces. The type of force-based interconnection among the substances is, furthermore, "the inverse-square relation of the distances" (1:24), an interconnection that also accounts for the three-dimensionality of space, although it is only one of the possible connections (and hence dimensions of space) that God could have established (1:23–25). In his later *New Elucidation* (1755), Kant adds that the inverse-square law of "Newtonian attraction" that holds among bodies is derived from the inverse-square connection among simple substances: "[i]t is...probable that this attraction [Newtonian attraction] is brought about by the same connection of substances, by virtue of which they determine space" (TP 1:415).[12]

One of the uncertainties related to Kant's early monadic philosophy in the *True Estimation*, however, pertains to the possible spatial presence of these substances after space emerges from their interconnections. As is evident in the long quotation provided above, "since there can be no location without external connections", it would seem to be the case that substances are located once these external connections are established. In the *Physical Monadology*, Kant sets out to address this issue, since it potentially leads to a problem that Newton had earlier confronted in *De grav*, i.e., if space is divisible and the foundational entities from which space

---

[12] There is an interesting problem concerning force given Kant's model: if force is prior to space, then how do we understand the distance relationships, such the $1/r^2$, that are incorporated into his conception of monadic force (as in NS 1:23 and TP 1:415 above). Kant does not address this issue, but his assertion that space "is the appearance of the external relations of unitary monads" (TP 1:479) suggests a possible response: any force manifest at the macrolevel of bodies that involves spatial distance, such as $1/r^2$, is an "appearance" (phenomena); i.e., the interconnections among monads are not spatiotemporal relations, but they nonetheless "bring about" matter, space, and the distance-based conception of force, at the material (phenomenal) level. See, also, §10.5.2, and J.J.C. Smart's (early) observations on an emergent spacetime theory.

emerges are situated in space, then these allegedly indivisible entities must be, on the contrary, divisible. Kant's solution, which parallels Leibniz' primitive/derivative force distinction (see Chap. 4), invokes a difference between a monad's non-spatial "internal determinations" and its "external determinations", or "sphere of activity", the latter giving rise to space. "[T]hough any monad, when posited on its own, fills a space", explains Kant, yet "the filled space is not to be sought in the mere positing of a substance but in its relation with respect to substances external to it....It must, therefore, be granted that the monad fills the space by the sphere of its activity" (TP 1:481). Described as an "accident" of the monad, he further identifies the sphere of activity with a body's impenetrability (1:482).[13]

Returning to the divisibility worry, Kant provides a detailed account of his version of monadic presence:

> But, you say, substance is to be found in this little space and is everywhere present within it; so, if one divides space, does not one divide substance? I answer: this space itself is the orbit of the external presence of its element. Accordingly, if one divides space, one divides the extensive quantity of its presence. But, in addition to external presence, that is to say, in addition to the relational determination of substance, there are other, internal determinations; if the latter did not exist, the former would have no subject in which to inhere. But the internal determinations are not in space, precisely because they are internal. Accordingly, they are not themselves divided by the division of the external determinations....It is as if one were to say that God was internally present to all created things by the act of preservation; and that thus someone who divides the mass of created things divides God, since that person divides the orbit of His presence—and than this there is nothing more absurd which could be said. (1:481)

By adopting this strategy, Kant has, in fact, forthrightly sanctioned FGS(prg)—and, quite importantly, he supports his conclusion by offering an analogy between a monad's sphere of activity and God's presence "to all created things by the act of preservation", a description that exactly matches Leibniz' account of repletive ubeity, whereby God "operates immediately on all created things, continually producing them" (NE II.xxiii.21; see §9.3.4). That is, just as Leibniz characterizes God's

---

[13] Friedman contends that, for Kant, space "is derivative from or constituted by the underlying non-spatial reality of simple substances", and is "metaphysically real", whereas, in contrast, Leibniz' "space is ideal because relations between substances are ideal" (Friedman 1992, 7–8). Needless to say, our investigation rejects a purely phenomenalist reading of Leibniz, since his metaphysics includes numerous realist descriptions of the manner by which material extension (and thus space, when generalized) is brought about by the monads (see Chap. 4): besides the primitive/derivative force distinction, there are many references to the "diffusion of materiality or antitypy" (e.g., AG 261). While there are no direct inter-monadic connections, Leibniz' plenum and the diffusion of resistance provides that same "metaphysically real" depiction that Friedman finds in Kant's pre-critical monadology. Likewise, Friedman contends that the appeal to God's role in fixing the inter-monadic connections (see below) heralds the transition from the Leibniz-Wolff doctrine of space to "the Newtonian doctrine of divine omnipresence" (1992, 7). While Kant does employ this terminology, such as by referencing the "infinite extent of divine presence" (NS 1:306), there are many different ways that God can be related to space (ubeity)—and, as will be disclosed, Kant follows Leibniz in sanctioning Leibniz' brand of repletive ubeity (i.e., presence by power), whereas Newton favored the circumscriptive variety. Hence, Friedman's singling out only Newton's form of ubeity is unwarranted.

repletive ubeity through the act of preserving the world, an hypothesis that rejects the presence in space of God's substance *per se*, so Kant posits a corresponding form of monadic repletive ubeity (or extension of power) doctrine that similarly denies the presence in space of a monad's internal determinations, where internal determinations are associated with the monad's substance or being *per se*, and the external determinations that give rise to space are accidents of the monad's substance (internal determinations). This interpretation of the internal/external determination distinction is corroborated in his discussion of the division problem: although one can divide a monad's sphere of activity into two parts, he concludes that "each [half of the monad's sphere of activity] is nothing but an external determination of one and the same substance, but accidents do not exist independently of their substances" (TP 1:482). As argued in §4.2.2, Leibniz also makes an analogy between a monad's operation/activity in space and the extension of power doctrine, the latter including God's continual preservation of the world (via repletive ubeity), but hesitates since he apparently associates the extension of power hypothesis for lesser beings with a physical influx, which he rejects (preferring his pre-established harmony thesis instead). Kant, as we have seen, openly supports a type of physical influx among monads via their external determinations, although he argues in the *New Elucidation* that these monadic interconnections do not constitute a physical influence "in the true sense of the term" because they are "outside the principle of substance, considered as existing in isolation" (1:415–416). In the language of the *Physical Monadology*, this would amount to the claim that the external determinations among monads do not involve or modify their internal determinations, even though the external determinations are accidents of internal determinations. And, since Kant claims that his form of inter-monadic connection incorporates a "reciprocal *dependency*", a dependency that he deems to be more robust than the mere "agreement among substances" that one finds in Leibniz' pre-established harmony doctrine (1:415), it hence follows that the *Physical Monadology* represents one the most elaborate attempts to defend the non-spatially situated component of Leibniz' original monadological hypothesis alongside a full-fledged notion of inter-monadic activity or interconnection.

Finally, the *New Elucidation* offers an hypothesis that implicates God's preservation of the world as the foundation of the inter-monadic connections among the individual, isolated monads, once again demonstrating that Leibnizian repletive ubeity underlies Kant's monadic philosophy:

> [T]he ground of their [monadic] reciprocal dependence upon each other must also be present in the manner of their common dependence on God. How that is brought about is easy for the understanding to comprehend. The schema of the divine understanding, the origin of existences, is an enduring act (it is called preservation); and in that act, if any substances are conceived by God as existing in isolation and without any relational determinations, no connections between them and no reciprocal relation would come into being. (TP 1:413–414)

This explanation, which appeals to God's "understanding" and "conceiving" to bring space into being (via the inter-monadic connections), can also be seen as anticipating the final destination of Kant's evolving conception of space: while divine cognition may no longer play this fundamental role in his later work, a sub-jectivist/idealist construction of space would remain, albeit limited to the human sphere (see Chap. 11).

9.3.6 Barrow and Descartes on Spatial Ontology

Finally, a brief glance at Barrow and Descartes will help to round out our analysis of seventeenth century spatial ontologies. In accordance with the prevailing theologically-based ontology of the period, Barrow also reckons that space is depen-dent on God: "there was Space before the World was created, and...there is now an Extramundane, infinite Space, (where God is present)" (LG 203). By declaring that there was "Space before the World", this passage likely prompted those assessments that group Barrow with the absolutists (substantivalists), such as Hall (1990, 210), a point first raised in Chap. 3. Nevertheless, Barrow actually follows Leibniz' nomi-nalism, for he explicates space's "existence" via the God-based capacity to receive bodies, i.e., a non-dimensional capacity to receive dimensional bodies (in keeping with the Scholastic imaginary space tradition). For instance, he explains that time "does not imply an actual existence, but only the Capacity or Possibility of the Continuance of Existence; just as space expresses the Capacity of a Magnitude contain'd in it" (LG 204). In his *Mathematical Lectures* (partially translated in UML), a more detailed discussion is provided: "Space is not any thing actually existent, and actually different from Quantity, much less that it has any Dimensions proper to itself, and actually separate from the Dimensions of Magnitude [i.e., bodily dimensions]" (UML 175), and he also rejects the "fictitious Vacuum endowed with real actual Dimensions such as Epicurus with his Followers" accepted (UML 179). Hence, while he insists that "*Space* is a thing really distinct from Magnitude" (UML 175), that "thing" ("Space before the World") is the mere capacity to receive bodily magnitude, a conception that he shares with Leibniz, as we have seen: in other words, SL(nom). In addition, since he claims on several occasions "that God is present to all *Space*" (UML 178, as well as UML 170 and LG 203), FGS(top) seems the more likely classification, but the evidence is consistent with FGS(prg) as well.

Descartes' conception of space shares many features with Leibniz' views, espe-cially the espousal of God's repletive ubeity (using Leibniz' term for the extension of power concept), or FGS(prg). In the correspondence with More, he explains that "the extension which is attributed to incorporeal things is an extension of power and not of substance. Such a power, being only a mode in the [corporeal] thing to which it is applied, could not be understood to be extended once the extended thing cor-responding to it is taken away" (CSMK 373). As such, this explanation leaves open the possibility that God may be situated in the points of space, FGS(top), but not extended; yet, a bit further in the same letter, he explicitly rejects More's contention that God "exists everywhere": "I do not agree with this 'everywhere'. You seem here to make God's infinity consist in his existing everywhere, which is an opinion I cannot agree with. I think that God is everywhere in virtue of his power; yet in virtue of his essence he has no relation to place at all" (373). That is, by insisting that God's essence "has no relation to place at all", it would seem to follow that a topological presence, FGS(top), is ruled out, thereby leaving FGS(prg) as the only option. Furthermore, since Descartes regards spatial extension as identical to matter (Pr II 10), it is highly unlikely that he would situate God in matter via FGS(top), for that maneuver may have the potential to threaten the coherence of his substance dualism, e.g., his claim that "God is not corporeal" (Pr I 23). More carefully, a point, as the boundary of a line, might qualify as a mode (or part) of extension, and thus qualify as corporeal if Descartes were to accept definitive ubeity (holenmerism). When later pressed by More to clarify his stance on the Scholastic relationship between God's essence and substantial presence, Descartes modifies his view but apparently retains his preference for a non-spatially situated God whose power can operate in space: "I said that God is extended in virtue of his power, because that power manifests itself, or can manifest itself, in extended being. It is certain that God's essence must be present everywhere for his power to be able to manifest itself everywhere; but I deny that it is there in the manner of extended being" (CSMK 381).[14] Turning to Descartes' understanding of this manifestation of power, he concludes that "all others [substances] can exist only with the aid of God's participation" (Pr I 51), hence it would seem to follow that God's continual conservation of the world (which is only conceptually distinct from the world's recreation; CSM II 33) qualifies as an instance of a divine action that is everywhere in space. Finally, like Leibniz, God is the foundation of space by way of upholding the existence of matter, and since matter is identical with space and a vacuum is impossible (Pr II 18), thus SL(nom).

---

[14] That is, in his last letter to More, Descartes now appears to interpret a presence by power as a form of presence by essence, but without a corresponding situation in space of God's being, perhaps following Aquinas' non-literal, analogical conception of God's presence (see, Funkenstein 1986, 50–57), and, of course, anticipating Leibniz' later stance (see §9.3.4). Reid and Pasnau argue that Descartes' concession to More concerning the presence of God's essence must mean that God "*has* to be spatially present" (Reid 2007, 105; and, Pasnau 2011, 333–338)—but this inference is quite problematic, since Descartes never claims that God's *substance* or *being* is actually present/situated in space, and it would raise havoc for his substance dualism (as argued above). On the whole, Descartes takes an analogical approach to God's extension, stating in his earlier letter that "if someone wants to say that God is in a sense extended, since he is everywhere, I have no objection"; and, "just as we can say that health belongs only to human beings, though by analogy medicine and a temperate climate and many other things also are called healthy, so too I call extended only what is imaginable as having parts within parts, each of determinate size and shape—although other things may also be called extended by analogy" (CSMK 361–362). Consequently, given this analogical approach to God's extension, it is best to interpret God's situation in space (via his essence) as equally analogical in nature. Yet, since Descartes, unlike Leibniz, does not explicitly deny God's actual substantive presence in space, an FGS(top) reading is consistent with his last letter to More. Reid (2012, 153–154) examines this issue among the later Cartesians, many of whom would take a definitive/holenmerist stance, although at least one Cartesian, Pierre Poiret, did embrace a non-spatial, non-situated God (which More would later call nullibism; see Chap. 2).

9.4   Reflections on Seventeenth Century and Contemporary Spatial Ontologies

One of the major themes of our investigation is that both Newton and Leibniz regard extension/space as requiring some form of foundation in a substance or entity, broadly construed, and that this aspect of their respective hypotheses strongly suggests that space functions, in an ontological sense, as some sort of property in the P(O-dep) vein (see §1.2). It may not be a property that is internal to, or "inheres" in, God for Newton and Leibniz, but they both claim that space is dependent on God; and Leibniz, additionally, views space much like a holistic property of the entire world's dynamical interconnections, ultimately grounded at the monadic level (see Chaps. 3 and 4). In short, both reject the notion that space is either an independent entity in its own right or that it can act upon things, thus it is not a substance. And, since both adhere to the Scholastic substance/property doctrine, any relationist construal of space as the extension *between* bodies would be reckoned "an attribute without a subject, an extension without anything extended" (L.IV.9), and thus outright rejected. It is in this sense that the modern attempts to appropriate Newton and Leibniz as would-be substantivalists or relationists go seriously awry (e.g., Sklar 1974, Friedman 1983). As suggested in Chap. 1, the modern substantival/relational debate has become inextricably intertwined with the larger scientific realism/anti-realism dispute that emerged in the second half of the twentieth century and an accompanying (positivist-influenced) disinterest in archaic non-empirical, metaphysical approaches to physics—in brief, the contemporary stalemate that afflicts the modern spacetime ontology debate is largely the result of this vain effort to graft seventeenth century metaphysics on to contemporary philosophy of science and physics; that is, the modern dichotomy strives to remain consistent to Newton and Leibniz but without utilizing their stock of metaphysical presuppositions.

These last observations shed light on some of the conclusions reached in Chap. 5. Given the recent emphasis on the dual role of g as both the metric and gravitational field in GR, the prospects for the modern spacetime ontology debate have faded even further, since the competing claims of g as either the unique spacetime substance or just another physical field are, it would appear, in principle irresolvable or conventional; i.e., there is no longer any relationist *unsupported* extension between matter if g is a physical field—it is "internal" on both the relationist and substantivalist construals of GR. The common refrain that the postulated entities in QG, whether strings, quantum loops, etc., can also support either substantivalism or relationism is, therefore, merely the subatomic analogue of this fruitless quarrel over the ontological status of such macroscopic "things" as the metric/gravitational field.

Finally, in order to better grasp how the platonism/nominalism distinction can assist the evaluation of spatial ontologies, it is worth examining Belot (2011), an important contribution to the spacetime ontology debate which also happens to reach some of the conclusions advanced in our investigation. As we have seen, Leibniz' conception of the geometric truths of space are both grounded by God and independent of matter: e.g., "[t]ime and space are of the nature of eternal truths, which equally concern the possible and the actual" (NE II.xiv.26); space's "truth and reality are grounded in God, like all eternal truths" (NE II.xiii.17); and, concerning how the world can be filled with matter, "there would be as much as there possibly can be, given the capacity of time and space (that is, the capacity of the order of possible existence); in a word, it is just like tiles laid down so as to contain as many as possible in a given area" (AG 151). Belot (2011, 2) has inferred from this evidence that Leibniz is a realist about geometry at the phenomenal level of matter. However, this quite justifiable observation overlooks the fact that Leibniz' realism about space at that level stems from God's immensity, and so his realism differs in only one significant way from Newton's similar God-grounded spatial realism; namely, at the secondary material macrolevel, Newton's virtual-platonism versus Leibniz' nominalism, a distinction that evades the contemporary substantivalism/relationism debate since that dichotomy conflates the ontological and geometrical/mathematical aspects of spatial theories, and thus it cannot track these more fine-grained distinctions. From a modern perspective, one might strive to equate Newton's virtual-platonism with a straightforward realism about geometric structure at the secondary material level, and Leibniz' nominalism with a modal relationist form of geometric realism at that level, but this approach fails to account for the rationale behind these different realist ascriptions, non-modal versus modal: (a) an underlying ontology (God) that is actually present in space (via circumscriptive ubeity), FGS(met), in conjunction with his virtual-platonism at the secondary level, SL(v-plt), thus explaining Newton's geometric realism at the secondary level; and (b), an underlying ontology (God, monads) that is not present in space (via repletive ubeity), FGS(prg), in conjunction with his nominalism at the secondary level, SL(nom), thereby explaining Leibniz' modal relationist geometric realism at the secondary level. Put differently, both Leibniz and Newton would deny the assumption that bodies/fields alone can ground geometric truths, since only God can—but, a macrolevel body/field foundation for space is, in fact, the motivation behind the modern approach to spacetime ontology (e.g., the sophisticated metric field versions of both substantivalism and relationism in GR), hence modal relationism does not capture Leibniz' spatial ontology.

It is worth delving into these last few points in a bit more detail. Since a realism about spatial structure is consistent with sophisticated modal relationism, one might therefore attempt to advance a sophisticated modal relationist hypothesis as the contemporary equivalent of Leibniz' nominalism, a strategy expertly developed in Belot (2011, 173–185). Yet, while Belot's efforts are quite informative, it nonetheless seems to strain the coherence of relationist doctrine. A true modal relationist must posit spatial (spatiotemporal) modality on actually existing matter/fields, or on the possibility of matter/fields coming into existence provided at least one preexisting body. Leibniz' hypothesis, on the contrary, fixes the truths of spatial structure, not on bodies or on the possibilities of bodies, but on God: "He is the source of possibilities and of existents alike, the one by his essence and the other by his will. So that space like time derives its reality only from him" (NE II.xv.2). Likewise, a complete and permanent vacuum state does seem plausible given Leibniz' additional claim that God could block the emergence of extended matter, and hence space: Leibniz states that a monad's primitive force is "a higher principle of action and resistance, from which extension and impenetrability emanate when God does not prevent it by a superior order" (A.I.vii.249; Adams 1994, 351). But, since they are secured by God's existence, the fixed Euclidean truths of spatial geometry remain inviolate, even in a void scenario, for they represent "the capacity of time and space (that is, the capacity of the order of possible existence)" and are "like tiles laid down so as to contain as many as possible in a given area" (AG 151). By claiming that space determines the configurations and positions of bodies, Leibniz has, accordingly, reversed the relationist's account of space, for the relationist holds that bodies and their relations determine spatial geometry. If forced to choose between substantivalism and relationism, one might reasonably draw the conclusion that Leibniz' spatial hypotheses fall more comfortably on the absolutist/substantivalist side of the debate, and not modal relationism. Yet, on second thought, a theory whose fixed spatial truths and structures at a given level do not depend on the matter/fields at that level—but instead posits that matter, and hence the instantiated truths of space, emerge from a quite different non-spatial layer of ontology (in this case, God/monads)—not only eludes modal relationist doctrine, but would seem to demand a separate classification beyond traditional substantivalism and relationism. That dichotomy, once again, grew out of the classical theories of nineteenth and twentieth century macrolevel physics, but struggles to find applicability and relevance both within the context of quantum gravity theories and, *if* their deep metaphysics is specified in full, seventeenth century spatial ontologies.

9.5   Platonism/Nominalism, Geometric Background, and Leibniz Shifts

This final section of the chapter highlights some of the important features and functions associated with the nominalist component of the new taxonomy, as well as examines a rival classificational scheme for spatial ontologies that is based on the standard dichotomy.

9.5.1   *Background Dependence/Independence*

While philosophers of space and time have largely ignored the significance of the platonism/nominalism distinction for interpreting the substantival/relational dispute, a close cousin of the SL(v-plt)/SL(nom) divide can be detected in one of the central themes connected to the search for QG, namely, background independence. Among the competing interpretations of this doctrine (see, Rickles 2008b, 352–355), background independence can be minimally construed as the "freedom from 'background structures', where a background structure is some element of the the- ory that is fixed across the models of the theory" (Rickles and French 2006, 1–2, n.1). The metric is normally the structural element at issue, hence the variable met- ric in GR marks that theory as background independent. Likewise, as Rickles and French go on to note, "background independence, when used in the context of quan- tum gravity, is usually meant in a restricted sense, covering the freedom from a background metric alone" (2006, 2, n.1). Conversely, if a theory's metric is back- ground dependent, then all formulations of the theory must conform to a predeter- mined, fixed geometry regardless of the matter configuration or other factors. This last stance is consistent with SL(v-plt), of course; but, since Euclidean geometry is standard throughout the seventeenth century, it would appear that all of these earlier theories are background dependent, including the SL(nom) cases. For instance, Leibniz defines "distance" in the *New Essays* as "the size of the shortest possible line that can be drawn from one [point or extended object] to another", and com- ments that "[t]his distance can be taken either absolutely or relative to some figure which contains the two distant things", and where "a straight line is absolutely the distance between two points" (NE II.xiii.3). Leibniz' choice of the term "absolute" to signify the Euclidean concept of a straight line (as the shortest distance) thereby calls into question those attempts, such as Smolin (2006, 201), to enlist Leibniz as an early advocate of the background independence cause.

In spite of these observations, Jean Buridan offered a different God-grounded, nominalist-inspired approach to the vacuum in the late Medieval period that hints at a peculiar form of background independence. If the terrestrial realm below the lunar sphere was transformed into a vacuum, Buridan concludes that, due to the absence of the dimensional quantity instantiated in matter, that vacuum would be without any measurable size. As Sylla notes, "[i]f so, then God could create many worlds there, even worlds much larger than what was destroyed. A body in motion inside the evacuated lunar sphere could move with high velocity for a long time and never get any closer to one side or further from the other" (Sylla 2002, 262). While this scenario might more plausibly constitute a sort of geometric anti-realism in the void sphere (or even a multiply connected space?), another possibility is that it resembles a limited case of background independence, since many different geometric "truths" are admissible within that sphere given the same God-grounded ontology; i.e., dif- ferent geometries compatible with the same theory.

9.5.2   The Shift Scenarios and Platonism/Nominalism

The difference between a platonist and nominalist construal of spatial geometry is also useful for understanding the static and kinematic shift cases, as well as the structuralism intrinsic to both Newton and Leibniz' respective approaches. Since structuralism is concerned with the structural relationships among things, as opposed to the things themselves, it is often regarded as more akin to relationism than substantivalism. Nevertheless, as disclosed in Chap. 6, Newton takes a structuralist line on the identity of the parts of space, so structuralism also enjoys support from a prominent seventeenth century thinker who is not associated with relationism. Given Newton's virtual-platonism at the secondary macrolevel, and the fact that an immobile God grounds the parts of space (via his incorporeal nominalism), it follows that an identical configuration of bodies can occupy many different positions in space, static shifts, as well as motions in space, kinematic shifts.

Leibniz' structuralism, on the other hand, is more directly tied to his nominalism, a point discussed at length in Chaps. 3 and 7. Because God's essence serves as the foundation of the Euclidean capacity of space (see §9.3.4), and space depends on bodies, nominalism entails that there can be only one instantiation of the same relative configuration of bodies, unlike on Newton's scheme. For example, if space "is nothing at all without bodies but the possibility of placing them, then those two states, the one such as it is now, the other supposed to be the quite contrary way [static shift], would not at all differ from one another"; L.III.5). Specifically, prior to the material world's instantiation, if space is only "the possibility of placing [bodies]", secured via God's essence, then to claim that the material world could have had a different spatial location while retaining its relative configuration (e.g., the material world has been uniformly repositioned five feet to the left) is to demand that space must exist both prior to, and independently of, the material world's instantiation of matter (since this prior and independent existence of space is the only way that the same relative configuration of bodies can obtain a different position in space). But, of course, the existence of space prior to its material instantiation is just what Leibniz' nominalism denies. The same argument applies in the case of kinematic shifts, since, on his unique conception of the force-space relationship, it is the dynamical interactions among bodies that is associated with space at the secondary macrolevel (see Chap. 3), and thus any uniform addition to, or subtraction from, the velocity (speed and/or direction) of all bodies does not alter the dynamical interconnections among those bodies.[15] In more detail, the primitive monadic forces that comprise monads at the deeper foundational level are manifest, at the secondary macrolevel, in the derivative force and motion of bodies (Lm 533). Consequently, on this unique conception of force, a uniform change in the velocity of all of the world's inhabitants is really just a uniform addition/subtraction of an identical amount of force to/from each body, and is thus another instance of a mere difference in scale or gauge.

To sum up, the platonist/nominalist divide helps to elucidate the different forms of structuralism intrinsic to both Newton and Leibniz' theories, as well as assists in understanding the rationale behind Leibniz' rejection of the shift arguments (although his rejection of kinematic shifts requires a nominalism that is conjoined with his belief in the dynamical interconnections of all bodies, which in turn stems

---

[15] In a work on the principle of the identity of the indiscernibles, Leibniz claims that "there are no purely extrinsic denominations [i.e., purely spatial differences], *because* of the interconnection of things", and, "if place does not itself make a change [i.e., no extrinsic denominations], it follows that there can be no change which is merely local" (MP 133, emphasis added; see, also, NE II. xxv.5). As argued in Chap. 3, a global, or holistic, basis for space is thereby linked to the dynamical interconnections among all bodies, and this holism, in turn, supports a unique form of spatial or geometric nominalism.

from the primitive force of monads). Yet, it is important to note that other nominalist approaches to space, similar to Leibniz' version, did admit uniform shifts of the entire material world, and this difference might be traceable to the difference between FGS(top) and FGS(prg) surveyed above, which in the seventeenth century stands for the difference between, respectively, definitive ubeity (or holenmerism), and repletive ubeity (or extension of power, nullibism). As first introduced in Chap. 7, Oresme treats an intra-world vacuum along the same lines as Leibniz, employing possible bodies, yet, unlike Leibniz, he reckons that God could move the entire world in a rectilinear kinematic shift (O 369). For Oresme, whose deep ontology of space is much closer to Gassendi's than Leibniz', God's immensity is both classified as "whole in every part" holenmerism,[16] and identified with all spaces/places, including the vacuum outside the world.[17] Hence, while Oresme's nominalism correlates with SL(nom), his ontology accepts FGS(top) by way of holenmerism (definitive ubeity). Given that holenmerism posits God's unextended but point-like presence in space (as is also the case with Gassendi's God), there is, accordingly, a consistent basis upon which to posit a kinematic shift of the world—i.e., God's actual presence in space, even though it is a non-dimensional, point-like presence. While not possessing any dimensional quantity, Oresme's God-grounded ontology of space can surely register the change of the material world's position from one point to another, and so a kinematic shift makes sense on this scheme. Leibniz, on the other hand, rejects holenmerism (definitive ubeity), and favors his form of repletive ubeity instead, so that God is not actually present in space *per se*. And, since only God's operation of conserving the world can be situated under the repletive ubeity hypothesis (see §9.3.4), a Leibnizian kinematic shift is thus not applicable on the grounds that it is inconsistent with Leibniz' conception of a non-situated God. Unlike the Oresme case, there is no spatial relationship between Leibniz' space-grounding entity (God, monads) and the supervening space, and so no basis upon which to meaningfully invoke a static or kinematic shift. Whether or not this additional line of argument played a role in Leibniz' decision to reject the shift scenarios remains unclear, needless to say, but it does pose an intriguing possibility since Oresme and Leibniz' respective spatial theories are nearly identical but diverge quite drastically on these two issues, i.e., Oresme's definitive ubeity (holenmerism), FGS(top), versus Leibniz' repletive ubeity, FGS(prg); and Oresme's embrace of kinematic shifts versus Leibniz' rejection of kinematic shifts.[18]

---

[16] Oresme argues: "Thus, outside the heavens, then, is an empty incorporeal space quite different from any other plenum or corporeal space.... Now this space of which we are talking is infinite and indivisible, and is the immensity of God and God Himself" (O 176).

[17] See, also, Grant (1981, 350, n.127; 349, n.123). Oresme claims: "Notwithstanding that He is everywhere, still is He absolutely indivisible and at the same time infinite... for the temporal duration of creatures is divisible in succession; their position, especially material bodies, is divisible in extension; and their power is divisible in any degree or intensity. But God's [duration] is eternity, indivisible and without succession....His position is immensity, indivisible and without extension" (O 721).

[18] Nevertheless, it is possible that other Scholastic or Early Modern natural philosophers may have accepted repletive ubeity, which (on our usage) equates with FGS(prg), and the shift scenarios. In short, even though God is not spatially situated, one might still claim that there is a residual or surrogate notion of "same position" that applies to things/entities in space, thus rendering the shift

9.5.3   Maudlin's Ontological Classification

At this point, it will be informative to examine aspects of the spacetime classification system put forward in Maudlin (1993), for there are limitations to his approach that validate our more complex two-part conception of spatial ontologies. On the whole, Maudlin's coupling of traditional substantivalism and relationism to different spacetime structures renders impressive results, and is directly motivated by Newton and Leibniz' theories of space, but his scheme nevertheless fails to track Leibniz' theory. One of his scenarios, dubbed "Newtonian relationism" (1993, 193), complements some of our earlier conclusions: provided Newtonian relationism, space retains the full Newtonian spacetime structure (absolute position, absolute velocity), but, since relationism is also in play, static shifts are ruled out, whereas kinematic shifts are possible (and, in all the cases examined in this section, a matter-filled plenum is presumed). That is, since matter instantiates Newtonian spacetime structure, static shifts (as different embeddings in a Newtonian spacetime) are not meaningful because they "would be interpreted as different representations of the same physical state" (193), a conclusion that exactly matches the nominalist rationale advanced above (§9.5.2). Yet, once matter is instantiated, the spacetime "rigging" that fixes absolute position, and hence velocity, entails that kinematic shifts are meaningful possibilities, albeit empirically inaccessible. Therefore, Maudlin's Newtonian relationism diverges from Leibniz' own theory, since Leibniz denies the possibility of kinematic shifts. Likewise, Maudlin's other relationist concept, "neo-Newtonian relationism" (1993, 193), is too weak to determine the individual states of motion that Leibniz insists are determinate in nature given a typical Galilean relationship between two bodies. That is, neo-Newtonian spacetime, which upholds the principle of Galilean relativity, cannot determine which body is really at rest or in motion given only a "change in distance" among the two bodies—but, Leibniz rejects this view and opts instead for individual states of motion via his notion of force (e.g., L.V.53; see Chap. 4). Consequently, Maudlin's scheme offers two possible interpretations of Leibniz' theory, Newtonian relationism and neo-Newtonian relationism, although both contravene important features of Leibniz' system: the first allows kinematic shifts, which runs counter to Leibniz' view, whereas the second rules out individual states of motion, also contra Leibniz.

The problem, from a spacetime perspective, can be put as follows: given Leibniz' concept of "universal place" (see Chap. 3), Newtonian spacetime structure holds between Leibniz' bodies (i.e., *within* the material world), since universal place records the positions of all bodies and their change of position; but, via his rejection of the shift scenarios (in the Leibniz-Clarke correspondence), it is also simultaneously true that neo-Newtonian spacetime structure applies to the *whole* material world! This contradictory outcome thus provides a further demonstration of the

---

scenarios meaningful. Descartes, who accepts FGS(prg), also holds that rest and motion are distinct states (Pr II 37), and that God can discern these states (CSMK 381), but he never (to the best of our knowledge) entertains the possibility of a static or kinematic shift (probably because it violates his definition of motion; see Chap. 1).

limitations of the spacetime approach when applied to Leibniz' theory, a point first examined in Chap. 3. Our investigation, in contrast, provides two separate grounds for rejecting the shift scenarios, and hence for avoiding the quandary just described for the spacetime approach: first, as noted above, Leibniz' nominalism precludes static shifts and his unique force-based conception of motion rules out kinematic shifts; second, the inherent difference in the foundational and secondary material levels of reality limits the domain of full Newtonian spacetime structure (i.e., universal place) to only those relationships *among* bodies at the secondary macrolevel—why?: because questions concerning the state of inertial motion of the entire material world can only be determined relative to a separate, non-comoving reference frame or entity, in this case, God/monads, that is/are *situated* in space. Hence, unlike Newton and Gassendi's God-grounded ontologies that do situate God in space (which thus allows the world's uniform motion to be measured relative to that entity), Leibniz' non-spatial, and thus non-situated, God obviates such universal states of uniform motion (i.e., since God is not in space, there is no fixed entity that can serve as a basis to determine the world's uniform motion). Both of these points, which are interdependent to some degree, stem from a nominalist reading of Leibniz' theory and our two-tiered ontological levels scheme, and have little connection with modern spacetime relationism. Overall, while Maudlin (1993) provides one of the most intricate treatments of the traditional substantival/relational dichotomy utilizing modern spacetime structures, it both fails to accommodate the complexities of many historically important theories, in particular, Leibniz, and also seems too wedded to modern macrolevel theories of physics (Newtonian mechanics and gravitation theory, special and general relativity), thereby forsaking both the deep metaphysical issues inherent in these historical theories as well as their latent similarities with modern QG approaches.

Chapter 10
A New Taxonomy Beyond Substantivalism and Relationism II: Some Philosophical Prehistory of Quantum Gravity

The analysis in the preceding chapter has set the stage for a closer examination of various quantum gravity (QG) hypotheses, background independence, pregeometry, spacetime emergence, and, ultimately, the deficiencies in the substantivalist/relationist dichotomy as it applies to the seventeenth century and QG. After a discussion of how the difference in historical context affects the new taxonomy, §10.1, analogues between seventeenth century spatial ontologies and QG hypotheses will be developed, §10.2, followed by a closer examination of nominalism and virtual-platonism in a QG setting, §10.3, the locality of beables, §10.4, and a concluding summary and evaluation of the new taxonomy and the competing third-way strategies, §10.5.

10.1 Introduction: Geometric Levels and Quantum Gravity

Returning to the division of spatial geometric levels, FGS, introduced in Chap. 9, there is a fascinating, and apparently natural, analogue of this distinction within the diverse array of QG hypotheses (albeit some QG hypotheses will pose various classificational difficulties due to their complex and hybrid construction). One might question the relevance of this exercise, of course, given that the seventeenth century's preoccupation with the theological underpinnings of space would seem to have little in common with the modern search for QG. Yet, as mentioned previously, the situation confronting both the seventeenth century and QG theorists is exactly the same: both are concerned with constructing an adequate theory of space, time, and the physical world based on a pre-given foundational entity or theory, specifically, the western God, on the one hand, and general relativity (GR), quantum mechanics (QM), or quantum field theory (QFT), on the other. Both "research programs", as it were, strive to retain the essential features of that underlying theory, but both recognize the need to adapt, revise, and sometimes overturn, various elements of the established system in the process of securing their respective goals; namely, a

© Springer International Publishing Switzerland 2016
E. Slowik, *The Deep Metaphysics of Space*, European Studies in Philosophy of Science, DOI 10.1007/978-3-319-44868-8_10

spatiotemporal theory grounded upon God, for the seventeenth century, and a spatiotemporal theory that successfully integrates GR and QM/QFT, for contemporary physics. Moreover, as first discussed in Chap. 1, the issue concerning the relationship between immaterial entities and space in the seventeenth century evolved into a new, but closely related issue, specifically, the relationship between Leibniz' force-based quasi-physical entities, monads, and space. Inspired by Leibniz' conception, a host of similarly structured hypotheses were advanced in the eighteenth century by the most prominent natural philosophers of the day, such as Wolff and the pre-critical Kant, and thus the emergence of space from a deeper force-based realm became a well-established theme in natural philosophy more generally. By this route, Leibniz' wide-ranging conception of force (dubbed "proto-energetics" in Bernstein 1984, 101) also helped to lay the groundwork for the development of the energy concept and the laws of thermodynamics in the nineteenth century. Indeed, Leibniz' view that force is the basis of matter and spatial extension can be seen, in retrospect, as the forerunner of the philosophical outlook that motivates the type of modern field theories that includes QG hypotheses; and, to demonstrate this point, a leading QG proponent, Lee Smolin, has even gone so far as to draw ideas for potential QG hypotheses directly from Leibniz' monadic hypothesis (see, Barbour and Smolin 1992). Consequently, there is a significant historical connection between Leibniz' analysis of the ubeity ("ways of being somewhere") of immaterial beings and the search for QG.

First, a few words are in order concerning pregeometry and spacetime emergence. As first discussed in Chap. 8, QG theories often posit a fundamental level of ontology that does not possess the 4-dimensional spacetime structure common to twentieth century field theories, most notably, GR. Pregeometry, in the context of QG, does not necessarily mean non-geometrical, yet these pregeometric structures are not the continuous, differential topological and metrical structures assumed in the standard constructions of GR or QFT (as will be elaborated further; see, also, Cao 1997, 111; Hedrich 2009, 14, n.31). A purely algebraic approach is also included in the pregeometric category, and, as mentioned previously, a pregeometric theory will also be referred to as a non-spacetime theory (since spacetime theories assume continuous metrical and topological structure). Furthermore, whereas virtual-platonism includes incorporeal nominalism in the context of the historical theories examined in Chap. 9, virtual-platonism as used in the modern setting of GR, QM, QG and other theories only pertains to the existence of the spatial geometry at a given level in the absence of the entities, energy/matter/fields, at that level.

Emergence is a notoriously difficult concept, but our analysis will use this term to include both of the strategies explored in Butterfield and Isham (2001) for going beyond the standard ingredients of QM (via QFT) and GR, i.e., a four-dimensional manifold and a classical, Lorentzian metric: (i) quantization, which is the quantizing of a classical structure "and then to recover it as some sort of classical limit of the ensuing quantum theory"; and (ii) emergence, where the classical structure is seen as "an approximation, valid only in regimes where quantum gravity effects can be neglected, to some other [more fundamental] theory" (2001, 35). Consequently, given that the relationship between the foundational and secondary levels in QG concerns *theories*, as opposed to the seventeenth century's *entities*, the form of third-way spatial property theory that is relevant to a QG setting is P(TL-dep), and not P(O-dep). That is, since an approximation and/or classical limit between the foundational level theory and the secondary level theory is the intended meaning of "emergence" in the modern QG case, it is more appropriate to view the secondary level theory as a property of the foundational level theory; and, as argued in §1.2, this approach can be applied to the Early Modern theories surveyed in our investigation as well. Accordingly, whereas our earlier interpretation of Leibniz' ontology in Chap. 4 sided with a supervenience interpretation over an emergence interpretation, a P(TL-dep) reading of the property theory in Leibniz' case would regard emergence and supervenience as synonymous in the general sense that they signify the manner by which foundational level theory (entities and processes at that level) bring about secondary level theory (entities and processes at that level).

10.2 Applying The New Taxonomy: Quantum Gravity Hypotheses and Their Seventeenth Century Analogues

10.2.1 QG Metric Structure and Newton

Turning to these analogies between seventeenth century and QG hypotheses, Newton and More's FGS(met) would correspond to the earliest geometrodynamic hypotheses (as a canonical quantization approach), as well the older covariant quantization techniques, since these approaches rely upon the general geometric structure employed by the foundational theory, which is, respectively, GR and QFT (which is the conjunction of QM and the flat Minkowskian spacetime of special relativity). In the (naive) covariant quantization strategy that flourished up through the early 1970s, the metric of the foundational theory, QFT, is split into two parts: the fixed background metric (usually Minkowskian), which "defines spacetime, namely it defines location and causal relations" (Rovelli 2004, 12), and a dynamical component that relies on perturbation techniques to secure the postulated graviton (and hence extend QFT to gravity). For these early covariant quantization hypotheses, matter and higher level physical fields, such as GR's metric and stress-energy field, would comprise the secondary level entities constructed from the foundational theory, QFT. In the old geometrodynamics (which is not technically a QG hypothesis, but similar), GR is the foundational theory, with the metric and the curvature of spacetime taken as the basic groundwork from which all other physical phenomena are presumed to be derived or constructed (i.e., as the secondary level entities).[1] The geometric outlook that motivates geometrodynamics also prompts Sklar's

---

[1] Here, we are assuming that a time parameter can be linked with the spatial 3-metric so as to approximate standard spacetime (an assumption that is by no means trivial). More accurately, however, FGS(met) in the case of geometrodynamics pertains to its 3-metric, so there is a sense in which 4-dimensional spacetime is emergent as well. Nevertheless, if the secondary level is interpreted as also possessing a 3+1 structure (3 space and 1 time), then FGS(met) is maintained.

well-known concept of supersubstantivalism: "not only does spacetime have reality and real structural features, but in addition, the material objects of the world, its totality of ordinary and extraordinary material things, are seen as particular structured *pieces* of spacetime itself" (Sklar 1974, 221). Newton's "determined quantities of extension" hypothesis, reviewed in §9.3.1, befits the supersubstantivalist definition quite nicely, that is, if one substitutes the term "spacetime" with the term "God's spatial (spatiotemporal) extension". If viewed within the context of the deep metaphysics of space, however, the various attempts to tie Descartes to geometrodynamics fail (e.g., Graves 1971, 87), since Descartes grounds space (= matter) on a non-spatial, non-extended (or non-geometric) conception of God, as we have seen.

In addition, the first phase of string theory, roughly up through the mid-1990s, would likely conform to the FGS(met) category as well. Despite invoking a number of compactified extra dimensions at the microlevel,[2] the perturbative method employed by these theories presupposes a classical spacetime backdrop and its metric:

> [T]he propagation of the [one-dimensional] string is viewed as a map $X : \to M$ from a two-dimensional worldsheet W to spacetime M (the 'target spacetime'). The quantization procedure quantizes X, but not the metric γ on M, which remains classical....[T]he classical spacetime metric γ on M satisfies a set of field equations that are equivalent to (the supergravity version of) Einstein's field equations for general relativity plus small correction of Planck size: this is the sense in which general relativity emerges from string theory as a low-energy limit. (Butterfield and Isham 2001, 71)

While the metric at the foundational microlevel of strings and the secondary macrolevel of GR is, approximately, the same in these string theories, hence FGS(met), the already significant topological differences at these two levels have evolved into a potentially more radical set of dissimilarities in the subsequent, second phase of string theory's history. As Butterfield and Isham note, concerning the possibility that there might exist a minimum spacetime length in these later approaches (via the duality symmetries), "these developments suggest rather strongly that the manifold conception of spacetime is not applicable at the Planck length; but is only an emergent notion, approximately valid at much larger length-scales" (73; and, Witten 1989, 350–351). Consequently, given the duality symmetries and the allied

---

[2] In more detail, Butterfield and Isham explain that "[i]n perturbative superstring theories, the target spacetime M is modeled using standard differential geometry, and there seems to be no room for any deviation from the classical view of spacetime. However, in so far as the dimension of M is greater than four, some type of 'Kaluza-Klein' scenario is required in which the extra dimensions are sufficiently curled up to produce no perceivable effect in normal physics [e.g., GR], whose arena is four-dimensional spacetime" (Butterfield and Isham 2001, 72). See, also, Rickles (2008b, 311–323). Given this topological difference between string theory, which is the foundational theory, and the secondary level theory, GR, it thus follows that early string theory does not precisely fit the definition of the FGS(met) category offered above (which requires identical metrical and topological structures between these two theories or levels). But, early string theory is quite close to FGS(met) due to its compactification of these extra dimensions. Overall, given our classificational system, the QG theories that precisely match Newton and More's FGS(met) would be the first two strategies mentioned above, i.e., the early, naive covariant quantization theories and early geometrodynamics (but not quantum geometrodynamics; see footnote 4 as well).

emphasis on non-perturbative formulations, the trajectory of this second phase in string theory development seems inclined to FGS(prg), where the foundational level of ontology exhibits entirely different geometric structures than at the secondary level: i.e., contra FGS(met), a discrete, non-continuous geometry at the foundational microlevel, but the standard differential geometry of the secondary macrolevel emergent theory (GR); and, contra FGS(top), a topology at the foundational level that likewise diverges from the topological structure at the secondary level.

10.2.2   QG Topological Structure and Gassendi

The well-known rival of string theory is LQG (loop quantum gravity), which, unlike string theory, does not rely upon a classical metric but does rely upon a classical topological manifold. Rather, LQG quantizes the metric of GR, incorporating a discrete quantum substructure at the foundational level but regaining the classical metric as an emergent phenomena at the secondary level. As a later variant of the canonical quantization program, a theory like LQG "uses a background dimensional manifold (but it uses no metric)", where this (spatial) manifold "becomes part of the fixed background in the quantum theory—so that...there is no immediate possibility in discussing quantum changes in the spatial topology" (Butterfield and Isham 2001, 76). Therefore, LQG upholds FGS(top). So, regarding the question that prompted our investigation at the start of Chap. 9, i.e., "Which seventeenth century philosophy of space best resembles the structure of LQG?", we are finally in a position to provide an answer: it is neither Smolin's choice of Leibniz, who accepts FGS(prg), nor Dainton's preference for Newton, who endorses FGS(met)—rather, it is Gassendi, since he endorses FGS(top)! As discussed in section §9.3.4, by declaring that "there is no absolute or spatial nearness or distance between monads" (LDB 255), it follows that any ascription of metrical properties to Leibniz' monadic ontology is ruled out, but so would most topological properties. A topological space involves a neighborhoods of points and their various non-metrical interrelationships, such as continuity and connectedness. Yet, since monads have "no position with respect to one another", and each monad is "a certain world of its own, having no connections of dependency except with God" (AG 199), even these weaker topological notions are apparently excluded. That is, if monads were situated in the points of Euclidean secondary macrolevel space, then classical topological structure would be applicable to Leibniz' ontology of monads, via FGS(top), but Leibniz consistently rejects this possibility. In addition to the LDB 255 quote provided above, there are many others: "If you examine the issue more carefully, perhaps you will try to say that a definite point at least can be assigned to a soul [i.e., monad]. But a point is not a definite part of matter, and an infinity of points gathered into one would not make extension....I do not think it appropriate to regard souls as though in points" (LDB 123–127).[3] In short, the structure of a continuous manifold at the foundational level, FGS(top), is incompatible with Leibniz' spatial ontology at that level, but not with respect to Gassendi's, for the latter situates God in the points of a Euclidean space and its corresponding continuous topology (see §9.3.2).

10.2.3   QG Pregeometric Structure and Leibniz

There is, however, a group of modern QG strategies that would seem analogous to the pregeometry of monads, namely, start with a mere set of points, M, without topological or differential structure, and build the continuous topological and metrical secondary macrolevel structures upon this non-spatiotemporal foundation. On Butterfield and Isham's estimation, "this set is formless, its only general geometrical property being its cardinal number", and is such that "there are no relations between the elements of M, and no special way of labeling any such elements [i.e., no topology]" (2001, 81). Butterfield and Isham's analysis is part of a larger discussion of alternative non-spacetime QG strategies, different from string theory and LQG, where the quantization is imposed "below the metric".[4] Quantum effects and

---

[3] As regards Leibniz' frequent analogies between points/monads and a line/matter (as in LDB 255), a few observations are in order. In short, points/monads are required for a line/entity, but they are not ingredients of lines/entities: "[O]ne must not infer that the indivisible substance enters into the composition of body as a part, but rather as an essential, internal requisite, just as one grants that a point is a not a part that makes up a line, but rather something of a different sort which is, nevertheless, necessarily required for the line to be, and to be understood" (AG 103). In addition, as discussed in Chap. 4, Leibniz often admits that monads have a derived situation in matter (i.e., while monads are not actually in space, we can attribute a derivate sense of situation to them via the bodies that result from the monads): "For although monads are not extended, they nevertheless have a certain ordered relation of coexistence with others, namely, through the machine which they control" (Lm 531). To recap, while one may be tempted to associate the structure of Leibniz' monadology with something akin to a set of points, with or without a topology, any such idea is misguided *if* it situates the monads in space (i.e., as the points of space). Matter is extended and in space, but matter is the "result" of non-spatial monads, and is thus akin to something like a supervenient or emergent property—but the spatial characteristics of the supervenient/emergent material bodies are not applicable to the ontology of monads (and ultimately, God).

[4] In Cao's history, the origins of pregeometry is correlated with Wheeler's later development of quantum geometrodynamics, in particular, the difficulties associated with reconciling quantum fluctuations in geometrodynamics with the concept of a multiply connected space: "According to quantum geometrodynamics, there are quantum fluctuations at small distances in geometry, which lead to the concept of multiple connected space….However, 'quantum fluctuations' as the underlying element of his geometrical picture of the universe paradoxically also undermined this picture. Quantum fluctuations entail change in connectivity. This is incompatible with the ideas of differential geometry, which presupposes the concept of a point neighborhood. With the failure of differential geometry, the geometrical picture of the universe also fails: it cannot provide anything more than a crude approximation to what goes on at the smallest distances. If geometry is not the ultimate foundation of physics, then there must exist an entity—Wheeler calls it 'pregeometry'—that is more primordial than either [differential] geometry or particles, and on the foundation of which both are built" (Cao 1997, 111).

structures can then be introduced at this lower level and associated, depending on the particular QG scheme, with a host of possible sub-metric structures, e.g., M, causal, algebraic, topological, differential, etc., in an ascending hierarchy (with different hierarchies erected based on the particular QG strategy). These sub-metric QG structures thus lie underneath, and can be said to generate or bring about, GR and QFT's common geometrical presuppositions as employed in their standard mathematical formalisms, i.e., a Lorentzian metric on a four-dimensional topological, differentiable manifold. Among these different strategies, the most Leibnizian would lie in the utilization of a discrete quantum substructure, similar to LQG's quantization strategy, but absent LQG's need for a differentiable manifold. Since the main goal of Leibniz' *analysis situs* is to provide an algebraic model of spatial situation (Lm 248–249), a theory of space that does not employ geometric structures (metric and manifold) as the foundational starting point is thus in keeping with his overall worldview, an approach that takes algebra/arithmetic as primary and geometry as derived.[5]

There are many non-spacetime QG theories that fit this general category: e.g., causal sets, computational universe, non-commutative geometries, etc. For example:

> Causal set theory arises by combining discreteness and causality to create a substance that can be the basis of a theory of quantum gravity. Spacetime is thereby replaced by a vast assembly of discrete "elements" organized by means of "relations" between them into a "partially ordered set" or "poset" for short. None of the continuum attributes of spacetime, neither metric, topology nor differentiable structure, are retained, but emerge it is hoped as approximate concepts at macroscales. (Dowker 2005, 446)

In what follows, however, we will concentrate on the quantum causal histories program (QCH). Hedrich (2009, 22) provides the colorful description, "geometrogenesis", for the process by which "spacetime emerges from a pregeometric quantum substrate" in QCH, with the "quantum substrate" correlated to the set M and a causal structure in Butterfield and Isham's account.

> [QCH's] basic assumptions are: There is no continuous spacetime on the substrate level. The fundamental level does not even contain any spacetime degrees of freedom at all. — Causal order is more fundamental than properties of spacetime, like metric or topology. — Causal relations are to be found on the substrate level in form of elementary causal network structures....[M]acroscopic spacetime is necessarily dynamical, because it results from a background-independent pregeometric dynamics. But, the dynamics of the effective degrees of freedom on the macro-level are necessarily decoupled from the dynamics of the substrate degrees of freedom. If they would not be decoupled, there would not be any spacetime or gravity on the macro-level, because there is none on the substrate level. In the same

---

[5] "[P]hysics makes use of principles from two mathematical sciences to which it is subordinated, geometry and dynamics....Moreover, geometry itself, or the science of extension, is, in turn, subordinated to arithmetic, since, as I said above, there is repetition or multitude in extension; and dynamics is subordinated to metaphysics, which treats cause and effect" (AG 251–252). Although a purely algebraic version of a QG theory would seem to fit the character of Leibniz' *analysis situs* best, we will utilize a pregeometric theory with a discrete geometric substructure as the basis for comparison with Leibniz' account of the rise of matter and space from monads. Furthermore, Leibniz' *analysis situs* incorporates both distance and orientation, and so his algebraic scheme allows one to naturally recover or derive these structures (see, Lm 667; and De Risi 2007 for a comprehensive examination).

way, causality on the macro-level, finding its expression in the macro-level interactions, is decoupled from causality on the substrate-level. And spacetime-locality on the macro-level, if it emerges from the dynamics of coherent excitation states, has nothing to do with locality on the substrate graph structure level. (Hedrich 2009, 22–23)

Given this strategy, FGS(prg) is the proper classification for QCH and the other non-spacetime QG hypotheses.

The analogue of these QG hypotheses will be readily evident to the Leibnizian devotee (see Chap. 4, in particular). First, monads and their intrinsic primitive forces correspond to the discrete elementary quantum events, which in the QCH program are excitation states in a finite-dimensional Hilbert space (as the discrete nodes in a graph structure).[6] For Leibniz, matter and space emerge from a hidden realm of constitutive entities that, like QM and QG theories, is more aptly described in terms of force: a monad is "endowed with primitive power" so that the "derivative forces [of bodies] are only modifications and resultants of the primitive forces" (AG 176). Derivative force, as a value of the primitive force, is also tied to material extension, which is described as "diffusion": "the derivative force of being acted upon later shows itself to different degrees in secondary [i.e., extended] matter" (AG 120); "the nature which is supposed to be diffused, repeated, continued, is that which constitutes the physical body; it cannot be found in anything but the principle of acting and being acted upon" (AG 179). Second, the derivative nature of the spatial and dynamical properties of bodies, as opposed to the intrinsic primitive forces of the non-spatial monads which bring about bodies, thus correlates with the term "decoupling"; i.e., the emergence of secondary macrolevel spatial and dynamical properties that are quite different from, and *seemingly* independent of, the foundational microlevel pre-spatial dynamical properties that generate those macrolevel properties ("spacetime-locality on the macro-level, if it emerges from the dynamics of coherent excitation states, has nothing to do with locality on the substrate graph structure level"). As discussed previously, Leibniz' monads (like his conception of God) are not in space *per se*, but they are the means by which God "brings about" matter and, hence, instantiates his nominalist account of space: "[c]ertainly monads cannot be properly in absolute place, since they are not really ingredients but merely requisites of matter" (Lm 607); and, "properly speaking, matter is not composed of constitutive unities [monads], but results from them" (AG 179). Much has been written on whether or not monads are really in space (see, e.g., Cover and Hartz 1994), a debate inflamed by these kinds of fairly opaque quotations; yet, QG theories provide a concrete example of the type of world view that Leibniz was likely striving to establish (see, Garber 2009, 383–384, who briefly mentions a particle physics-inspired interpretation of Leibniz as well).

---

[6] "The basic structure [at the foundational microlevel] is a discrete, directed, locally finite, acyclic graph. To every vertex (i.e. elementary event) of the graph, a finite-dimensional Hilbert space (and a matrix algebra of operators working on this Hilbert space) is assigned. So, every vertex is a quantum system. Every (directed) line of the graph stands for a causal relation: a connection between two elementary events; formally it corresponds to a quantum channel, describing the quantum evolution from one Hilbert space to another. So, the graph structure becomes a network of flows of quantum information between elementary quantum events. Quantum Causal Histories are informational processing quantum systems; they are quantum computers" (Hedrich 2009, 22).

A third similarity between QCH and Leibniz relates to one of the major themes of our investigation, namely, nominalism:

> But what are these coherent, propagating excitation states, resulting from the substrate dynamics and leading to spacetime and gravity? And how do they give rise to spacetime and gravity? —The answer given by the *Quantum Causal Histories* approach consists in a coupling of geometrogenesis to the genesis of matter. The idea is that the coherent excitation states resulting from and at the same time dynamically decoupled from the substrate dynamics are matter degrees of freedom. And they give rise to spacetime, because they behave as if they were living in a spacetime. (Hedrich 2009, 23).

By its "coupling of geometrogenesis to the genesis of matter", i.e., that the emergence of space at the secondary level is coupled to the emergence of matter at that level, QCH can truly claim a lineage with Leibniz' brand of nominalism, as opposed to LQG, the latter permitting possible states that are absent matter at the secondary level but which retain topological structure at that level. That is, a vacuum state occurs in LQG when the foundational s-knots, which constitute the discrete structure of space (by means of equivalence classes of spin networks formed by spatial diffeomorphisms), lack the requisite quantum excitations needed for the existence of matter (see Rickles 2005, 426–427). In short, using the new taxonomy, LQG's version of space at the secondary level upholds FGS(top) even in the complete absence of matter. For this reason, LQG is closer to Gassendi's theory, where a matter-less topological space is possible at the secondary level as well, and unlike the strategy employed by Leibniz and QCH, where the emergence/actualization of space (spacetime) at the secondary level is linked to the emergence of matter at that level. However, as will be explained in §10.3, LQG actually supports nominalism at the secondary level since the metric field counts as a secondary level entity in addition to matter, whereas we have argued that Gassendi supports virtual-platonism at the secondary level since a matter-less void is possible (with matter being his secondary level entity). Of the two theories, LQG and QCH, there are other reasons for preferring QCH as more comparable to Leibniz' monadic system. The result of LQG's quantization of the metric of GR is an array of spin networks with finite area and volume, which essentially constitutes "quantum chunks" of space (Rovelli 2001, 110). The QM-rooted spin networks are therefore inconsistent with the non-spatial, non-geometric character of monads, and the same is true of the spatial diffeomorphisms required to form the s-knots from the spin networks (since diffeomorphisms are geometric transformations on a differential, hence continuous, manifold, contra Leibniz' FGS(prg)). Moreover, given the direct quantization of GR's metric, gravity is rendered a fundamental interaction for LQG; but gravity is emergent for both Leibniz and QCH, since it is tied to the existence of matter at the secondary macrolevel. Rather, the spin networks in LQG, which are contiguous discrete chunks of a quantum field, are much closer to Leibniz' conception of contiguous discrete chucks of matter, as opposed to the non-contiguous discrete objects that comprise his pregeometric monadic metaphysics. This last point is nicely captured by Rovelli as it pertains to LQG:

> This discreteness of the geometry, implied by the conjunction of GR and QM, is very different from the naive idea that the world is made by discrete bits of something. It is like the discreteness of the quanta of the excitations of a harmonic oscillator. A generic state of spacetime will be a continuous quantum superposition of states whose geometry has discrete features, not a collection of elementary discrete objects. (Rovelli 2001, 110)

Once again, this description more accurately tracks Leibniz' material plenum, as opposed to the deep metaphysical entities, monads, that underlie his material realm.

Nevertheless, there is one significant issue on which Leibniz' monadic system diverges from QG theories, namely, causal or dynamical structure at the foundational level, like QCH's quantum channels (the lines of the graphs that connect the vertices). Because "monads have no windows through which something can enter or leave" (AG 214) there is as an absence of any inter-monadic causal mechanism at that level (which we will dub the "no windows" stipulation). What is needed is a mechanism that could connect or interrelate the monads so as to generate matter and space, a process that would roughly correlate with the dynamics of the pregeometric substrate in QCH.[7]

On the other hand, even granting the legitimacy of the "no windows" criticism (as regards the analogy between QCH and Leibniz' monadic system), a counter-reply might reside in the fact that the type of connection that binds the elementary quantum events in QCH is quantum information (see footnotes 6 and 8). The nature of quantum information within QM is difficult to assess, but it would seem to be a type of physical property that, for lack of a better description, is situated near the material/immaterial divide (see, e.g., Clifton et al. (2003), and Bub (2010), for an overview of the information approach to QM). Consequently, since the lines that connect the nodes of the graph structure in QCH constitute quantum information, and, since quantum information evokes "immaterialist" connotations, it could be interpreted by the Leibnizian as an acceptable surrogate for an inter-monadic connection at the monadic level. Whether or not this strategy is plausible is open to question, of course, but there are precedents for utilizing a graph structure to model Leibniz monads (see §10.2.3).[8]

---

[7] One might be tempted to interpret Leibniz' "substantial chain" idea, which is tentatively proposed in the late correspondence with Des Bosses, as a remedy for the lack of an inter-monadic connection, where a substantial chains is defined as "something substantial which is the subject of [the monads'] common predicates and modifications, that is, the subject of the predicates and modifications joining them together" (AG 203). As first discussed in Chap. 4, the substantial chain seems tantamount to a continuous space, since "[r]eal continuity can arise only from a substantial chain" (AG 203). Nevertheless, like matter, "monads aren't really ingredients of this thing [substantial chain] which is added [to the monads], but requisites for it" (AG 198). That is, the substantial chain is a property that results from the monads, as is also the case with matter (see, e.g., Lm 607). Therefore, the substantial chain arises at the secondary macrolevel, and so it does not constitute the sought-after foundational level connection. See Chap. 4 for further analysis of the substantial chain as regards the issues of supervenience and emergence.

[8] Nevertheless, Julian Barbour (1994, 2003), who was one of the first to associate a graph structure with Leibniz' monadology, seems to interpret the lines between the nodes (monads) as purely perceptual or mind-based, for he concludes that these graph strategies constitute "mathematical

10.3 Quantum Gravity and Nominalism

This section explores the nominalist component of the new taxonomy as applied to QG, with background independence and further analogies with Leibniz included in the discussion.

10.3.1 Background Independence, Nominalism, and the Void

From the modern QG perspective, Smolin has argued that the substantival/relational dichotomy converts to a dispute over background dependent (fixed geometry for all models) or background dependent methods (more than one geometry possible for these models), thus it follows that the early string theories that employed a fixed background metric and manifold are more substantivalist (absolutist) than the alternative QG options that only rely on a point manifold, such as LQG (Smolin 2006, 199; see, also, §9.5). However, the utilization of the manifold's topological, dimensional, and differential structure can still be deemed to violate a fully background independent scheme, and it is for these reasons that Smolin declares LQG to be only partially relational (2006, 215). Nevertheless, whether it is metric or manifold structure that counts as a more substantivalist orientation is itself a contentious issue; see, Earman (1989, chaps. 8 and 9), for a defense of manifold structure as the better candidate. If one accepts Earman's argument, then LQG and covariant quantization theories are *both* equally substantivalist, contra Smolin, since both employ a point manifold. A further obstacle for Smolin's attempt to link the substantivalist/relationist dichotomy to background dependence/independence is that one of the most substantivalist leaning theories, geometrodynamics, which Sklar associates with supersubstantivalism, is background independent, and thereby should count as relationist given Smolin's conjecture (see §10.2.1).

---

realizations of idealism", and adds that he has since abandoned this method for exploring QG theories (2003, 56). Interestingly, Barbour developed his ideas on a graph structure representation of Leibniz' monadic system in conjunction with Smolin (Barbour and Smolin 1992), who hoped to use the results of their investigation to develop a potential QG strategy that relies on a hidden-variables approach (see, Barbour 2003, 57). In §10.3.2 below, we will briefly investigate Smolin's hidden-variables concept, although he does not specifically associate that strategy with Leibniz' monadic system. It is possible that Barbour's idealist interpretation of the graph model may have dissuaded Smolin from invoking any Leibnizian analogy as regards his hidden-variables idea, and, consequently, this may also explain Smolin's preference, in his (2006) article, for LQG as the better comparison with Leibniz' philosophy. Nevertheless, the unique status of quantum information might overcome these reservations, since quantum information, interpreted as physical information associated with the complete quantum state vector, is not a mental or cognitive attribute of quantum systems (i.e., quantum information would exist even if no minds existed, so Barbour's reticence is all the more curious). In fact, as argued above, since quantum information involves the interconnectedness of entangled quantum states, there is a ready-made analogy with the interconnectedness of monads.

Leaving aside the problem of which geometric component structure should be identified with substantivalism, manifold or metric, a somewhat different way of explaining the inadequacy of using the background dependence/independence divide as a substitute for the substantival/relational dichotomy connects with our earlier analysis of virtual-platonism and nominalism. In short, whether the background geometry is, or is not, fixed is not a crucial factor in the platonist/nominalist distinction; rather, it is the presence of geometric structure at a given ontological level in the absence of all physical entities or processes at that level that is crucial, since that possibility would refute nominalism at that level. For these reasons, the argument that LQG is in conflict with relationism, because it allows vacuum solutions (see, Rickles 2005, 425; Earman 2006, 21), gains no traction against our new system of classifying spatiotemporal ontology. LQG does not violate nominalism because the secondary level theoretical entities—i.e., the entities postulated by GR, in particular, GR's metric field (which are all emergent phenomena or properties of the foundational theory's s-knots)—instantiate the spatial structures required at the secondary level. Additionally, as first discussed in Chap. 5, not only is there no void or vacuum state of the metric field (i.e., a region of the manifold lacking a metrical value), but since the metric is also the gravitational field, which carries energy, the metric/gravitational field can thus legitimately claim to be a nominalist-friendly physical thing (see, once again, Earman and Norton 1987, 519, who detail some of the physical consequences of gravity waves). Accordingly, given our new taxonomy, there is no conceptual room to invoke a further distinction between sophisticated substantivalist and sophisticated relationist interpretations of GR's metric field—at the secondary level, both of these rival interpretations fall under nominalism (as opposed to virtual-platonism), and hence are identical as judged by our new ontological scheme (see also, Chap. 5, and Slowik 2005a, for a less detailed version of this line of argument).

One might dispute this last point by drawing attention to the difficulties that surround the very notion of emergence in QG theories (or in any theory, for the matter). The "emergence of GR" from LQG, as explained in §10.1 above, is really about approximations or limiting processes, whereby the higher level theory, GR, is recovered from the deeper QM-based processes at work in LQG, and the same likely holds true of most other QG theories that embrace spacetime emergence as well. There are, to put it differently, no emergent *entities* at the secondary level, such as the claim that GR's metric field is an emergent entity, but only the emergence of a higher level *theory* and its *theoretical* entities from the more fundamental quantum processes via these approximation strategies, with GR's metric field comprising a theoretical component of that recovered higher level theory. So, in contrast to the definitions provided at the outset of our new taxonomy, nominalism versus virtual-platonism at the secondary level in QG is really a question about nominalism and virtual-platonism at the foundational level. On the seventeenth century ontologies that we have examined, conversely, the secondary level is inhabited by actual secondary level entities, namely, matter, which are not mere approximations or limiting cases of foundational entities and processes.

Yet, even if we concede the above argument, and envision emergence in QG as more theoretical or phenomenal than ontological, it does not alter the conclusions reached utilizing our new classificational scheme, since this difference has been factored into the switch to the P(TL-dep) version of the property theory of space in the modern physics setting, as opposed to the more straightforwardly ontological P(O-dep) utilized for the pre-twentieth century spatial theories examined in our investigation. Returning to the issue of LQG's vacuum state will help to demonstrates this point. As we have seen, Leibniz' nominalist spatial geometry is only instantiated by matter (in conjunction with the truths of space grounded in God), and so a matter-less world is absent spatial geometry. LQG, on the other hand, can admit scenarios which retain the s-knot spatial structure, and hence the spatial manifold, in the absence of matter (see §10.2.2, where it is argued that Gassendi's theory more closely resembles LQG over Leibniz' theory for these reasons). Nevertheless, if we leave aside the issue of GR's metric field as either an emergent secondary level ontological or theoretical entity, respectively, P(O-dep) or P(TL-dep), the allegation that LQG permits a vacuum state is somewhat misleading due to the fact that the underlying quantum processes at the foundational microlevel would still remain even in the absence of matter. Although traditional matter-based conceptions of spatial relationism and nominalism are indeed undermined by these vacuum solutions, the finite value of the vacuum energy and its effects in QFT (virtual particles, Casimir effect, Higgs field) upholds a field-based form of nominalism at the foundational microlevel, since there are no voids totally absent of energy at that level in LQG. Likewise for the energy of the metric field in GR (as noted above), whether conceived as an emergent secondary level QG entity or phenomenal feature, or as the foundational entity in the standard interpretation of GR.[9] That is, for both the vacuum energy in QFT and the energy of the metric field in GR, the "fullness of space", so to speak, is not a contingent fact, but a built-in feature of these *field* theories.[10]

---

[9] In this context, a "matter-based" platonist/nominalist distinction places the distinction on the existence of matter, i.e., whether or not spatial geometry requires matter (nominalism) or does not (platonism). "Field-based" platonism/nominalism places the same distinction on the existence of physical fields, which need not involve matter.

[10] In the context of a universe entirely filled with energy/matter, Friedman (1983, 222–223) argues that the mere possibility of a true, energy-less vacuum is enough to separate absolutism (substantivalism) from relationism, and he offers the example of the cosmic background radiation as an example. Yet, leaving aside the question of the alleged contingency of the background radiation left over from the Big Bang, the vacuum energy of QFT and the energy of the metric field in GR are entirely different cases, with an energy-filled space (spacetime) a non-contingent feature of these theories (QFT and GR). In addition, QFT's vacuum energy may contribute to the cosmological constant, Λ, the latter constituting another alleged physical field that permeates all of space, and which may play a role in the observed expansion of the universe. Thus, as a physical field with potentially measurable effects, Λ is in keeping with both a property theory of space and a sophisticated field version of relationism, and not substantivalism (contra Baker 2005). The recent discovery of the Higgs boson, and hence Higgs field, also supports the contention that there is no true, energy-less vacuum in space.

More carefully, recalling the distinction first introduced in §9.2.2, if the platonism/nominalism question is pushed to the foundational level of ontology, FL(v-plt) and FL(nom), then all of our examined theories, whether from the seventeenth century or contemporary physics, align with nominalism, FL(nom). As revealed in our investigation, God is the foundational entity required for the existence of space in the seventeenth century, thereby securing nominalism at the foundational level via that unique (immaterial) entity. In the same way, modern QG theories are not committed to a virtual-platonist background structure given the complete nonexistence of the relevant QG entities and processes at the foundational level. So, just as nominalism does not discriminate between sophisticated substantivalist and sophisticated relationist interpretations at the secondary level of GR's emergent metric (whether taken ontologically or phenomenally, as above), the same holds true at the foundational level in QG (or if the metric is taken as the foundational entity in standard GR). For these reasons, our new system also provides an insight as to why Earman and Rickles' arguments are only effective against a relationist (or nominalist) interpretation of LQG *if* confined to matter at the secondary level.

Nominalism at the foundational level, FL(nom), furthermore, is well-documented, both for seventeenth century philosophers of space and modern QG theories. At this foundational level, seventeenth century philosophers only required a sort of congruence of the domain of God's substance or operation and the extent of space, since they all reject the inherence conception that treats space as an internal accident or attribute of God (with the exception of the later More and possibly other Cambridge Neoplatonists, such as Raphson; see EM 56–57). For all of the seventeenth century thinkers surveyed in Chap. 9, space is not "external" to God. To be exact, space cannot exceed either the bounds of God's own extension (Newton and More) or God's non-extended immensity, where that non-extended immensity takes the form of either definitive ubeity (Gassendi and most Scholastics) or repletive ubeity (Descartes and Leibniz). As previously disclosed, Newton denies "that a dwarf-god should fill only a tiny part of infinite space" (TeL 123), and Gassendi claims that "since it follows from the perfection of the divine essence that it be eternal and immense, all time and space are therefore connoted" (RIV 94). For Leibniz, "[t]he immensity and eternity of God are things more transcendent than the duration and extension of creatures", yet, "[t]hose divine attributes do not imply the supposition of things extrinsic to God, such as are actual places and times" (L.V.106). This dependence of space on God is further revealed in his rejection of the idea that space is God's place, since that would imply that "there would be a thing [space] coeternal with God and independent of him" (L.V.79). Hence, while God is not situated in space for Leibniz, God's powers or operations are situated (by way of repletive ubeity) and thus congruent with the world's actual or possible spatial extension; if not, space would be independent of God. Finally, since monads generate the matter that instantiates space (and are a sort of interface between Leibniz' God and the material world), it naturally follows that space is not independent of the monads and is congruent with monadic affects.

In a similar fashion, there is a sort of congruence of the physical quantum states and their Hilbert spaces, or the field in QFT and its Minkowski spacetime, in that the

QM-based QG theories do not sanction void spaces entirely devoid of energy, where an absolute void would imply that the geometry at this foundational level exceeds the bounds of, or is not congruent with, the physical entities/fields and their associated states at that level. The same holds true for standard GR, since there are, once again, no metrical voids that would undermine a nominalist interpretation of that theory. One could even go so far as to claim a certain analogy between God's grounding the possibilities of bodies at the macrolevel in Leibniz' spatial ontology (as above, NE II.xv.2), and, for a physical system in QM, the state vectors grounding the probability of the physical observables in a Hilbert space. In many of the pregeometric QG hypotheses, in fact, it is often claimed that space emerges from "internal" QM processes, a description that upholds the nominalist ban on entirely void spaces (virtual-platonism) at the foundational level: e.g., in the model of Kaplunovsky and Weinstein (1985), "the distinction between 'geometric' and 'internal' degrees of freedom can be seen as a low-energy artifact that has only phenomenological relevance. Space is finally nothing more than a fanning out of a quantum mechanical state spectrum" (Hedrich 2009, 16).

Turning to string theory, the story becomes a bit more complex. Given the intuitive picture derived from the early versions of string theory, where strings vibrate against a fixed background geometry, it might seem possible that there could exist (energy-less) void spaces among the strings, a conclusion that would favor FL(v-plt) since there is no longer a coincidence, or mutual overlap, of strings and background geometry. This state-of-affairs would recall the type of ontology favored by (at least some of) the ancient atomists, such as Epicurus, where atoms and void are separate and, presumably, independent of one another. Likewise, FL(v-plt) is likely the concept that undergirds the definition of absolutism (and later substantivalism) from the mid-eighteenth century onward, after the decline of the God-based conception of space and the rise of view that space is an entirely independent entity.[11] However, since string theory posits many curled-up dimensions at each point in space, with a corresponding potential energy associated with these extra dimensions at each point, an entirely energy-less empty space is thus not sanctioned by the theory, and hence FL(v-plt) would not apply. Furthermore, the second phase in the development of string theory may sanction an emergent spacetime scenario; i.e., just like the non-spatiotemporal pregeometric QG theories described in §10.2.3, if spacetime emerges from the complex QM-based interconnections that are posited at the pregeometric level in these later string theories, then the intuitive picture of a fixed background geometry populated by strings and interspersed with energy-less void gaps is, needless to say, no longer applicable. More generally, as long as any string theory does not sanction energy-less void spaces, even the fixed background

---

[11] Unlike Newton, the spatial ontology seemingly implicit in at least some of the ancient atomists, such as Epicurus, accepts FL(v-plt) and FGS(met), since the spatial geometry is, presumably, fully metrical and exists independently of the atoms. This type of spatial ontology, which may epitomize the purest form of substantivalism in the minds of many contemporaries, is, ironically, quite rare in the history of spatial ontologizing, and only became part of the standard dichotomy in the latter half of the eighteenth century. See, Sorabji (1988, Part II), Algra (1995), on these ancient atomist theories.

structure of the older forms of string theory can be seen as siding with nominalism over virtual-platonism. Furthermore, given the rise of the braneworld scenarios in M-theory (a further development of string theory in its second period), another potential nominalist strategy is to link the existence of three-dimensional space to three-dimensional branes (where strings are one-branes), and thereby regain a congruence of the physical (branes) and the spatial. For example: "If a three-brane is enormous, perhaps infinitely big,…[a] three-brane of this sort would fill the space we occupy.…Such ubiquity suggests that rather than think of the three-brane as an object that happens to be situated within our three spatial dimensions, we should envision it as the very substrate of space itself" (Greene 2011, 113–114).

10.3.2 Pre-established Harmony and QG

To round out our investigation of background structure, it should be noted that Smolin's quest for a completely background independent QG theory provides a unique Leibnizian twist to this principle, for he employs a hidden variables conception of QM as a key component in his scheme. In response to the query, "Can there be a fully background-independent approach to quantum theory?", he states, "I believe that the answer is only if we are willing to go beyond quantum theory, to a hidden variables theory" (Smolin 2006, 232). In more detail, he argues:

> We know from the experimental disproof of the Bell inequalities that any viable hidden variables theory must be non-local. This suggests the possibility that the hidden variables are relational. That is, rather than giving a more detailed description of the state of an electron, relative to a background, the hidden variables may give a description of relations between that electron and the others in the universe. (232)

While not directly mentioning Leibniz' monadic system, Smolin's hidden variables approach to a fully background independent QG theory not only evokes the holistic, pre-established harmony of Leibniz' monadic metaphysics, but, in fact, was inspired by it (Barbour and Smolin 1992, see also footnote 8). Although "the monad's natural changes come from an *internal principle,* since no external cause can influence it internally" (AG 214), their pre-established harmony mimics the holistic interconnections of a hidden variables theory: "This interconnection or accommodation of all created things to each other, and each to all the others, brings it about that each simple substance [monad] has relations that express all the others, and consequently, that each simple substance is a perpetual, living mirror of the universe" (AG 220). Smolin characterizes his hidden variables strategy as "relational"; but the relational aspect of these entities, whether a monad or a hidden variables electron, is not *spatial* relationism, but the non-spatial interrelatedness of intrinsic metaphysical (monad) or physical (electron) *properties* — and this demonstrates, once again, the inability of the traditional substantival/relational distinction to probe the conceptual depths of the foundational realm of ontology, whether in the seventeenth century or in the context of modern QG strategies.

10.4 Alternative Interpretations: Ubeity and Local Beables

There is a further analogy that can be drawn between seventeenth century and QG theories that involves the "local beables" issue first discussed in Chap. 8. As noted, various commentators have argued against those QG theories that posit an underlying non-spatiotemporal ontology on the grounds that the "locality" of these fundamental entities either remains a mystery or may be incoherent in some manner. Leaving aside the evident flaw in these arguments (namely, that they beg the question), the issues that pertain to the locality of QG entities would seem to naturally invoke a set of historical parallels with the seventeenth century's version of the same dilemma, namely, ubeity, which Leibniz describes as "ways of being somewhere", and includes the circumscriptive, definitive, and repletive categories surveyed in §9.2. Up to this point, our investigation of ubeity in its seventeenth century setting has concentrated on the grounding relationship between the fundamental entity, God (or, for Leibniz, God and monads), and the secondary level of reality, matter and Euclidean space, that emerge from the foundational level. These concepts from seventeenth century ontology were then utilized as the basis of the FGS classification developed above, with analogies drawn to geometric structures that are also identical at the foundational and secondary levels in modern QG theories. Nevertheless, Leibniz' inquiry is additionally concerned with the ubeity of lesser non-fundamental entities, such as angels and souls, finite immaterial entities that, like each individual monad, are not congruent with the whole of space. However, as first discussed in Chap. 9, Leibniz' analysis of circumscriptive and definitive ubeity focuses on the hypotheses put forward by other natural philosophers, and hence the finite immaterial entities at issue in this portion of the *New Essays* does not include his own monadic conception, i.e., it does not involve entities that are the ontological foundation of matter and extension. It is with respect to these non-monadic finite immaterial entities that further comparisons can be made to various hypotheses concerning the locality of QG's non-spatiotemporal beables (theoretical entities).

To recap Leibniz' discussion in NE II.xxiii.21, circumscriptive ubeity maps an entity to space directly, point by point, so that "measuring them depends on being able to specify points in the located thing corresponding to points in space". In definitive ubeity, "the located thing lies within a given space without being able to specify exact points or places which it occupies exclusively", i.e., it is not "possible to specify an exact point such that the soul or something pertaining to it is there and at no other point". Lastly, as regards repletive ubeity, Leibniz explains that God "operates immediately on all created things, continually producing them, whereas finite minds cannot immediately influence or operate upon them". Immediate operation is also linked to the entity's absence of spatiotemporal situation: "God is not present to things by situation but by essence; his presence is manifested by his immediate operation" (L.III.12). Yet, as argued in §9.3.4, there are strong parallels between Leibniz' concept of God and his concept of monads—both are posited as non-spatiotemporal foundations of the spatiotemporal material world—hence one night possibly infer lessons on the ubeity problem as it pertains to monads from the analysis in the *New Essays* quotation, NE II.xxiii.21.

Intriguingly, in an attempt to address the local beables issue for those QG hypotheses that embrace spacetime emergence, Huggett and Wüthrich put forward a detailed analysis of locality that mirror Leibniz' three forms of ubeity, as well as match the results of our earlier FGS taxonomy as regards the specific QG theories that fit in each of the FGS categories. Starting with circumscriptive ubeity, Huggett and Wüthrich's comment that, on a simple reading of string theory, "it looks exactly as if strings are local beables, bits of stuff describing worldsheets in a classical spacetime", but they go on to add that the duality structure of the later versions of string theory undermines that conclusion (2013, 280). (In §10.2.1, the second phase of string theory's history was likewise seen as favoring the pregeometric, FGS(prg), as opposed to the metric, FGS(met).) Nonetheless, Huggett and Wüthrich's discussion would seem to concede that the earliest string theories, as well as any other QG strategy with a classical background space (spacetime), such as the naive covariant quantization techniques and geometrodynamics, would not violate the demand for local beables, and hence these theories would meet Leibniz' circumscriptive designation and our FGS(met).

Turning to definitive ubeity, there is close parallel between the problems associated with the locality of LQG's beables, the discrete QM-based spin networks from which spacetime emerges, and Leibniz' analysis of the locality (situation) of finite immaterial beings. A critical obstacle in the development of LQG has centered on its inability to preserve some notion of adjacency or neighborhood among spin networks at the emergent level of spacetime. As described by Huggett and Wüthrich, "[t]he problem is that any natural notion of locality in LQG—one explicated in terms of the adjacency relationship encoded in the fundamental structure—is at odds with locality in the emerging spacetime. In general, two fundamentally adjacent nodes [i.e., of two spin networks] will not map to the same neighborhood of the emerging spacetime" (279). Since locality and adjacency in spacetime are topological notions, LQG's local beables quandary correlates with our FGS(top) category— but, focusing strictly on the locality of finite immaterial beings as mentioned in the passage from Leibniz' *New Essays*, the inability to localize these immaterial beings in space is analogous to the inability to localize LQG's spin networks, e.g., "the located thing [angel, soul] lies within a given space without being able to specify exact points or places which it occupies exclusively" (NE II.xxiii.21). In both cases, Leibniz' finite immaterial beings and LQG's spin networks, the difficulty is topological in nature and specifically concerns the adjacency (next to) relationship: under Leibniz' definitive ubeity concept, there is no determinate adjacency (next to) relationship among the "parts" of a soul or angel in space; whereas in the case of LQG, adjacent spin networks fail to remain adjacent in the emergent spacetime.

The third form of ubeity is repletive, where only an entity's actions or effects can be situated in space, but not the entity itself. Repletive ubeity, consequently, is the seventeenth century equivalent of the absence of local beables; and, under our interpretation (see, once again, §9.2.1), repletive ubeity has been linked to pregeometry, our FGS(prg) category. Given all of the parallels disclosed thus far between ubeity and local beables, it probably should not be surprising that Huggett and Wüthrich's proposal for a surrogate notion of locality for those QG theories with the most thoroughgoing non-spatiotemporal (pregeometric) ontology, such as causal set theory, bears an uncanny similarity with Leibniz' repletive ubeity: "take localization in causal terms, and argue that it is causal nexus [among the non-spatiotemporal basal elements], rather than spatiotemporally understood locality, which supplies the condition relevant for empirical coherence [of the theory]" (2013, 278). Just as Leibniz' God and monads are not situated in space, but God's actions and the monads' "results" (i.e., matter) are situated, so it would seem that the pregeometric elements of causal set theory are not spatially located, but one can obtain a proxy notion of locality via their causal structure. It is important to point out that the term "causal" as used in this context pertains to a structural relationship among the basal elements, and not to the more familiar philosophical notion of causality that is a subject within metaphysics. In particular, causal structure does not include metric structure ("[t]here simply is nothing on the fundamental level corresponding to lengths and durations"), and it cannot "identify 'space', in the sense of a spacelike hypersurface", or, "[i]n other words, nothing but a difference analogous to that between spacelike and timelike remains at the fundamental level" (278). At a minimum, maintaining a difference of this sort would guarantee that, at the foundational level, two basal elements are distinct and do not overlap, and this restricted form of structural relationship finds a parallel in the structural relationship among God's conservation of the world's material occupants and the structural relationship among monads. That is, since God's effort to maintain each material object would naturally impose a discrete order among these actions (i.e., for each object), and the monadic forces that bring about matter (and thus space) emanate from *individual* monads, the only type of quasi-spatial relationship that can be attributed at the foundational level to God's conservation actions and the monads themselves would likely amount to a non-overlapping or discreteness criterion analogous to the quasi-spacelike separation requirement mentioned by Huggett and Wüthrich with respect to causal set theory's structure.

In summary, Huggett and Wüthrich's analysis of the local beables issue not only correlates with the ubeity system introduced in §9.2, but their proposals as regards particular QG theories, e.g., string theory, LQG, and causal sets, also fall under the specific categories of ubeity, and with respect to the same geometrical structures, as referenced in Leibniz' approach to ubeity: circumscriptive/metric with (early) string theory or naive covariant quantization; definitive/topological with LQG; and repletive with pregeometric QG hypotheses, such as causal set theory. For the same reasons, Huggett and Wüthrich's suggestions concerning the locality of beables in, respectively, string theory, LQG, and causal set theory thus fit the exact FGS classificational categories (metric, topological, pregeometric) that were assigned to those same theories in previous sections of this chapter. There is much more work that can be done on these issues, of course, but the results of our investigation offer a notable example of a rich, and largely unexplored, set of ontological themes that should inspire further investigations by historians and philosophers.

10.5   Overview and Conclusion

In this final section of the chapter, the advantages of the new taxonomy in comparison with the standard dichotomy will be elaborated, alongside an examination of potential relationist criticisms of this new method, and, finally, a summary and assessment of the different third-way theories that have comprised our investigation.

First, however, it would be useful to summarize the earlier defense (see §10.1) of the analogy between Leibniz' seventeenth century ubeity concept and QG theories developed in Part III. It is understandable that philosophers may have misgiving about employing Leibnizian ubeity in an analogy with contemporary non-spatiotemporal hypotheses, since the former seems pseudo-scientific and far removed from contemporary physics. Nevertheless, the best way to counter the impression that these seventeenth century non-spatiotemporal hypotheses are akin to magic or astrology is to historically connect them with current, widely-accepted scientific (as opposed pseudo-scientific) hypotheses that are similar in structure and purpose as regards physical emergence, i.e., leaving aside the mind/soul aspect of these Early Modern theories while focusing exclusively on the material and spatiotemporal components of their emergence concepts. As revealed in this chapter, there are close similarities between the Early Modern and QG hypotheses on specific spatiotemporal properties or features that obtain at the foundational and secondary levels of reality, thus our detailed investigation is justified on these grounds alone. Likewise, while the ubeity concept first developed in §9.2.1 may seem, from our current perspective, to be outside the realm of natural philosophy, it was not judged so by Leibniz and Newton, a realization that demonstrates the historically-situated, contingent nature of any attempt to label an hypothesis as pseudo-scientific. Furthermore, many of Leibniz ideas, in particular, that force is essential to matter, have been embraced by modern physics, and so perhaps he was right about the significance of his ubeity distinction as well.

10.5.1   Assessing the New Taxonomy

To recap, the main goal of Chaps. 9 and 10 has been to expose the limited capacity of substantivalism and relationism to assess spatial ontologies by offering an alternative, and more successful, classificational system. The evidence for the weakness of the substantivalist/relationist dichotomy resides in the ambiguity and uncertainty that characterizes any application of the distinction, whether in the seventeenth century or in the modern context of QG and GR. In its place, a different set of distinctions has been advanced that concern (i) the different levels of spatial geometry at the foundational and secondary levels, and (ii) the platonist/nominalist divide in spatial geometry at these levels—and these new dichotomies, which more accurately track the content of seventeenth century and modern QG theories both conceptually and historically, do not naturally align with the substantival/relational distinction, as we have seen. Consider substantivalism: with respect to (i), some alleged substantivalists embrace a similarity of geometric structure at the material secondary and God/monad foundational level (Newton, later More), but some do not (Gassendi, early More); as regards (ii), some alleged substantivalists favor virtual-platonism (later More, Newton, Gassendi), but some do not (Barrow). And, while both of the alleged relationists in our investigation (Descartes, Leibniz) are in the same camp concerning (i) and (ii), i.e., both posit a complete difference in geometric structure at the foundational and secondary levels, as well as accept geometric nominalism at the secondary level, modern substantivalism and relationism cannot adequately explain these similarities since they do not take account issues (i) and (ii). For instance, while issue (i) is clearly not a factor in the modern dichotomy, if the possibility of a vacuum were invoked as a surrogate for nominalism, our issue (ii), and hence as a means of separating substantivalists from relationists, then Leibniz would now count as a substantivalist since he admits the possibility of a vacuum. Therefore, despite the obvious similarities between Descartes and Leibniz' respective theories of space, the substantivalist/relationist dichotomy simply cannot pair them together in a natural way, and this demonstrates, once again, that it is far too crude and erratic an instrument to assess spatial ontologies. Moreover, since modern relationism rejects the idea that a fixed geometric structure accounts for the relative configuration of actually existing bodies, but Leibniz does invoke this type of explanation (see §9.4), modern relationism is simply inapplicable to Leibniz' spatial ontology.

In this chapter and the previous one, a more accurate set of dichotomies on the deep ontology of space has been offered that does accomplish a number of important goals that naturally align with the third-way emphasis of our investigation. First, it successfully groups together seventeenth century spatial ontologies that are indeed similar on specific issues, but it also accounts for their differences concerning other issues: specifically, Newton and the later More, but not Gassendi, with FGS(met); Gassendi and the early More with FGS(top); Newton, the later More, and Gassendi with SL(v-plt), Leibniz, Descartes, and the pre-critical Kant with FGS(prg); Leibniz, Descartes, Barrow, and the pre-critical Kant with SL(nom). In addition, FL(nom) is apparently upheld by all of the theories surveyed in our examination, whether in the Early Modern period or as regards contemporary QG hypotheses, as well as QM and GR (but see footnote 11 on ancient atomism). Second, our two-part dichotomy, (i) and (ii), not only successfully partitions the various QG approaches into natural categories, but, more importantly, it also provides a basis for drawing successful analogies with seventeenth century theories, e.g., Leibniz with the pregeometry of QCH, Gassendi with the continuous topological structure required for LQG, and Newton with the fixed background metric in early string theory. Moreover, it offers a consistent explanation for why Leibniz' nominalism does not support the shift scenarios, whereas Oresme's nominalism does, since the former holds FGS(prg), while the latter accepts FGS(top). Ironically, our system also successfully accomplishes some of the goals that have eluded previous assessments that rely upon the substantival/relational dichotomy to draw historical analogies: it links Newton, but not Descartes, with geometrodynamics, and Leibniz and Descartes with a pregeometric subvenient entity which lacks any continuous degrees of freedom. The new taxonomy advanced in this essay has a further advantage in that it does not utilize nor sanction the apparently arbitrary and unconstructive ontological distinction between the sophisticated substantivalist and sophisticated relationist interpretations of LQG and the metric field in GR (whether as an emergent entity or phenomenal feature of a QG theory, or in standard GR). Finally, as should be readily apparent by this point, the new taxonomy is also in harmony with the third-way theories of spatial ontology explored in Part II, especially the property theory of space, of either the P(O-dep) or P(TL-dep) sort.

Besides presenting a host of more fine-grained distinctions than the standard dichotomy allows, another appealing feature of our new system is that it does not retrospectively single out the historical winners and losers: in conjunction, both of these consequences of our new taxonomy are quite useful provided the greater complexity and ongoing nature of the search for a viable QG theory. For instance, our new taxonomy, by correlating seventeenth century spatial theories with the modern QG hypotheses, leaves open the possibility that either Newton, Gassendi, or Leibniz, may have foreshadowed the correct theory regarding the geometric structure—metric, topological, or pregeometric—which the foundational level of QG shares with the secondary level. That is, concerning issue (i), perhaps the non-spatiotemporal pregeometric hypotheses will never pan out, nor the latest non-perturbative string theories, nor LQG, but an older naive covariant quantization hypothesis or a variant of the early string theories will provide the link between QM and GR, thereby vindicating Newton's metric foundationalism, FGS(met). Or, perhaps not, and the same holds for the modern theories that espouse FGS(top) and FGS(prg). Turning to issue (ii), while nominalism at the foundational level, FL(nom), has a decided advantage over the virtual-platonist option, FL(v-plt), in contemporary physics, it is always possible that some form of energy void at the foundational level may yet gain ascendency, and thereby vindicate a form of platonism that even Newton's own spatial ontology rejected; e.g., as discussed in §10.3.1, perhaps string theory is actually compatible with a scenario that admits an energy-less void space. On the other hand, virtual-platonism as the secondary level, SL(v-plt), seems a more plausible outcome of the search for QG than FL(v-plt), and SL(v-plt) is a feature of Newton's spatial ontology. Given issues (i) and (ii) and their many different combinations of structures, there are, accordingly, a host of potential outcomes as regards the specific aspects of spatial ontology that any single QG hypothesis may satisfy. For example, a single QG hypothesis may favor a Leibnizian approach regarding one aspect of the spatial ontology, say, SL(nom), while favoring a Newtonian conception on another aspect, such as FGS(met)—and this stands in sharp contrast to the traditional dichotomy's prevailing method of debate, where a simplistic, all or nothing, vindication of the entire spatial scheme of either Newton or Leibniz is the usual goal of contemporary substantivalist and relationist interpretations of modern theories in physics. This more simplistic and undifferentiated approach may explain the historical and conceptual confusion that has long attended the substantival and relational classificational scheme, especially when applied in hindsight from the perspective of our currently accepted physical theories, such as GR. As the earlier chapters have disclosed, not only do the alleged chief representatives of that dichotomy, Newton and Leibniz, fail to meet many of the key tenets of, respectively, substantivalism and relationism, but, as just recapped, the contemporary sophisticated versions of those two ontological positions are practically indistinguishable in the context of our best current physical theories. Given these confusions and uncertainties, clear-cut historical comparisons and the evaluation of conceptual content have been rendered nearly impossible, and so commentators are often at a loss to provide a coherent case for singling out, say, Newton or Leibniz as GR's genuine progenitor.

In addition, it should be noted that the new classificational scheme developed in this chapter does not claim to be final word on the ontology and structure of spatial (spatiotemporal) theories. There are, at least potentially, other taxonomies that might be devised that can supersede the substantival/relational dichotomy, or even adjustments to the system advanced above that may prove beneficial in various ways. For example, if FGS(met) is confined to metrical structure alone, and does not include shared topological structure between the levels, then string theory would likely fit the FGS(met) category more successfully (see footnote 2). Likewise, a purely spatial version of FGS(met), as opposed to a spacetime construal, would better harmonize with geometrodynamics (see footnote 1); and a totally algebraic formulation of a foundational theory would seem to be the best match for Leibniz' *analysis situs* (see footnote 5). On the other hand, separating the metric and topology in FGS(met) would diverge from Newton and More's spatial ontologies, and, since it is doubtful that any mathematical structure can capture Leibniz' real ontology, it appears more reasonable to associate Leibniz' theory with both the purely algebraic and discrete spatial QG substructures. Besides, the main point of the "pregeometry" label is to disclose the difference between the foundational level's structure and the secondary level's metric and topological structure, and both the algebraic and discrete spatial structures accomplish that task. In short, there are many possibilities, and associated pros and cons, for any classificational system that improves upon the standard substantival/relational division. All of these potential taxonomies, however, like the one offered in this chapter, would aim to reveal more information and insights, both historically and conceptually, about the ontology of space than the standard dichotomy can provide. For instance, one could divide the FGS(top) category utilized in our investigation into two more specific sub-categories: a continuous structure with non-dimensional points, as Gassendi accepts; or the conception that posits a minimal length to points, as supposedly Charleton and Patrizi favor. This further distinction would augment the accuracy of the taxonomy developed above (although, for the sake of simplicity, it has not been employed in Part III).

As a concluding synopsis, while it can be argued that the substantival/relational distinction remains somewhat serviceable in the context of macrolevel Newtonian mechanics, it has become practically dysfunctional in the debates on the status of the metric field in GR and in the assessment of QG hypotheses. The deep ontology of space, which is a paramount concern for seventeenth century thinkers and QG theorists alike (but not necessarily macrolevel Newtonian mechanists), may now hopefully prompt a much needed recalibration of the tools used for ontological appraisal by philosophers of space and time.

10.5.2   The Sophisticated Relationist Counter-Offensive

One of the principal objections that might be raised against the new taxonomy would almost certainly focus on the close similarity between sophisticated relationism and the property theory, of either the P(O-dep) or P(TL-dep) type. If a sophisticated relationist can embrace a host of seemingly non-relational, absolute structures, such as inertial structure, in the same way that a property theorist can, then what difference remains between these two spatial ontologies? As first acknowledged in Chap. 5, there may indeed be little or no difference between sophisticated relationism and the property theory of space—but, so much the worse for sophisticated relationism! The property theory, whether conceived along the lines of ontological dependence, P(O-dep), or the dependence of a higher level theory upon a foundational level theory, P(TL-dep), or even P(loc), sanctions spatial structures that contravene the more austere eliminative brands of relationism, whether inertial structure or a primitive location predicate. This feature of the property theory is, by the way, their primary motivation and chief advantage, namely, that space is not an independent entity but that it can nonetheless include absolutist-leaning spatial structures that transcend eliminative relationism. Indeed, the differences between a sophisticated relationism that accepts inertial structure, and an eliminative relationism that does not, would appear so vast as to call into question a common designation for these theories—hence, a more adequate response to the overlap of sophisticated relationism and the property theory would be to identify the former as an instance of the latter. Put simply, modality is a property of bodies, and not a relation among bodies, a point revealed most clearly in Teller (1991), a sophisticated modal relationism that suggests that a *single* body may instantiate inertial structure (see Chap. 5). Therefore, the property theorist's counter-reply is that sophisticated relationism is usurping the role that the property theory had been designed to fill in the larger spatial ontology debate. Put differently, since the property theory has been largely ignored in the debates on spatial ontology, unlike relationism, the latter view has unwittingly appropriated the place and function of the former view within the larger landscape of potential spatial ontologies. As revealed in Chap. 1, the common conception of relationism in the latter half of the seventeenth century (after Descartes) normally associated that doctrine with external relations among existing bodies (e.g., Newton and Clarke's criticism of the alleged inability of a relationism to account for metrical differences given an order of situations among bodies; see Chaps. 2 and 6). Consequently, the introduction of possible bodies, inertial structure, etc., into relationist doctrine would seem to signal the transition to a quite different hypothesis—an hypothesis that, in fact, fits the description of a property theory of space.

The same argument holds for those spatial ontologists who would venture to classify an emergent spacetime theory, whereby space arises from the entities and processes at a non-spatiotemporal microlevel, as relationist. Interestingly, in J. J. C. Smart's popular anthology, *Problems of Space and Time*, a work that coincided with the rise of the contemporary movement in the philosophy of space and time, one also finds the suggestion that a non-spatiotemporal fundamental ontology violates relationism, as well as absolutism (substantivalism), although the context is theoretical physics alone (and not any grand analogy between the seventeenth century's metaphysics of space and contemporary physics):

> It is also possible that science will develop in such a way that the simple notion of space as a system of relations between particles as well as that of space as absolute will have to be given up. It is just possible that we shall come to regard space and time as statistical properties on the macroscopic level only—just as, for example, temperature is a statistical property on the macrolevel, which has no meaning in micro-physics. In this case, the particles of microphysics will be related only by relations which are not spatio-temporal, and so these particles will bear a remarkable likeness to Kant's "things-in-themselves". (Smart 1964, 16–17)

Whether a successful QG theory would envision an emergent space as a statistical property of the underlying ontology is debatable, as is the reference to Kant's noumenal world, but the analysis is, without a doubt, prophetic. Smart argues that an emergent space does not qualify as absolute (substantival), presumably due to the fact that it depends on a material microlevel, but nor does it count as relational, since the relations among these microlevel particles "are not spatio-temporal". As demonstrated above, many prospective QG hypothesis retain geometric structure of some sort at the microlevel, although they often diverge radically from, or fall short of, the continuous metrical and topological spatial structures employed by classical macrolevel physics. Nevertheless, even in these cases, since the continuous geometric structures at the secondary macrolevel space *are not grounded on the relations between the entities at the secondary level*, it is difficult to categorize a theory of this type as relational, whether of the strict eliminativist or sophisticated modal variety, and for the same reasons offered with respect to Belot's appraisal of Leibniz' spatial hypotheses discussed in §9.4. Instead, the relations that are responsible for the continuous spatial structures at the secondary macrolevel are, as Smart correctly notes, among non-spatial entities at the lower foundational microlevel, with continuous macrolevel space emerging at the secondary level as a direct result of these non-spatiotemporal (non-continuous) lower level interconnections—and it is a property theory of space, of either the P(O-dep) or P(TL-dep) type, that best captures this spatial ontology, not the standard dichotomy. Moreover, as explored in §10.3.1, some QG theories admit the existence of secondary level space in the absence of matter at the secondary level, a scenario that seriously undercuts the plausibility of most relationist interpretations (except the field version of relationism, of course, but the property theory is a better classification than relationism in the case of fields, too; see §5.2).

Of course, contemporary philosophers of physics, for want of a better taxonomy, may still strive to invoke relationism to describe these kinds of QG theories (e.g.,

Norton 2008); but, once again, the uncertainty that marks the extension of the old dichotomy into these new conceptual domains only exposes its shortcomings: that is, substantivalists will also try to obtain ownership of these emergent space theories, contra the relationists, by contending that an emergent space is an entity, which thereby upholds substantivalism (e.g., Dainton 2010, 405–406, as well as Pooley's and Earman's substantivalist reading of Newton's references to space as God's emanative effect in §2.5). In essence, this is a more general manifestation of the conflicting substantivalist and relationist interpretations of LQG that we observed at the outset of Chap. 9. To sum up: QG theories, from the pregeometric to the more familiar LQG and string theory (especially in the latter's second phase of development), are neither relationist nor substantivalist, but much more akin to a property theory of space (spacetime). This divergence from the standard dichotomy is, in fact, implicitly confirmed by the recent introduction of the new description, "emergent spacetime" theories, that has sometimes been used to describe these types of QG proposals.

Finally, before leaving the topic of counter-arguments to the new taxonomy, it is worth pointing out that the committed relationist may accept the general line of argument advanced in this section, but nonetheless defend the relevance and merits of the traditional dichotomy. The relationist can simply declare that, first, while it is true that Leibniz' natural philosophy is not really relationist, at least in the modern sense of the conjunction of eliminative relationism and relational motion, there was at least one "true" relationist in the seventeenth century, namely, Huygens. Hence, citing Leibniz or Descartes (who frequently diverges from modern relationism as well) was just an historical oversight in the quest to establish the relationist ancestry. Second, since Huygens did embrace eliminative relationism and relative motion in a fairly straightforward and unrestricted manner (unlike Leibniz and Descartes), it follows that only those hypotheses that likewise accept eliminative relationism and relative motion should count as relationist—and, in fact, the modern conception that best captures those two doctrines is Leibnizian spacetime structure, which forsakes the absolutist-leaning inertial structure (see §1.1.1). Yet, restricting relationism to only those hypotheses that reject inertial structure would seriously reduce the field of prospective relationist candidates, possibly to just a handful of recent attempts (such as Barbour and Bertotti's efforts, and probably few before the twentieth century). On the other hand, one can plausibly maintain that the very concept and function of relationism has only recently been clarified, and thus it is not surprising that many purported relationists have fallen short of producing a truly relationist account, especially as regards pre-twentieth century hypotheses. This would explain, for instance, why even Huygens and Mach seem to rely at times upon a spacetime structure stronger than Leibnizian (i.e., neo-Newtonian) in modeling their physics, despite their reliance on a material framework to determine motions (which they apparently assumed was sufficient to overturn absolutism). In brief, both the strict eliminative relationist and the property theorist (or any other third-way approach) can thereby agree that their respective theories were not clearly formulated and distinguished from one another in the past, for instance, by demonstrating that many alleged relationists of the Early Modern period simply conflated issues (such as nominalism and eliminative relationism), when their real intent was to deny sub-stantivalism, which a property theory can achieve on its own without need of rela-tionism. Hence, the hardcore relationist can happily concede that Leibniz, and possibly many other past natural philosophers, were really property theorists about space, or nominalists, etc., and not "true" relationists after all, and the same holds for the mischaracterization of emergent spacetime theories as relationist. On the whole, this form of response is not only perfectly reasonable, but, more importantly, it is consistent with the historical evidence.

Furthermore, having taken this purist stance, the strict eliminative relationist can now side with the property theorist's case against the sophisticated brands of both relationism and substantivalism, declaring that the former is really a property theory in disguise (as argued above), and that the latter must revert to its traditional roots as well (i.e., manifold substantivalism, which permits the static and kinematic shifts implicit in Newton's conception of absolute space; see §1.1.1). By this means, the relationist can reclaim the relevance of the traditional dichotomy, but it does require jettisoning sophisticated relationism, such as metric field relationism and any modal relationist hypothesis that tacitly accepts inertial structure, and this may strike some as too high a price to pay.

10.5.3 Synopsis of General Themes

In the remainder of this chapter, a survey of the principle third-way spatial ontolo-gies, and a review of the three major parts of the work, will bring together many of the separate topics and arguments that have been featured in our investigation. In Part I, the most straightforwardly historical portion of the study, the many non-relationist and non-substantivalist components of both Newton and Leibniz's spatial ontologies were explored, and these findings served as the groundwork for the new taxonomy developed in Part III. While material from Part II was also incorporated into the new classificational system, especially as regards platonist and nominalist conceptions, the major share of that portion of work was dedicated to examining several contemporary third-way approaches to spatial ontology that forsake the standard dichotomy, substantivalism versus relationism; namely, the definitional conception, spacetime structural realism, and the property theory of space. What is the relationship among these different approaches, and what are their relative strengths and weaknesses?

With respect to the definitional (or dynamical) interpretation of space advanced by DiSalle and (arguably) Stein, examined in Chaps. 2 and 5, the chief obstacle is their reticence to discuss the ontological implications of their approach. While their rejection of the standard dichotomy is clear, the claim that space and time are needed for a system of mechanics, or that they comprise basic facts that pertain to existing things, does not in itself provide any answer to our ontological inquiry. As DiSalle states in the preface of his (2006) investigation, "[w]hat this book attempts to show is that the best philosophy of space and time—the part that has been decisive in the evolution of physics—has been a connected series of arguments that began with Newton, arguments about how physics must define its conceptions of space and time in empirical terms" (2006, xii). This empiricist or positivist project does fall within the purview of a third-way conception of space, it should be noted once again, since, on our interpretation, any conception of space that rejects the standard dichotomy for another account merits a third-way classification. A similar verdict might seem in store as regards the version of spacetime structural realism championed in this investigation, i.e., the liberal form of epistemic structural realism, or ESR-L (see Chap. 8). Despite their common epistemological focus, ESR-L is a species of scientific realism, hence an underlying ontology is responsible for the observed invariance of theoretical structure, such as spatiotemporal structure, across theory change, although the specific details of that ontology are currently unknown (but perhaps knowable in the limit of scientific theorizing). Nevertheless, there is an undoubtedly close resemblance between ESR and some of Stein's comments on scientific realism, and this may suggest that a hidden ontology also undergirds his version of the definitional method: "our science comes closest to comprehending 'the real', not in its account of 'substances' and their kinds, but in its account of the 'Forms' which phenomena 'imitate' (for 'Forms' read 'theoretical structures', for 'imitate', 'are represented by')" (Stein 1989, 57). On the positive side, ESR-L does not fall afoul of the pessimistic meta-induction, nor the many underdetermination problems, thus it presents many advantages over rival forms of scientific realism, such as ontic structural realism.

Since the definitional approach and epistemic structural realism stem from, respectively, the empiricist/positivist and scientific realist wings of the philosophy of science, it is probably not a surprise that their engagement with questions that pertain to the metaphysics of space is, to say the least, quite muted—and, in the case of ESR-L, that it treats the sophisticated spatial ontologies (sophisticated substantivalism, sophisticated relationism, and the property theory) as the same ontology (see, §5.2.2). For these reasons, the third-way hypothesis that constitutes the best *ontological* option is the property theory of space. As argued in Part I, Newton and Leibniz' theories of space are much closer to the property theory than the standard dichotomy, and, as revealed in Parts II and III, the theories in modern physics that are most closely associated with questions of spatial ontology, specifically, GR, QFT, and QG, can be successfully modeled as property theories, whether under the ontological dependence or theoretical levels reading, but without the various drawbacks associated with either a substantivalist or relationist interpretation (see, once more, Chap. 5). In fact, a property theory of space would seem to be the best choice among the competing ontological hypotheses to complement the decidedly empirical strategy embraced by the definitional approach and ESR-L, but it also suits the ontology implicit in OSR. For these reasons, the property theory offers more advantages for a prospective spatial ontologist than the other third-way contenders, since it appears to embody the common ontological presupposition among those competing strategies.

Chapter 11
Epilogue: The Post-Seventeenth Century Evolution of the Standard Dichotomy

By way of conclusion, this final chapter will briefly explore the development of the standard dichotomy, substantivalism (absolutism) versus relationism, in the period after Newton, specifically, the eighteenth and nineteenth centuries. Only a cursory synopsis of this rather intricate history can be offered, but many of the spatiotemporal concepts and strategies that would shape the future course of the standard dichotomy were in play during the late seventeenth century as well, and thus a full accounting of the spatiotemporal ontology of Leibniz and Newton's time merits an assessment of their content, function, and evolution. In §11.1, the decline of the seventeenth century's God-infused spatial metaphysics will be assessed, along with a survey of both the empiricist-centered replacement concepts and the transformation of the ontological dependence version of the property theory during the eighteenth century. Finally, Kant's unique relationist interpretation of Newtonian physics, along with its implications for the standard dichotomy, will round out the investigation in §11.2.

11.1 Spatial Ontology in the Eighteenth Century

In this section, an outline of the development of the standard dichotomy in the wake of the demise of the seventeenth century's God-based spatial ontologies will be attempted. After a summary of the relevant material from Chap. 1 and some preliminary suggestions, Berkeley's metaphysics and natural philosophy will be offered as an exemplar of the empiricist trends that were to rapidly change the tenor of spatial theorizing over the course of the eighteenth century. Lastly, the fate of the alternative property conception of space, P(O-dep), in the centuries after Newton and Leibniz will be discussed.

© Springer International Publishing Switzerland 2016

E. Slowik, *The Deep Metaphysics of Space*, European Studies in Philosophy of Science, DOI 10.1007/978-3-319-44868-8_11

11.1.1   Who Banished God from Space?

In Chap. 1, an overview of the development of the standard dichotomy in the decades prior to the introduction of the *Principia* revealed the central role that Huygens' work played in fusing the debate concerning absolute and relational conceptions of space and motion with the interpretation of various impact rules and conservation laws in mechanics. As we have seen, Descartes provided the blueprint for these collision rules and conservation laws, but his treatment of motion eludes a relational classification given its many absolutist elements and various Scholastic complexities. Huygens, in contrast, wholeheartedly embraced a strict eliminativist relational conception of motion and conjoined it with a relational interpretation of the Cartesian collision rules via the center-of-mass frame. Newton, like many other natural philosophers in the period, adopted the absolute/relational terminology but rejected the novel relationist construal of mechanical quantities favored by Huygens for an orthodox absolutist interpretation, just as Borelli, Pardies, and many others had in the 1660s and 1670s. As explained in Chap. 1, it is important to place the standard dichotomy in its proper historical context, for there remains a pervasive belief that this conceptual duality first appeared in Newton's *Principia*, an erroneous assumption that obscures the historical contingencies that shaped its development.

Yet, how did the standard dichotomy evolve in the period after Newton, i.e., in the eighteenth and nineteenth centuries, prior to the contributions of the neo-Kantians and positivists? While a topic of such vast proportions cannot be covered in depth given the constraints of this investigation, a brief synopsis can be provided nonetheless. In short, there were several rapidly growing trends in the philosophical milieu of the late seventeenth century that would ultimately conspire to overthrow the God-based approach that served as the foundation of Newton and Leibniz' respective spatial ontologies. One of these developments has already been discussed (with a more elaborate treatment in Chap. 1), i.e., the use of the standard dichotomy to assess the motions of bodies, true versus apparent, as required by collision rules and conservation laws, a purely mechanical strategy that, at best, remained neutral on the alleged theological foundation of space. The other pivotal element in the rise of a "de-theologized" standard dichotomy can be traced to the ascent of the empiricist tradition, with its attendant reliance on subjective perceptual and cognitive states to arbitrate ontological questions. It is the conjunction of the burgeoning empiricist movement with the newly developed mechanics-based appropriation of the absolute/relational dichotomy that would, in a fairly short span of time, topple the God-based spatial ontologies that had held sway for over a millennium. Grant concludes that "[n]ot until the eighteenth century did space achieve independence from the divine omnipresence so that it might serve only the needs of physical science" (1981, 263),[1] adding that "[b]y 1750, it would appear that the defenders of Newton and Clarke had been vanquished" (416, n.425).

Yet, in addition to the influence of Huygens' work on impact and the growing empiricism of the late seventeenth century, a further factor in the demise of the standard God-based spatial ontology of the period can be found in the particular type of God-based ontology favored by Leibniz, namely, his version of repletive ubeity, where only God's actions are situated in space. As Grant reflects in the concluding paragraph of his ground-breaking history:

> Before God could be removed from space, a general realization had to develop that the various mechanisms devised over the centuries to explain His omnipresence in infinite space were not only unsatisfactory but ultimately unintelligible....After centuries of debate and controversy, this stage was finally reached in the eighteenth century, helped to be sure, by Berkeley and Leibniz. In a curious sense, one might argue that John Dun Scotus and Gottfried Leibniz triumphed over Thomas Aquinas and Isaac Newton. It was better to conceive God as a being capable of operating wherever He wished by His will alone rather than by His literal and actual presence. (264)

In other words, by situating God's actions in space rather than God's being, the elimination of this foundational element from the spatial ontology is but a short step away (although one must also reject Leibniz' repeated claims that the truths of space are founded in God, etc.; see Chaps. 3 and 4). His attack on the concept of a spatially situated God, either of the circumscriptive or definitive variety, takes up a considerable part of the Leibniz-Clarke correspondence, of course, and it was through this widely read work that Leibniz' persuasive critique of the Newton/ More brand of Neoplatonist spatial ontology gained currency, even as his own account of space proved more controversial and was largely misunderstood (as argued in Part I).

But, returning to the passage quoted above, how does Berkeley fit into Grant's picture? While he does not provide a rationale, the juxtaposition with Leibniz would seem to implicate Berkeley's equally non-spatial conception of God (as will be discussed below), but there are many additional features of his natural philosophy that might count as well. In the ensuing discussion, these aspects of Berkeley's empiricist worldview will be shown to be a harbinger of the currents that would soon transform the debate on spatial ontology in the early eighteenth century into the modern form of the dichotomy that is still in operation today.

---

[1] Grant continues: "Until then, nonscholastics grappled with a scholastic problem and often employed the same terminology and concepts. For all of these reasons, the history of spatial doctrine must include the faceless scholastics as well as the major and minor figures who directly shaped the Scientific Revolution. The exclusion of scholastics from previous histories of space has limited our perspective and prevented genuine comprehension of the developments that eventually produced the fundamental frame of the Newtonian universe" (1981, 263). This investigation wholeheartedly endorses this assessment.

11.1.2   The Rise of the Empiricist Approach to Space: Berkeley as Exemplar

Unlike Locke, whose empiricist philosophy did not prevent his ultimate, but tentative, endorsement of an absolutist-leaning, God-based spatial ontology,[2] Berkeley enthusiastically embraced a relational conception of motion and space from his earliest major works. The *Principles of Human Knowledge* (1710) provides the details of his particular form of empiricist or phenomenalist metaphysics, a system that, as it were, derives spatiotemporal relationism from a thoroughgoing idealism (where, in Berkeley's case, idealism is synonymous with immaterialism, such that only immaterial beings, i.e., God, angels, souls, and their mental content, e.g., ideas and perceptions, exist). After rejecting the primary/secondary property distinction (as did Leibniz; see §4.1.3), Berkeley concludes that the source or cause ("archetype") of the idea of extension must come from another mind, since, just as color can only exist in minds, so the same must be true of extension: "[W]hen we attempt to abstract *extension* and *motion* from all other qualities, and consider them by themselves, we presently lose sight of them, and run into great extravagances" and thus "all sensible qualities are alike *sensations* and alike *real*; that where the extension is, there is the colour, too, to wit, in his mind, and that their archetypes can exist only in some other *mind*...none of all which can be supposed to exist unperceived" (WGB I 123). The archetype of extension and all other ideas, needless to say, is

---

[2] Locke, like Barrow, ultimately develops a type of imaginary space conception, an approach that traditionally exhibits both relationist and absolutist tendencies: relational in that imaginary space does not have dimensions and relations absent extended bodies, and absolutist in that imaginary space is a God-based (and not matter-based) ontological capacity to receive bodies (see, e.g., E II.xv.7-8 for both tendencies). Locke's spatial ontology rejects relationism, moreover, for he *contrasts* a relationist interpretation of space with his God-based alternatives (E II.xiii.27), and he clearly prefers the latter: in a later section he states that we cannot understand "the boundless invariable oceans of duration and expansion, which comprehend in them all finite beings, and in their full extent belong only to the Deity" (E II.xv.8). Likewise he admits a non-relational notion of the universe's place (E II.xiii.10), even though his epistemology of space is decidedly relational, much like Newton's (save for Newton's appeal to the non-inertial effects of rotation). Interestingly, Locke also appears to ultimately side with Newton's circumscriptive ubeity over holenmerism (definitive ubeity), although he hedges his bets by introducing the term "expansion" to refer to the manner by which *both* space and God/spirits are extended, whereas "extension" refers straightforwardly to matter (see, E II.xiii.26): e.g., "this present moment is common to all things, that are now in being, and equally comprehends that part of their Existence, as much as if they were all but one single Being;.... Whether Angels and Spirits have any Analogy to this, in respect of Expansion, is beyond my Comprehension;...'tis near as hard to conceive Existence, or have any *Idea* of any real Being, with a negation of all manner of Expansion; as it is, to have the Idea of any real Existence, with a perfect negation of all manner of Duration: And therefore what Spirits have to do with Space, or how they communicate in it, we know not" (E II.xv.11). In other words, God may be dimensionally extended or "whole in every part" (and thus not really extended), but God "fills" (occupies, is situated throughout, etc.) space in some fashion; Leibniz' repletive ubeity, in contrast, would perhaps amount to "a negation of all manner of Expansion", and thus be rejected. Overall, by claiming that space and immaterial beings are expanded, while matter is extended, Locke seems to be offering another clue that he reckons God to be the ontological foundation of space.

Berkeley's immaterial, unextended God, for "there is *not any other substance than Spirit*, or that which perceives" (89), and hence only God can be source of our ideas: "*there is a mind which affects me every moment with all the sensible impressions I perceive*" (188).

From Berkeley's, so to speak, "phenomenalization" of extension, he goes on to rebuff absolute motion and absolute space, since "all the absolute motion we can frame an idea of…[is] at bottom no other than relative motion thus defined" (130), and hence "it follows that *the philosophic consideration of motion doth not imply the being of an absolute Space*, distinct from that which is perceived by sense and related to bodies: which that it cannot exist without the mind, is clear upon the same principles that demonstrate the like of all other objects of sense" (131). The Neoplatonist concept of an extended space-constituting God is likewise dismissed: "[t]he chief advantage" of his idealist rendering of space, Berkeley contends, is that it dispels "that dangerous *dilemma*…of thinking either that real space is God, or else that there is something beside God which is eternal, uncreated, infinite, indivisible, immutable. Both which may justly be thought pernicious and absurd notions" (131).

In an ironic twist, it thus follows that both Leibniz and Berkeley posit a non-spatial, unextended God as the ontological foundation of space, albeit within the framework of entirely different spatial ontologies. That is, unlike the idealism that prompts Berkeley to dismiss the existence of matter and non-mental extension (i.e., extension that exists in an external world apart from minds/souls), Leibniz, as we have seen, only denies that matter and spatial extension constitute independent substances or real unities, but he does accept that they possess a sort of derivative reality, namely, as aspects of the material world that result from the non-spatial monads (see Chap. 4). As the means by which the monads bring about extension, Leibniz likewise supports the reality of force. Berkeley, on the other hand, regards force as equally suspect as matter, a point discussed at length in the later *De Motu*, where the force "attributed to bodies", and "used as if it signified a known quality, distinct as well from figure, motion, and every thing sensible, as from every affection of the animated life", is reckoned to be "nothing else than an occult quality" (WGB II 86).[3]

---

[3] Berkeley does admit, in the *Principles*, that while "in every motion it be necessary to conceive more bodies than one, yet it may be that one only is moved, namely that on which the force causing the change in the distance is impressed, or in other words, that to which the action is applied" (WGB I 129). This might be construed as sanctioning a Leibnizian interpretation of force over the purely kinematic approach of Huygens, but the later *De Motu* spells out more clearly the content of Berkeley's appeal to force/action. He states, that "no power can be known unless by action, and is measured by the same; but we cannot abstract the action of a body from its motion" (WGB II 87). The positivist streak in Berkeley's thought is clearly evident in his reduction of force/action to motion: "Action and reaction are said to be in bodies; and such expressions are convenient for mechanical demonstrations. But we should be on guard not therefore to suppose in them some real virtue which may be the cause or origin of motion" (91); that is, "physics contemplates the series or succession of the objects of sense, by what laws they are connected, and in what order; observing what precedes as a cause, what follows as effect. And in this way we say that a moved body is the cause of motion in another, or impresses motion on it" (103). This interpretation of the cause of motion, which is assigned from the perspective of mechanical laws and the past motions of the system, is (as argued in Chap. 4) Leibniz' view, too—although Leibniz additionally holds, contra

Leibniz' own conception of force is especially singled out for criticism, moreover: "Leibnitz also maintains that effort is every where and always in matter. It must be allowed that these things are too abstract and obscure, and of the same sort as substantial forms and entelechies" (89).

In retrospect, by concentrating on the kinematic component as opposed to the dynamic (force), Berkeley's approach to mechanics is much closer to Huygens' than Leibniz', as is his analysis of relational motion, which, like Huygens' treatment (see §1.3.3), explicitly denies that a lone body in an empty world can move: "[I]t doth not appear to me, that there can be any motion other than *relative*: so that to conceive motion, there must be at least conceived two bodies, whereof the distance or position in regard to each other is varied. Hence if there was one only body in being, it could not possibly be moved" (WGB I 129). On the other hand, Berkeley's relationism and kinematical account of mechanics are both consequences of his unique idealist worldview, whereas Huygens' relationism likely stems from a broadly empiricist, metaphysics-avoiding conception of science that strives to remain limited to observable quantities (unlike absolute space, for instance). Nevertheless, a dedicated idealist and a cautious empiricist would seem united in their antipathy to absolutism, although Newton's rotating bucket experiment complicates this assessment (since the seemingly non-relational character of the centrifugal force effects is observable).[4] Berkeley's phenomenalist proclivities are also responsible for one of the more forward-looking components in his system, namely, his claim that space is not derived from the visual sense, but from the tactile sense allied with bodily motions:

> When I excite a motion in some part of my body, if it be free or without resistance, I say there is *space*: but if I find a resistance, then I say there is *body*: and in proportion as the resistance to motion is lesser or greater, I say the *space* is more or less *pure*. So that when I speak of pure or empty space, it is not to be supposed that the word *space* stands for an idea distinct from, or conceivable without body and motion…When therefore supposing all the world to be annihilated besides my own body, I say there still remains *pure space*: thereby nothing else is meant, but only that I conceive it possible for the limbs of my body to be moved on all sides without the least resistance: but if that too were annihilated, then there could be no motion, and consequently no space. (131)

As Popper noted long ago (Popper 1953), there are many affinities between Berkeley and the logical positivist movement, but no single item in Berkeley's metaphysics reflects his empiricism or phenomenalism better than his body-centered conception of space. Yet, in contrast to Popper's historical emphasis, it is not the Mach of *The*

Berkeley and Huygens, that this assignment of individual motions can be further correlated with the existence of a real force at the simple substance/monad level (and hence his disagreement with Huygens on this point concerns the metaphysical grounding of individual motions, and not the physical assignments of motion from the perspective of the mechanical hypothesis).

[4] Incidentally, Berkeley joins both Leibniz and Huygens in rejecting Newton's argument, and in striving to account for the non-inertial force effects via rectilinear inertial motions determined relative to bodies (WGB II 100–101). Berkeley's argument is, nevertheless, rather muddled (see Earman 1989, 73–76).

Science of Mechanics that the quotation above evokes, but the physiological and psychological investigation of space in the latter's *Space and Geometry* (1906), as well as the phenomenology of Husserl's *Thing and Space* (1997).

A possible further consequence of the increasingly empiricist outlook of the period, whether inspired by the work of Hobbes, Locke, Berkeley, or others, is the option to eliminate physical space as a proper subject of study within those philosophical investigations that strive to remain limited to human perception and cognition. Hume, at least, would appear to have reached this conclusion in the *Treatise*: in response to the idea that a vacuum is required to allow bodily motion, he states that "I shall not enlarge upon this objection, because it principally belongs to natural philosophy, which lies without our present sphere" (T 1.2.5.4); and, commenting on the Newtonian conception of void, he adds that "[n]othing is more suitable to that philosophy [i.e., Newtonian], than a modest scepticism to a certain degree, and a fair confession of ignorance in subjects, that exceed all human capacity" (T 1.2.5.n). On the topic of empty space, Hume's stance has prompted Frasca-Spada to observe: "God's absence and the consequent apparent lack of balance in Hume's discussion of empty space—one may at this point suggest—depend on the shift of focus from natural philosophy to human nature and on a drastic concomitant change of metaphysical mood" (Frasca-Spada 2002, 189). Yet, an earlier instance of this change in mood can be found in Hobbes' natural philosophy, where a similar non-theological emphasis is placed on subjective sensory and cognitive functions, in conjunction with a language-based conceptual system, as a means of resolving various problems related to space. For example, unlike the Scholastics, who often appealed to God to guarantee the immobility of place (see Chap. 6, and Leijenhorst 2002, 102–127), Hobbes reckons that sensation, imagination, and a proper understanding of the role of concepts and language, can secure space's immobility: "For whilst one affirms that place is therefore said to be immovable, because space in general is considered there; if he had remembered that nothing is general or universal besides names or signs, he would easily have seen that that space, which he says is considered in general, is nothing but a phantasm, in the mind or memory, of a body of such magnitude and such figure" (EW 106). The upshot of this section of *De Corpore* (1655), arguably, is that space/place is a name that signifies a cognitive process or item (phantasm) obtained from extended bodies via sensation and memory, but, since names and/or cognitive features do not literally move, the problem of space's immobility is therefore resolved.[5] In short, and leaving aside the plausibility of this line of thought, there were seventeenth century precedents for using empirical and cognitive resources to handle the problems of space prior to Berkeley and Hume, and

---

[5] See, Slowik 2014, for a more detailed discussion of this aspect of Hobbes' philosophy of space. Among the natural philosophers whose work should also be factored into this account of the move towards an empiricist conception of space is Bayle and Malebranche, although it is beyond the bounds of this investigation.

some of these hypotheses, such as Hobbes', did not invoke the standard space-supporting God that had been the mainstay of spatial ontologies during the reign of Scholastic natural philosophy.[6]

11.1.3 The Fate of Third-Way Spatial Ontologies After Newton and Leibniz

One of the often overlooked lessons of Grant (1981) is God's ubiquitous role as the ontological foundation of space for nearly all natural philosophers from roughly the early middle ages up through the mid-eighteenth century. Although the God-based, third-way conception of spatial ontology fell out of favor by the dawn of the nineteenth century, while the standard dichotomy continued to ascend, the core idea of the former approach—namely, the ontological dependence of space on some grounding entity, P(O-dep)—retained its influence among subsequent generations of spatial ontologists, but often at a subliminal level of theorizing. With the demise of this theological basis, the material world, at the macrolevel, found its "support" in such doctrines as Leibniz-Wolffian monads, Boscovichian forces, and ultimately, the electromagnetic aether. The true heir to the God-grounded ontology that both Newton and Leibniz accepted was, therefore, already in play at the beginning of the eighteenth century, namely, in Leibniz' force-based monadic hypothesis. Needless to say, Leibniz' system, where an unseen monadic realm gives rise to the material world, dovetails with the plenum and atomic conceptions of nature, both ancient and in the versions practiced in his time. Yet, Leibniz sowed his plenum and microlevel entities with force, and envisioned the material macrolevel as a manifestation of force, thus presaging the future course of physics at the foundational level of reality. Leibniz' approach, in turn, would serve as the template for the similar hypotheses advanced by Wolff and his school, the pre-critical Kant, Boscovich, Herbart, and many others. Augmented by new discoveries and developments in physics in the ensuing centuries, this same conception can be seen, as noted previously, as the direct forebear of the electromagnetic aether in the nineteenth, as well as GR's metric/gravitational field and quantum field theory in the twentieth.

In brief, the P(O-dep) concept lives on in those interpretations of contemporary field theories that do not insist on a thoroughgoing relativity of motion (i.e., Leibnizian spacetime or weaker, see §1.1) but which simultaneously reject the idea that space (spacetime) is an independently existing entity. The number of theorists

---

[6]While popular among various natural philosophers of the period, such as Keill, Raphson, and MacLaurin, it was Samuel Clarke, John Clarke (the brother of Samuel), and John Jackson who were the most outspoken defenders of Newton's brand of God-based spatial ontology (conjoined with absolutism) in Britain during the first half of the eighteenth century. Their chief opponents, besides Berkeley and Hume, included Joseph Butler, Edmund Law, Joseph Clarke, and Isaac Watts (see, Ferguson 1974, Frasca-Spada 2002). These critics of absolute space, such as Law, often accepted a metaphysics of divine presence along the same lines as their seventeenth century precursors and contemporary opponents, but they also appealed to the same types of empirical or phenomenal arguments against absolute space that one finds in Berkeley.

that fit this description is quite large, and many do not explicitly state a preference for relationism, but their common refrain is the rejection of void space for a field or aether concept—in other words, space must have some type of material/physical grounding, a stance that is best captured by P(O-dep) as opposed to any of the other ontological positions surveyed in this investigation. Among this class of scientists and philosophers, some of the most notable are Faraday, Ørsted, Kelvin, Poynting, Planck, and Hilbert, with August Föppl going so far as to compare "the possibility of space without ether to the contradictory notion of a forest without trees" (Kragh 2011, 70–71). In the context of GR, an excellent candidate for a P(O-dep) classification might be Weyl's "world-structure" hypothesis, which explicitly criticizes both absolutism and relationism but maintains that spacetime structures have objective meaning and are real features of the world (1949, 70–74).[7] Einstein's late reflections on the proper historical analogue of GR's conception of space, i.e., Descartes' plenum as opposed to Newton's absolute space (see §5.2), also supports P(O-dep), since Einstein had rejected a Machian relationist construal of GR by that point. His remark, that "[s]pace-time does not claim existence on its own, but only as a structural quality of the field", even employs a key property theory term to characterize space's ontological status, i.e., as a "quality" (= property) of a physical field (Einstein 1961, 155).

Finally, even when shorn of its divine tethering in the seventeenth century, one can detect traces of the P(O-dep) outlook in the style of the defense of absolute space that gained prominence in the latter half of the eighteenth century. Euler (1748), to take a notable example, would defend absolute space much as Newton had, by insisting that relationism could not adequately explain the Newtonian laws of motion; but, unlike Newton, Euler did not ground his conception of absolute space on God, nor was he keen to defend the notion of absolute position in explicit terms. Rather, Euler moved the discussion to inertial structure, i.e., the sameness of direction of inertially moving bodies over time, a distinction that in modern nomenclature signals the replacement of Newtonian spacetime with neo-Newtonian spacetime. This change in emphasis—from a postulated immovable space to the inertial structure required for dynamics—is presaged in Newton's *Principia*, where the earlier *De grav*'s direct appeal to absolute position to counter relationism is dropped for a more indirect line of argument based on relationism's inability to coherently explicate rotational motion (see DiSalle 2013). An anti-relationist maneuver of this type is, moreover, consistent with the Stein-DiSalle approach to space, i.e., as definitional or structural components of a viable dynamics, and this view can be included within the P(O-dep) classification on the grounds that space becomes inextricably linked to material phenomena (in a manner that is comparable, although not identical, to sophisticated relationist construals). In short, the ontological status of a void

---

[7] Weyl focuses on the inertial (or guiding) field, i.e., the metric/gravitational field, in discussing world-structure. As explored in Korté (2006), Weyl's "inertial field as space" ontology is open to both a substantivalist and relationist construal; however, as argued in Chap. 5, a property theory interpretation of GR's metric field, such as P(O-dep), seems more appropriate than sophisticated relationism.

space is no longer one of the chief determinative factors that separates substantivalism from relationism once inertial structure is singled out as the key component of
the former view—and, since inertial structure can only be ascertained, and is only
meaningful, in the context of bodily behavior (as first explored in Chap. 5), this
change in emphasis on the part of the substantivalists inevitably draws their side of
traditional dichotomy closer to a property conception of space. If, as noted above,
Newton only hinted at this approach to absolutism, Euler forthrightly advanced and
defended it, although both would surely insist that inertial structure is a feature of
absolute space, and not any matter-based version of the P(O-dep) account of space.

In fact, like the three-dimensional void of the ancient atomists, Euler can be interpreted as accepting a conception of space that is both metrical and entirely independent of all other entities, whether at the foundational or derived levels; i.e., employing
the new taxonomy developed in Chaps. 9 and 10: FGS(met), FL(v-plt) and SL(v-
plt)—and it is the conjunction of these categories that constitutes modern spatial
absolutism (substantivalism). In contrast, modern eliminative relationism would
endorse nominalism at the foundational and secondary levels, and, under the FGS
category, invoke the geometric structure (whether, pregeometric, topological, or
metric) employed by the foundational level physical theory. Consequently, one can
justifiably trace the origins of the modern absolutist/substantivalist versus relationist
dichotomy to the mid-eighteenth century, when an influential non-theological version of the spatial ontology that Newton had championed (i.e., metrical, virtual-
platonist) was finally juxtaposed with the relationism that Huygens had pioneered
(i.e., metrical, nominalist). What is truly ironic, and supports Grant's interpretation
examined above, is that Euler's non-theological conception of space is coupled to a
conception of God that supports Leibniz' version of repletive ubeity: "My soul, then,
does not exist in a particular place, but it acts there, and as God possesses the power
of acting upon all bodies, it is, in this respect, we say, He is every where, though his
existence is attached to no place" (Euler 1761, 355). Put simply, whereas Leibniz
grounds his spatial ontology on God's being or essence by means of repletive ubeity,
Euler interprets repletive ubeity as a means of separating his theology from his natural philosophy, and thus from his spatial ontology (i.e., Euler never links his spatial
ontology to God's being, although he references God's actions in space, as the quote
above attests). As mentioned previously, the repletive ubeity hypothesis allows one
to, as it were, banish God from space, whereupon God's actions in space can be
given a purely metaphorical interpretation. Euler may have been one of the first
natural philosophers of consequence to have taken this momentous path to a purely
natural, as opposed to theologically-grounded, ontology of space.

11.2   The Kantian Synthesis of Relationism and Newtonian
Physics

In Chap. 1 and the previous section, several developments in the natural philosophy
of the late seventeenth and early eighteenth centuries have been posited as the chief
contributing factors in the formation of the modern absolutist (substantivalist)

versus relationist dichotomy. Above all, Huygens' relationist interpretation of Descartes' collision rules focused the attention of natural philosophers on the distinction between real and apparent motion within mechanical systems that employ quantitative laws, thereby introducing a basis for the absolutist/relationist distinction in the newly developed science of mechanics (whereas the distinction had earlier remained largely within the province of a deeply metaphysical, and often theological, natural philosophy). When coupled to the growing empiricism of the time period, whether of the non-theological sort (Huygens, Hobbes), or theologically inclusive and radically idealist variety (Berkeley), the result was the gradual decline of the type of spatial ontology advocated by Newton and Leibniz, namely, where space's metaphysical status is roughly akin to a property of God (albeit in different ways that depend on the precise details of the God-grounded ontology at hand).

The evolution of Kant's philosophy of space and motion exemplifies these changes, and it also prefigures the difficulties that nineteenth century natural philosophers would face in reconciling the prevailing Newtonian theory with the challenges imposed by the absolute/relational dichotomy. In many ways, the transition from Kant's early conceptions of space and motion to his later views can be viewed as a sort of conduit between the seventeenth and nineteenth centuries on these issues, and hence a brief examination provides a fitting coda to the historical and conceptual scope of this investigation.

11.2.1 The Center-of-Mass Frame as Absolute Space

As explored in §9.3.5, Kant's earlier pre-critical work follows the Leibniz-Wolff tradition by postulating a material macrolevel that arises from a non-extended monadic realm, hence space still meets the P(O-dep) classification, i.e., as a property of monads, with God retaining a constitutive role as well. Turning to the critical period, both the monadic microlevel and theological basis of space are no longer emphasized, but Kant's treatment of spatial hypotheses at the material macrolevel relevant to his physics is continuous with the approach that he had espoused in various pre-critical works, specifically, in that he utilizes a Huygens-style center-of-mass (or center-of-gravity) frame to construct a relationist interpretation of the Newtonian distinction between absolute and relative space (as will be explained below). In the first few pages of his most elaborate treatment of physics in the critical period, the *Metaphysical Foundations of Natural Science* (1786), Kant lays out the major themes of his unique relationist-inspired interpretation of absolute space:

[A]ll motion that is an object of experience is merely relative; and the space in which it is perceived is a relative space, which itself moves in turn in an enlarged space, perhaps in the opposite direction, so that matter moved with respect to the first can be called at rest in relation to the second space, and these variations in the concept of motions progress to infinity along with the change of relative space. To assume an absolute space, that is, one such that, because it is not material, it can also not be an object of experience, as given in itself, is to assume something, which can be perceived neither in itself nor in its consequences (motion in absolute space).... Absolute space is thus in itself nothing, and no object at all, but rather signifies only any other relative space, which I can always think beyond the given space, and which I can only defer to infinity beyond any given space, so as to include it and suppose it to be moved.[8] (MF 4:481)

In short, all motion is relative and perceived in a relative space, which is the empirical (sensible) space of material bodies.[9] Absolute space, on the other hand, is not an "object of experience" and "is thus in itself nothing", but is simply an "idea of reason" that assists in the construction of a series of ever larger empirical relative spaces that proceed to infinity. Since one of Kant's main objectives in this work is to supply his own interpretation of Newtonian gravitation theory, the application of the conceptual apparatus outlined above to the celestial realm, as Friedman explains, ultimately "indicates how the earth's state of true rotation can nonetheless be empirically determined, and concludes by considering the cosmos as a whole, together with the 'common center of gravity of all matter', as the ultimate relative space for correctly determining all true motion and rest" (Friedman 2004, xiii). Consequently, it is "the common center of gravity of all matter", using Kant's phrase (MF 4:563), i.e., the center-of-mass frame of all matter, and not the inertial structure of space *per se*, that constitutes absolute space (see, Friedman 2013b, 503–509, on this issue).

One of the truly novel features of Kant's system, which also ties into the discussion in §1.3, is that he envisions his center-of-gravity approach as an instance of a larger strategy for interpreting all bodily interactions that also includes within its scope the center-of-mass frame collision model first pioneered by Huygens (although Kant may have been unaware of this history). As regards impact, Kant provides an example involving two bodies, A and B, that approach from opposite directions along the same rectilinear path, collide, and reverse their motion:

[T]he change of relation (and thus the motion) between the two is completely mutual; as much as the one body approaches every part of the other, by so much does the other approach every part of the first....On this basis, the motion of a body A with respect to

---

[8] In the final pages, he reiterates these points: "Absolute space is therefore necessary, not as a concept of an actual object, but rather as an idea, which is to serve as a rule for considering all motion therein merely as relative; and all motion and rest must be reduced to absolute space, if the appearance thereof is to be transformed into a determinate concept of experience (which unites all appearances)" (MF 4:560).

[9] "*Matter*, as opposed to *form*, would be that in the outer intuition which is an object of sensation, and thus the properly empirical element of sensible and outer intuition, because it can in no way be given a priori. In all experience something must be sensed, and that is the real of sensible intuition, and therefore the space, in which we are to arrange our experience of motion, must also be sensible—that is, it must be designated through what can be sensed—and this, as the totality of all objects of experience, and itself an object of experience, is called *empirical space*. But this, as material, is itself movable. But a movable space, if its motion is to be capable of being perceived, presupposes in turn an enlarged material space, in which it is movable; this latter presupposes in precisely the same way yet another; and so on to infinity" (MF 4:481). As the last few sentences indicate, the relativity of perceived motion is an integral part of Kant's conception.

another body B at rest, in regard to which it can thereby be moving, is reduced to absolute space; that is, as a relation of acting causes merely related to one another, this motion is so considered that both have an equal share in the motion which, in the appearance, is ascribed to body A alone. And the only way this can happen is that the speed ascribed in relative space to body A alone is apportioned between A and B in inverse ratio to their masses. (MF 4:546)

In short, the center-of-mass frame is the position where the "speed ascribed in relative space" is "apportioned between A and B in inverse ratio to their masses". As briefly noted above, Kant had earlier employed the center-of-mass frame in this same manner in his pre-critical (1758), *New Doctrine of Motion and Rest* (NS 2:18–25), alongside the same eliminative relationist conception of place and motion that one finds in Huygens and Berkeley: before introducing his impact model, he argues that "the place of a thing is known by its position, situation, or by its external relationship to other objects around it", which he dubs a "relative space" (2:16), and he insists that the terms "motion and rest" should never be used "in an absolute sense but always relatively" (2:17). The relationist version of absolute space that he would later develop in the critical period is absent in the *New Doctrine*, however.[10]

Returning to the *Metaphysical Foundations*, the assimilation of gravitation and impact using the center-of-mass frame strategy falls under his third law of mechanics, which stipulates that "[i]n all communication of motion, action and reaction are always equal to one another" (MF 4:544). After detailing his impact model, he comments that "the communication of motion through *impact* differs from that through *traction* [gravitation] only in the direction in which the matters resist one another in their motions. It follows, then, that *in all communication of motion* action and reaction are always equal to one another" (4:546–547). The basis of Kant's third law of mechanics has, furthermore, distinctly Leibnizian, or (more accurately) Leibniz-Wolffian, roots, since Wolff and his followers were the likely source of Kant's dynamical notions.[11] In short, and leaving aside the obvious influence of Newton's third law of motion (N 71), the attempt to merge a kinematical treatment of the motions and quantities conserved in collision with a dynamical action/reaction principle employing the center-of-mass frame can be found in Leibniz' "Specimen Dynamicum" (AG 117–138), and, more generally, the action/reaction principle was taken up and developed by Wolff and his school. In the pre-critical *New Doctrine*, the action/reaction principle is also introduced (NS 2:19), but it is the *Metaphysical*

---

[10] In Kant's first published work from 1747, the *Thoughts of the True Estimation of Living Forces* (examined in §9.3.5), there are a few references to absolute rest and motion (NS 1:90, 125–126, 158), but the context suggests that he is using "absolute" to indicate real versus apparent motion, since he contrasts an absolute motion with a body being "at rest with regard to all things", which is consistent with a relationist conception (1:126), and there are no corresponding references to absolute space.

[11] On the action/reaction principle and the center-of-mass frame method in Leibniz, see Slowik (2006), and for the action/reaction principle among the Wolffians in general, and other related themes, see Watkins (1997, 2003, 2005) and Stan (2012, 2013, 2014a, b). Stan's work explores the inadequacy of Kant's use of the center-of-mass frame to support a consistent relational account, although that topic will not be addressed below.

Foundations that incorporates the impact model and the action/reaction principle with gravity under the same center-of-mass (center-of-gravity) scheme, as well as introduces the absolute/relative space distinction surveyed above.[12]

Besides the replacement of the orthodox interpretation of absolute space (as an independent, fixed world space, etc.) with a materially-based reference frame, the relationism inherent in Kant's *Metaphysical Foundation* becomes all the more evident once the details are specified. First of all, Kant's approach to motion matches the standard relationist conception that one finds in, say, Leibniz (at the level of well-founded phenomena, excluding force) and Huygens, where the individual states of motion assigned to the bodies are perspectival but the invariance of the relative change in distance *among* the bodies is emphasized: "all motion of material things...count as merely relative with respect to one another, as alternatively mutual, but none as absolute motion or rest" (MF 4:559–560). Second, Kant's explication of rotation, a form of motion that had stymied so many earlier accounts of relational motion, appeals to the dynamic (force) effects among the bodies undergoing the rotation, thus providing an empirical means of distinguishing these cases from an identical, non-rotating configuration. In short, Kant converts one of Newton's empirical arguments for absolute space, via the rotating globes thought-experiments (N 68–70), into a form that, allegedly, upholds relational motion among the material parts of the rotating system: "circular motion, although it in fact exhibits no change of place in the appearance,...exhibits nonetheless a continuous dynamical change, demonstrable through experience, in the relations of matter within its [relative] space, for example, a continual diminution of attraction in virtue of a striving to escape" (MF 4:561). Kant's account, in effect, mimics the types of hypotheses offered earlier by Huygens (H 39–47), Leibniz (AG 135–136), and Berkeley (WGB II 100–101), i.e., where one appeals to a set of resting external bodies as a backdrop, or the relative motion among a body's parts or a pair of bodies, to explicate rotational motion and its effects. Concerning the rotation of the earth, for instance, he states:

> [T]his motion, even though it is no change of relation to the empirical space, is nevertheless not absolute motion, but rather a continuous change in the relations of matters to one another, which, although represented in absolute space, is thus actually only relative, and, for just that reason, is true motion—this rests on the representation of the mutual and continuous withdrawal of any part of the earth (outside the axis) from any other part lying diametrically opposite to it at the same distance from the center. (4:561–562)

Third, Kant rejects as "utterly impossible" (4:563) a scenario wherein the entire cosmos moves uniformly and rectilinearly through space (i.e., a kinematic shift), and likewise denies a potential cosmic rotation, but concedes that "it is always possible to think such a [rotational] motion, although to suppose it would, so far as one can see, be entirely without any conceivable use" (4:563). Since a uniform rectilinear or rotational motion of the entire cosmos would not be relative to another body, a relationist must forbid, or deem as useless, these scenarios. Unlike Euler (1748),

---

[12] It should also be noted that the *Theory of the Heavens* (1755) provides an early model of the concentrically arranged rotating celestial systems that would form the basis of his later approach to gravity in the *Metaphysical Foundations*.

whose conception of absolute space would sanction the motion of a lone body in an otherwise empty universe, Kant reasons that "absolute motion, thought without any relation of one matter to another, is completely impossible" (MF 4:559). Fourth, Kant reflects on various meanings that can be ascribed to empty space, either within or outside the material world, and he ultimately concludes that these possibilities are, at a minimum, "not *necessary*", and potentially impossible on dynamic or physical grounds (4:563–564). By taking this stance, Kant is thereby relieved of the burdensome task of explicating the ontological status of a vacuum, a possible state of the world that many natural philosophers of the time period would have interpreted as supporting absolutism over relationism.

11.2.2   Kant's Relationist "Applied Metaphysics" of Motion and Its Aftermath

While there have been many notable exceptions over the past few decades, such as Carrier (1992) and Stan (2014b), to name only a few, the relationist orientation of Kant's natural philosophy of space and motion is—somewhat remarkably—largely unknown outside perhaps a small subsection of Kant scholars. Part of the reason for this oversight may lie in the tendency to treat Kant's transcendental idealist notion of space in the critical years as an alternative ontological conception, different from substantivalism and relationism, with the *Critique* and the incongruent counterparts argument (more on this below) forming the primary supporting evidence. Allison (1983, 25) has denounced this general outlook, claiming that Kant's idealist standpoint should be understood as epistemic in character, rather than as an ontological solution. Yet, a more prosaic reason for the curious oversight of Kant's relationism may stem from the simple fact that the *Metaphysical Foundations* and the relevant pre-critical works that advance relationism are themselves largely overlooked, and that previous investigations of the center-of-mass frame strategy, such as Friedman (2013b), have not addressed the connection with Huygens' similar strategy, or with relationism in general.[13] DiSalle (2006), on the other hand, focuses his attention on Newton and Euler's contribution, concluding that Kant's "mature concern was not to establish one of two opposing metaphysical positions [absolutism versus relationism]" (2006, 66), but to demonstrate that "[t]he metaphysical concepts that occur in physics—body, force, motion, space, time—become intelligible to us precisely, and only, as they are constructed by physics itself; physics provides us with the only intelligible notions we have on these matters" (60). Therefore, since the physics of Kant's day was Newtonian, "the metaphysical concepts underlying the

---

[13] Freidman does note that "[f]or Kant,...space, motion, and rest are always relative concepts" (2013b, 43), which he describes in the context of the "Copernican revolution in astronomy" (41). The point is, however, that a discussion of the distinctly relationist conclusions that Kant reaches utilizing his system, and how it relates to prior and past relationist hypotheses, is not a part of Freidman's ground-breaking investigation.

sensible world first become intelligible", for Kant, "in the framework of Newtonian physics" (64); and, "Kant's analysis of absolute space, accordingly, is an effort to clarify its place within the system of Newtonian principles" (67).

Nevertheless, the evidence of the texts presented in §11.2.1 indicates that various non-Newtonian principles played a role in Kant's system at least as important as Newton's, with his rejection of a uniform inertial motion of the material world presenting the most conspicuous example. If perchance a body were located outside the cosmos, Kant reasons that the mutual gravitational interaction between the cosmos and the body (which falls under the "the law of antagonism in all community of matter through motion") would "shift the common center of gravity of all matter, and thus the entire cosmic system, from its place", but "then the motion would already be relative" (MF 4:562–563). As it stands, this hypothesis is perfectly consistent as well as relational, and it is in accordance with the *Principia*'s stipulation that "the centre of the system of the world is immovable" (Newton 1962b, 419). Yet, it is the manner by which Kant reaches this conclusion that demonstrates his divergence from the *Principia*, especially when the proposed hypothesis is conjoined with his second law of mechanics, which holds (following Newton's first law of motion) that "[e]very body persists in its state of rest or motion, in the same direction, and with the same speed, if it is not compelled by an external cause to leave this state" (MF 4:543). This law figures prominently in Kant's assessment of the observed non-inertial force effects of rotation: "according to the law of inertia…the body, at every point on [a] circle (according to precisely the same law), is striving, for its own part, to proceed in the straight line tangent to the circle" (4:556). Consequently, by stipulating a decidedly Newtonian concept of inertia, a uniform inertial motion of the entire cosmos should be a possible state-of-affairs, rather than rejected out-of-hand as "utterly impossible".[14] Or, to put this point in specifically Newtonian terms, Kant's hypothesis (where only a gravitational interaction with an outside body can cause a rectilinear unison motion of the cosmos) is tantamount to claiming that the world's inertial motion, which comes under Newton's first law of motion, would violate Newton's third law of motion, the latter holding that every action has an equal and opposite reaction (and which also comprises Kant's own third law of mechanics, as explained above). This maneuver essentially constitutes a fundamental reconstruction of Newtonian physics, a revaluation of basic principles that just so happens to fall in line with a Huygens-style center-of-mass frame version of relationism and an action/reaction principle that, as Stan concludes, Kant "developed by constructive engagement with post-Leibnizian dynamics, rather than Newton's *Principia*" (2013, 503).[15] Accordingly, while Newton undoubtedly played

---

[14] That is, given Kant's acceptance of a Newtonian form of inertial motion, something like Newton's Corollary 4 should be in effect, namely, that the center-of-mass of the world is either at rest or moves uniformly in a straight line; as well as Corollary 5 (Galilean relativity), that one cannot distinguish a state of rest from a state of uniform rectilinear motion (see, N 76–79; and Friedman 2013b, 443–445, who explores the absence of these corollaries in Kant's work).

[15] The other noteworthy facet of Kant's system that favors relationism is his reluctance to embrace the possibility that the entire cosmos rotates, even though each sequence in his series of ever larger center-of-gravity systems apparently does (in order to preserve the stability of the order of bodies

a significant role in Kant's mature natural philosophy, perhaps it would be more accurate to infer that, for Kant, "the metaphysical concepts underlying the sensible world first become intelligible" in the framework of Huygens-Leibniz-Wolffian physics.[16]

Indeed, once the details are taken into account, it is rather difficult to avoid the conclusion that one of the chief goals of the *Metaphysical Foundations* is to provide an anti-substantivalist interpretation of Newtonian physics that generally follows the relationist precedent set by such thinkers as Huygens and Berkeley, as well as his own earlier *New Doctrine*. While the Huygensian spirit of Kant's system has been detailed above, via his relationist interpretation of the center-of-mass frame, it is the (transcendental) idealist foundation of the critical period that evokes Berkeley's approach to absolute space via its link with motion. Philosophical reflection, for Berkeley, "*doth not imply the being of an absolute Space*, distinct from that which is perceived by sense and related to bodies" (WGB I 131), a conclusion echoed in Kant's assertion that absolute space cannot "be an object of experience", and "can be perceived neither in itself nor in its consequences (motion in absolute space)" (MF 4:481). Likewise, Kant's common refrain, that "absolute motion, thought without any relation of one matter to another, is completely impossible" (4:559), is in harmony with Berkeley's insistence that "all the absolute motion we can frame an idea of [is] at bottom no other than relative motion thus defined" (WGB I 130). In the *Critique*, as is well known, Kant rebuffs the metaphysical doctrines of *spatial absolutism* and *relationism* (respectively, as an entity or the relations among entities; CPR A23/B38) in favor of space as an a priori intuition (and the same for time), i.e., as a subjective feature of the mind's operation. But, as uniquely revealed in the *Metaphysical Foundations*, the upshot of this critical period doctrine when applied to physics is a distinctly Berkeleyan approach to absolute space and motion—with Berkeley's outlook, in turn, amounting to a sort of "phenomenalized" Huygensianism

---

within each concentric system via a balance of gravitational and centrifugal forces; MF 4:557–563). According to Friedman (2013b, 501–502), this restriction stems from the *Critique*'s first antinomy (in particular, CPR A429/B457), which rules out the completion of an infinite sequence of this sort on epistemic (or transcendental idealist) grounds, hence the rationale behind Kant's belief that a rotation of the entire cosmos is "without any conceivable use". But this reasoning, which is quite dubious in its own right, is purely metaphysical, and not grounded in the physics at hand.

[16]More carefully, the upshot of the historical interpretation offered by DiSalle, and possibly Friedman, is that Kant's critical period works, and the *Metaphysical Foundations* in particular, rely on a conception of space and dynamics that more naturally fits the type of world view championed by Newton (or, for DiSalle, Newton and Euler) as opposed to the Leibniz-Wolff school. And, since Newtonian dynamics is anti-relationist, it must thereby follow that Kant's later system has more in common with Newtonian absolutism than Leibniz-Wolff relationism—although, to be fair, this last inference is not specifically drawn by Freidman, but it would appear to be an acknowledged consequence of DiSalle's reading. Indeed, as DiSalle comments: "Kant had started from a Leibnizian view of the world as constituted of monads, and consequently a relationalist view of space; he was moved in the direction of Newton's view largely by his reading of Euler. Evidently Kant was impressed by the argument...that dynamics must assume certain aspects of space and time—above all, the idea of a privileged state of uniform motion—that cannot be squared with Leibniz's relationalism" (2006, 60–61).

(see §11.1.2 and §1.3.3). To be more precise, it is not space as a non-empirical intuition that is at issue. Like Leibniz, Kant is quite clear that "[s]pace is not an empirical concept that has been drawn from outer experiences" (CPR A23/B38), and thus he differs from Berkeley (and Locke and Hume, for that matter) by offering an a priori interpretation of space that is tied to his own brand of subjectivist/idealist cognition. Yet, Kant holds that motion *is* an empirical concept (A41/B58), and it is in the *Metaphysical Foundations* that this topic is addressed at length:

> [S]ince the *movability* of an object in space cannot be cognized a priori, and without instruction through experience, I could not, for precisely this reason, enumerate it under the pure concepts of the understanding in the *Critique of Pure Reason*; and that this concept, as empirical, could only find a place in a natural science, as applied metaphysics, which concerns itself with a concept given through experience, although in accordance with a priori principles (MF 4:482).

To summarize, while the contention that Kant did not try to establish either an absolutist or relationist spatial ontology in the critical period is thus technically correct, he did offer a devoutly relationist construal of the phenomenal world of bodily motion, which he categorizes as "applied metaphysics" in the passage above (see, also, Buroker 1981, 123–130).

Consequently, it would seem that the relationism that Kant espoused in his precritical period was never really abandoned, but simply transformed into a version more amenable to his newly developed species of subjectivism/idealism.[17] The ontological form of eliminative relationism that one finds in the *New Doctrine* has been dropped, of course, but a subtle form of relationism is still operative that prefigures the sophisticated strains that would be developed in subsequent centuries — that is, in conjunction with the relative space of actual bodies, Kant's "absolute space" signifies the *possibility* of constructing ever larger relative spaces, and thus the function of the relative/absolute distinction in Kant's *Metaphysical Foundations* strongly resembles the sophisticated modal varieties of relationism currently in vogue, albeit at the subjective level of phenomena. In the *Directions in Space* (1768), where Kant first launches his criticisms of spatial relationism and invokes absolute space by way of the incongruent counterparts argument (see below), there is no corresponding sanction of the *ontology* of spatial absolutism. Rather, Kant seems already inclined towards an idealist interpretation, for he concludes that "absolute space is not an object of outer sensation; it is rather a fundamental concept which first of all makes possible all such outer sensation" (TP 2:383). In addition, the *Directions in Space* puts forward a critique of Euler's well-known (1748, 329–330) argument for absolute space (see §11.1.3), concluding that it "does not quite achieve its purpose", rather, "[i]t only shows the difficulties involved in giving a determinate meaning to the universal laws of motion if one operates with no other

---

[17] Kant's late *Opus Postumum* suggests that his pre-critical conception of a force-based material world, upheld by God, is still in play as well, although it is now presented in the form of an aether (see, Friedman 1992, 50–53). A subjective force-oriented account of space, predicated on the human body, and which is vaguely reminiscent of Berkeley's body-centered approach (as above, WGB I 131), is also included.

concept of space than that which arises from abstraction from the relation between actual things" (TP 2:378). Kant's evaluation of Euler's argument, in effect, antici-pates the rationale that contemporary sophisticated modal relationists offer to jus-tify their rejection of strict eliminative relationism; specifically, that a spatial concept abstracted from the relations among *actual* bodies is incapable of meeting the demands of physics. Unlike modern sophisticated relationists, however, Kant does not regard the requisite modality as a primitive ontological fact grounded in actually existing bodies, but as an a priori contribution of the mind that secures the unity (or holism) of spatial geometric structure.

Why did Kant reject the ontology of absolute space? While his Leibniz-Wolff background likely predisposed him against this notion, the "incongruent counter-parts" argument in the *Directions in Space* may contain important clues. Incongruent counterparts are objects with a spatial asymmetry, such that the object and its mirror image cannot be superimposed; e.g., left- and right-handed gloves and screws. Kant argues that the relations among the parts of a left- or right-handed object cannot account for this asymmetry (since the internal relations among parts of each object are identical), but Kant is quite clear that, if one were to appeal to the parts of space to account for this asymmetry, then the same problem would arise for the parts of space: after concluding that "[t]he direction...in which this order of parts [of the body] is orientated, refers to the space outside the thing", he adds, "[t]o be specific: it refers not to places in this space—for that would be the same thing as regarding the position of the parts of the thing in question in an external relation—but rather to universal space as a unity, of which every extension must be regarded as a part" (TP 2:378; Earman 1971 discusses this point as well). Yet, in its ontological form, i.e., as an entity, absolute space would assuredly have parts, thus raising the same incongruent counterparts problem for the parts of space as for the original incongru-ent counterpart bodies. However, Kant seems to have accepted as a general principle by this point that the infinite divisibility of space implies that it cannot be a compos-ite of simple parts, unlike real entities (i.e., "things in themselves", which do have parts).[18] Hence, given these difficulties, Kant may have reasoned that only an idealist route could provide the unity or holism of absolute space required to explicate the difference between incongruent counterparts, but without necessitating the standard part-whole structure possessed by space if regarded as an entity. By the time of the *Prolegomena* (1783), the idealist lesson that Kant draws from the incongruent coun-

---

[18] In a Leibnizian vein, he reasons that "the *composite of things in themselves* must certainly consist of the simple, for the parts must here be given prior to all composition" (MF 4:507), and thus, "if matter is divisible to infinity then (concludes the dogmatic metaphysician) *it consists of an infinite aggregate of parts;* for a whole must already contain in advance all of the parts in their entirety, into which it can be divided. And this last proposition is undoubtedly certain for every whole *as thing in itself.* But one cannot admit that matter, or even space, *consists of infinitely many parts* (because it is a contradiction to think an infinite aggregate, whose concept already implies that it can never be represented as completed, as entirely completed)" (4:506). See, Falkenstein (2004, 293–300), for an informative discussion. Friedman (1992, 29) hints that Kant may have accepted an idealist conception of absolute space (such that it only exists in minds) in the late 1760s transi-tional period.

terparts argument is made clear (PFM 4:286), but the transformation to space as an a priori intuition, which is tentatively suggested in *Directions in Space*, is largely complete in the *Inaugural Dissertation* penned two year later (1770): "Space is not something objective and real, nor is it a substance, nor an accident, nor a relation: it is, rather, subjective and ideal; it issues from the nature of the mind in accordance with a stable law as a scheme, so to speak, for co-ordinating everything which is sensed externally" (TP 2:403). Nevertheless, as demonstrated above, this "scheme" becomes overtly relational in the *spatiotemporal* context of the experience of bodily motion, although it is confined to the "subjective and ideal".

Kant's journey to his transcendental idealist solution to the problem of space can be viewed, in retrospect, as emblematic of the evolution of the spatial ontology debate from the seventeenth century to the contemporary scene. The immaterialist ontologies that Newton, Leibniz, and Kant's own pre-critical writings had accepted as the foundation of space—i.e., God, as well as interconnected monads—were no longer deemed tenable, but another "immaterialist" item of sorts might provide the same grounding and holistic function, namely, the human mind. When released from the theological and the more radical immaterialist implications associated with Berkeley's equally mind-based spatial theory, and augmented by a constructivist process that (allegedly) leans towards the logical/mathematical at the expense of the metaphysical, the result was an approach to space that mirrors the "duality" of contemporary scientific theorizing: a subjectivist spatiotemporal construction (theory), on the one hand, and a hidden world of "things in themselves" (reality), on the other, with the former somehow standing as a representation, interpretation, model, etc., of the latter. Viewed in this light, the confluence of empiricist and constructivist elements in Kant's transcendental idealist treatment of space does appear remarkably modern, but, no doubt, only if this structural similarity is admitted on very broad terms.

Finally, it should be noted that, after Kant, the utilization of the world's center-of-mass frame as a substitute for the standard Newton/Euler conception of absolute space would find many advocates, most notably, Mach (SM 287), although one should also include other material-based reference frame schemes, such as Neumann's "body Alpha" (Neumann 1976). In fact, the center-of-mass frame strategy remains to this day the default relationist approach, e.g., the Barbour-Bertotti program, which employs a least-action principle in order to secure a non-rotating universe (and thereby recover the standard Newtonian results; Barbour and Bertotti 1977, 1982). And, much like the quandary over the proper categorization of Kant's own achievement in the *Metaphysical Foundations*—as either relationist, absolutist, or something else—there were equivalent disagreements among the nineteenth century advocates of a materially-conceived surrogate for absolute space. Mach, for instance, forthrightly sides with a relationist classification for his center-of-mass frame method, and he even suggests, contra Newton's spinning bucket thought-experiment, that a fully relational account of rotation may be correct (i.e., the water would rise up the bucket's sides if the bucket is at rest and the universe undergoes the rotation; SM 284). Neumann, in contrast, argues that his body Alpha scheme, which stipulates a single inertially-moving rigid body to which all other motions are referred, undermines a relational account of motion since there is a privileged bases for determining all motions, and hence motion relative to the body Alpha is absolute (1976, 127). Russell (1938, 491) finds these types of strategies implausible, for the question regarding the state of motion of any privileged material frame or body naturally arises, and hence only a non-material (and hence non-empirical) conception of absolute space is acceptable. Regardless of these disputes, Kant's relationist appropriation of the term "absolute space" can be viewed, in hindsight, as having foreshadowed the future direction of the absolute versus relational debate, a controversy that would grow ever more ambiguous and arbitrary as fundamental disagreements over the meaning and scope of the dichotomy's basic concepts and terminology gradually engulfed the rival camps.

11.2.3   Conclusion: A Note on Kant as Precursor of Mach

It would be worthwhile at this point to summarize the relationship, as presented in §11.2 of this chapter, between Kant critical period natural philosophy and Newtonian physics. While Newtonian physics is, needless to say, central to the *Metaphysical Foundations*, is it due to: (i) the established status that Newtonian theory had obtained by Kant's later years, such that it had to be accommodated in any system of natural philosophy; or, is it that (ii), for Kant, "Newtonian physics is in itself a philosophical critique of metaphysics as traditionally practiced" (DiSalle 2006, 57)? On the basis of the evidence presented above, i.e., the many non-Newtonian and relationist features in Kant's theory, option (i) appears to be the better interpretation. In essence, Kant's chief innovation in the *Metaphysical Foundations*—the incorporation of gravity within the same bodily action/reaction center-of-mass frame model as used for impact, and based on the same relationist interpretation of that strategy for gravity as for impact—is thoroughly Huygensian in character. Newton also employed the center-of-mass frame, but he did not sanction a relational interpretation of that system. Returning to the discussion in §1.3.3, Newton's conception of gravitational attraction is identical to the interpretation of impact favored by Pardies, Borelli, Mariotte, et al. in that they all distinguish an absolute (or world) space from the merely relative space of bodies. Kant, on the other hand, sides with Huygens in sanctioning a relative space interpretation alone, while simultaneously rejecting the reality of absolute space—and, as revealed in §11.2.2, Kant even goes so far as to reinterpret fundamental Newtonian principles in order to uphold his relationism, contra (ii). That is, by specifically rejecting both a uniform inertial motion and rotation of the world, Kant has exposed his true allegiance to the non-Newtonian relationist cause, notwithstanding the many Newtonian trappings of his approach, especially the (somewhat cynical) use of the term "absolute space" to signify a scheme for constructing relative spaces. Put simply, Kant's non-trivial deviations from Newtonian orthodoxy undercut the claim that he came to see Newtonian physics as a philosophical critique of metaphysics as traditionally practiced; rather, the evidence points in the opposite direction: a relationist "applied metaphysics" serves as the basis of a philosophical critique of Newtonian physics as traditionally practiced. Kant was deeply impressed by the Newtonian accomplishment, needless to say, but his goal in the *Metaphysical Foundations* would appear to have been a reconstruction of Newtonian theory grounded on a strict relationist conception of space and motion. Kant's achievement, accordingly, places him in a line of development that starts with Huygens and Berkeley and extends forward to Mach (and even Barbour and Bertotti), namely, a center-of-mass frame relationist mechanics. Kant, in fact, stands at the exact midway point in the evolution of this approach to physics, incorporating both the earlier Huygensian impact version of this strategy as well as the later Machian gravitation type. Historians and philosophers of science should not allow themselves, therefore, to be (with apologies to Pope) blinded by Newton's light in seeking a full accounting of Kant's strategy and innovations in the *Metaphysical Foundations*.

References

Adams, R. 1983. Phenomenalism and corporeal substance in Leibniz. In *Midwest studies in philosophy*, vol. 8, ed. P.A. French et al., 217–257. Minneapolis: University of Minnesota Press.

Adams, R. 1994. *Leibniz: Determinist, theist, idealist*. Oxford: Oxford University Press.

Albert, D. 1996. Elementary quantum metaphysics. In *Bohmian mechanics and quantum theory: An appraisal*, ed. J.T. Cushing, A. Fine, and S. Goldstein, 277–284. Dordrecht: Kluwer.

Algra, K. 1995. *Conceptions of space in Greek thought*. Leiden: Brill.

Allison, H. 1983. *Kant's transcendental idealism*. New Haven: Yale University Press.

Aquinas, T. 1964. *Summa Theologiae*, Blackfriars edition, 61 volumes. New York: McGraw-Hill.

Ariew, R. 1999. *Descartes and the last Scholastics*. Ithaca: Cornell University Press.

Aristotle. 1984. In *The complete works of Aristotle*, ed. J. Barnes. Princeton: Princeton University Press.

Aristotle. 1996. *Physics*. Trans. R. Waterford. Oxford: Oxford University Press.

Armstrong, D.M. 1988. Can a naturalist believe in universals. In *Science in reflection*, ed. E. Ullmann-Margalit, 103–115. Dordrecht: Kluwer.

Armstrong, D.M. 1997. *A world of states of affairs*. Cambridge: Cambridge University Press.

Arntzenius, F. 2012. *Space, time, and stuff*. Oxford: Oxford University Press.

Arthur, R. 1994. Space and relativity in Newton and Leibniz. *British Journal for the Philosophy of Science* 45: 219–240.

Arthur, R. 2012. *Leibniz on the relativity of motion*. Unpublished manuscript, 6th Annual Conference of the Leibniz Society of North America, Montreal.

Arthur, R. 2013. Leibniz's theory of space. *Foundations of Science* 18: 449–528.

Auyang, S. 1995. *How is quantum field theory possible?* New York: Oxford University Press.

Auyang, S. 2000. Mathematics and reality: Two notions of spacetime in the analytic and constructionist views of gauge field theories. *Philosophy of Science* 67: 482–494.

Auyang, S. 2001. Spacetime as a fundamental and inalienable structure of fields. *Studies in History and Philosophy of Modern Physics* 32: 205–216.

Azzouni, J. 2004. *Deflating existential consequence*. Oxford: Oxford University Press.

Bain, J. 2006. Spacetime structuralism. In *The ontology of spacetime*, vol. 1, ed. D. Dieks, 37–66. Amsterdam: Elsevier.

Bain, J. 2009. Motivating structural realist interpretations of spacetime. *Metaphysics of Science*, Melbourne, July 2–5, 2009, Philsci-archive.

Baker, D. 2005. Spacetime substantivalism and Einstein's cosmological constant. *Philosophy of Science* 72: 1299–1311.

Balaguer, M. 2009. Platonism in metaphysics. In *The Stanford encyclopedia of philosophy*, ed. E. Zalta. http://plato.stanford.edu/archives/sum2009/entries/platonism/.

Balaguer, M., E. Landry, S. Bangu, and C. Pincock. 2013. Structures, fictions, and the explanatory epistemology of mathematics in science. *Metascience* 22: 247–273.

© Springer International Publishing Switzerland 2016
E. Slowik, *The Deep Metaphysics of Space*, European Studies in Philosophy
of Science, DOI 10.1007/978-3-319-44868-8

Bannier, U. 1994. Intrinsic algebraic characterization of spacetime structure. *International Journal of Theoretical Physics* 33: 1797–1809.

Barbour, J. 1982. Relational concepts of space and time. *British Journal for the Philosophy of Science* 33: 251–274.

Barbour, J. 1989. *Absolute or relative motion?* The discovery of dynamics, vol. 1. Cambridge: Cambridge University Press.

Barbour, J. 1994. On the origin of structure in the universe. In *Philosophy, mathematics and modern physics*, ed. E. Rudolph and I.O. Stamatescu. Berlin: Springer.

Barbour, J. 1999. *The end of time*. Oxford: Oxford University Press.

Barbour, J. 2003. The deep and suggestive principles of leibnizian philosophy. *The Harvard Review of Philosophy* 11: 45–58.

Barbour, J., and B. Bertotti. 1977. Gravity and inertia in a Machian framework. *Nuovo Cimento* 38B: 1–27.

Barbour, J., and B. Bertotti. 1982. Mach's principle and the structure of dynamical theories. *Proceedings of the Royal Society* 382: 295–306.

Barbour, J., and L. Smolin. 1992. *Extremal variety as the foundation of a cosmological quantum theory*. http://arXiv.org/abs/hep-th/9203041.

Barrow, I. 1734. *The Usefulness of Mathematical Learning Explained and Demonstrated*. Trans. J. Kirkby. London: Stephen Austin.

Barrow, I. 1860. *The mathematical works of Isaac Barrow D. D.*, ed. W. Whewell. Cambridge: Cambridge University Press.

Barrow, I. 1976. *Lectiones Geometricae*, 4–15. Trans. M. Capek. In *The concepts of space and time*, ed. M. Capek, 203–208. Dordrecht: Reidel.

Belkind, O. 2007. Newton's conceptual argument for absolute space. *International Studies in the Philosophy of Science* 21: 271–293.

Bell, J.S. 1987. *Speakable and unspeakable in quantum mechanics*. Cambridge: Cambridge University Press.

Belot, G. 2000. Geometry and motion. *British Journal for the Philosophy of Science* 51: 561–595.

Belot, G. 2006. The representation of time and change in mechanics. In *Handbook of philosophy of physics*, ed. J. Butterfield and J. Earman. Dordrecht: Kluwer.

Belot, G. 2011. *Geometric possibility*. Oxford: Oxford University Press.

Belot, G., and J. Earman. 2001. Pre-Socratic quantum gravity. In *Physics meets philosophy at the Planck Scale*, ed. C. Callender and N. Huggett, 213–255. Cambridge: Cambridge University Press.

Berkeley, G. 1843. *The works of George Berkeley, D. D.*, 2 vol., ed. G. Wright. London: Thomas Tegg.

Bernstein, H. 1984. Leibniz and Huygens on the 'relativity' of motion. *Studia Leibnitiana* 13: 85–101.

Bird, A. 2007. *Nature's metaphysics*. Oxford: Oxford University Press.

Blackwell, R. 1966. Descartes' laws of motion. *Isis* 57: 220–234.

Bolton, M. 1998. Universals, essences, and abstract Entities. In *The Cambridge history of seventeenth century philosophy*, vol. 1, ed. D. Garber and M. Ayers, 178–211. Cambridge: Cambridge University Press.

Borelli, G. 1667. *De Vi Percussionis*. Bologna: Ex typographia Iacobi Montii.

Boscovich, R. 1966 [1764]. *A Theory of Natural Philosophy*. Trans. J.M. Child. Cambridge, MA: MIT Press.

Boylan, M. 1980. Henry More's space and the spirit of nature. *Journal of the History of Philosophy* 18: 395–405.

Brading, K., and A. Skiles. 2012. Underdetermination as a Path to Structural Realism. In *Structural realism: Structure, object, and causality*, ed. E. Landry and D. Rickles, 99–116. Dordrecht: Springer.

Brading, K., and E. Landry. 2006. A minimal construal of scientific structuralism. *Philosophy of Science* 73: 571–581.

Bricker, P. 1990. Absolute time versus absolute motion: Comments on Lawrence Sklar. In *Philosophical perspectives on Newtonian science,* ed. P. Bricker and R.I.G. Hughes, 77–91. Cambridge, MA: MIT Press.

Brickman, B. 1943. On physical space, Francesco Patrizi. *Journal of the History of Ideas* 4: 224–225.

Brighouse, C. 1994. Spacetime and holes. In *PSA 1994,* vol. 1, ed. D. Hull, M. Forbes, and R. Burian, 117–125. East Lansing: Philosophy of Science Association.

Broad, C.D. 1981 [1946]. Leibniz's Last Controversy with the Newtonians. In *Leibniz: Metaphysics and philosophy of science,* ed. R.S. Woolhouse, 157–174. Oxford: Oxford University Press.

Brown, H. 2005. *Physical relativity.* Oxford: Oxford University Press.

Brown, H., and O. Pooley. 2006. Minkowski space-time: A glorious non-entity. In *The ontology of spacetime,* vol. 1, ed. D. Dieks, 67–92. Amsterdam: Elsevier.

Bub, J. 2010. Quantum probabilities: An information-theoretic interpretation. In *Probabilities in physics,* ed. C. Beisbart and S. Hartmann, 231–262. Oxford: Oxford University Press.

Bueno, O. 1997. Empirical adequacy: A partial structures approach. *Studies in History and Philosophy of Science* 28: 585–610.

Bueno, O. 2005. Dirac and the dispensability of mathematics. *Studies in History and Philosophy of Modern Physics* 36: 465–490.

Bueno, O. 2012. Structural empiricism, again. In *Scientific structuralism,* ed. P. Bokulich and A. Bokulich, 81–104. Dordrecht: Springer.

Buroker, J. 1981. *Space and incongruence: The origins of Kant's idealism.* Dordrecht: Reidel.

Burtt, E.A. 1952. *The metaphysical foundations of modern science.* Atlantic Highlands: Humanities Press.

Butterfield, J. 1989. The hole truth. *British Journal for the Philosophy of Science* 40: 1–28.

Butterfield, J., and C. Isham. 2001. Spacetime and the philosophical challenge of quantum gravity. In *Physics meets philosophy at the Planck Scale,* ed. C. Callender and N. Huggett, 33–89. Cambridge: Cambridge University Press.

Cao, T.Y. 1997. *Conceptual development of 20th century field theories.* Cambridge: Cambridge University Press.

Cao, T.Y. 2006. Structural realism and quantum gravity. In *The structural foundations of quantum gravity,* ed. D. Rickles, S. French, and J. Saatsi, 40–52. Oxford: Oxford University Press.

Carrier, M. 1992. Kant's relational theory of absolute motion. *Kant-Studien* 83: 399–416.

Carriero, J. 1990. Newton on space and time: Comments on J. E. McGuire. In *Philosophical perspectives on Newtonian science,* ed. P. Bricker and R.I.G. Hughes, 109–134. Cambridge, MA: MIT Press.

Casey, E. 1997. *The fate of place.* Berkeley: University of California Press.

Cassirer, E. 1952 [1910/1921]. *Substance and Function and Einstein's Theory of Relativity.* Trans. W.C. Swabey and M.C. Swabey. New York: Dover.

Cassirer, E. 1979 [1945]. Reflections on the concept of group and the theory of perception. In *Symbol, myth, and culture,* ed. D.P. Verene. New Haven: Yale University Press.

Chakravartty, A. 2011. Scientific realism and ontological relativity. *The Monist* 94: 157–180.

Charleton, W. 1654. *Physiologia Epicuro-Gassendo-Charletoniana.* London: Johnson Reprint Corporation, 1654.

Chihara, C. 2004. *A structural account of mathematics.* Oxford: Oxford University Press.

Churchland, P.M. 1988. *Matter and consciousness.* Cambridge, MA: MIT Press.

Clagett, M. 1959. *The science of mechanics in middle ages.* Madison: University of Wisconsin Press.

Clifton, R., J. Bub, and H. Halvorson. 2003. Characterizing quantum theory in terms of information-theoretic constraints. *Foundations of Physics* 33: 1561–1591.

Colyvan, M. 2001. *The indispensability of mathematics.* Oxford: Oxford University Press.

Cook, J.W. 1979. A reappraisal of Leibniz's views on space, time, and motion. *Philosophical Investigations* 2: 22–63.

Copenhaver, B. 1980. Jewish theologies of space in the scientific revolution: Henry More, Joseph Raphson, Isaac Newton and their predecessors. *Annals of Science* 37: 489–548.

Cover, J.A. 1989. Relations and reduction in Leibniz. *Pacific Philosophical Quarterly* 70: 185–211.

Cover, J.A., and G. Hartz. 1994. Are Leibnizian Monads spatial? *History of Philosophy Quarterly* 11: 295–316.

Da Costa, N., and S. French. 2003. *Science and partial truth*. Oxford: Oxford University Press.

Dainton, B. 2010. *Time and space*, 2nd ed. Montreal: McGill-Queen's University Press.

De Risi, V. 2007. *Geometry and monadology: Leibniz's analysis situs and philosophy of space*. Basel: Birkhäuser.

De Risi, V. 2012. Leibniz on relativity: The debate between Hans Reichenbach and Dietrich Mahnke on Leibniz's theory of motion and time. In *New essays in Leibniz reception: Science and philosophy of science 1800–2000*, ed. R. Krömer and Y.C. Drian, 143–185. Basel: Springer.

De Risi, V. 2015. *Francesco Patrizi and the new geometry of space*, unpublished manuscript.

Debs, T.A., and M. Redhead. 2007. *Objectivity, invariance, and convention: Symmetry in physical science*. Cambridge, MA: Harvard University Press.

Demopolous, W., and M. Friedman. 1985. Critical notice: Bertrand Russell's *The Analysis of Matter*: Its historical context and contemporary interest. *Philosophy of Science* 52: 621–639.

Des Chene, D. 1996. *Physiologia: Natural philosophy in late Aristotelian and Cartesian thought*. Ithaca: Cornell University Press.

Descartes, R. 1976. In *Oeuvres de Descartes*, ed. C. Adams and P. Tannery. Paris: J. Vrin.

Descartes, R. 1983. *Principles of Philosophy*. Trans. V.R. Miller and R.P. Miller. Dordrecht: Kluwer Academic Publishers.

Descartes, R. 1984. *The Philosophical Writings of Descartes, Vol. 2*. Ed. and Trans. J. Cottingham, R. Stoothoff, and D. Murdoch. Cambridge: Cambridge University Press.

Descartes, R. 1985. *The Philosophical Writings of Descartes, Vol. 1*. Ed. and Trans. J. Cottingham, R. Stoothoff, and D. Murdoch. Cambridge: Cambridge University Press.

Descartes, R. 1991. *The Philosophical Writings of Descartes, Vol. 3, The Correspondence*. Ed. and Trans. J. Cottingham, et al. Cambridge: Cambridge University Press.

Dieks, D. 2000. Consistent histories and relativistic invariance in the modal interpretation of quantum mechanics. *Physics Letters A* 265: 317–325.

Dieks, D. 2001a. Space-time relationism in Newtonian and relativistic physics. *International Studies in the Philosophy of Science* 15: 5–17.

Dieks, D. 2001b. Space and time in particle and field physics. *Studies in History and Philosophy of Modern Physics* 32: 217–241.

DiSalle, R. 1994. On dynamics, indiscernibility, and spacetime ontology. *British Journal for the Philosophy of Science* 45: 265–287.

DiSalle, R. 1995. Spacetime theory as physical geometry. *Erkenntnis* 42: 317–337.

DiSalle, R. 2002. Newton's philosophical analysis of space and time. In *The Cambridge companion to Newton*, ed. I.B. Cohen and G.E. Smith, 33–56. Cambridge: Cambridge University Press.

DiSalle, R. 2006. *Understanding space-time*. Cambridge: Cambridge University Press.

DiSalle, R. 2013. The transcendental method from newton to Kant. *Studies in History and Philosophy of Science* 44: 448–456.

Dobbs, B.J.T. 1991. *The Janus Face of Genius*. Cambridge: Cambridge University Press.

Dorato, M. 2000. Substantivalism, relationism, and structural spacetime realism. *Foundations of Physics* 30: 1605–1628.

Dorato, M. 2008. Is structural spacetime realism relationism in disguise?: The supererogatory nature of the substantivalism/relationism debate. In *The ontology of spacetime II*, ed. D. Dieks, 17–38. Amsterdam: Elsevier.

Dowker, F. 2005. Causal sets and the deep structure of spacetime. In *100 years of relativity: Space-time structure: Einstein and beyond*, ed. A. Ashtekar, 445–464. Hackensack: World Scientific.

Ducheyne, S. 2008. J. B. Van Helmont's *De Tempore* as an influence on Isaac Newton's Doctrine of absolute time. *Archiv für Geschichte der Philosophie* 90: 216–228.

Dugas, R. 1958. *Mechanics in the Seventeenth Century*. Trans. F. Jaquot. Neuchatel: Griffon.

Earman, J. 1971. Kant, incongruous counterparts, and the nature of space and SpaceTime. *Ratio* 13: 1–18.

Earman, J. 1977. *Perceptions and relations in the monadology*, 212–230. IX: Studia Leibnitiana.

Earman, J. 1989. *World enough and space-time*. Cambridge: MIT Press.

Earman, J. 2006. The Implications of general covariance for the ontology and ideology of space-time. In *The ontology of spacetime*, vol. 1, ed. D. Dieks, 3–23. Amsterdam: Elsevier.

Earman, J., and J. Norton. 1987. What price space-time substantivalism?: The hole story. *British Journal for the Philosophy of Science* 38: 515–525.

Eddington, A. 1939 [1958]. *The philosophy of physical science*. Ann Arbor: University of Michigan Press.

Einstein, A. 1923. The foundations of the general theory of relativity. In *The Principle of Relativity*. Trans. W. Perrett and G.B. Jeffery, 109–164. New York: Dover Publications.

Einstein, A. 1961. Relativity and the problem of space. In *Relativity: The special and the general theory*. New York: Crown Publishers.

Elzinga, A. 1972. *On a research program in early modern physics*. New York: Humanities Press.

Esfeld, M. 2013. Ontic structural realism and the interpretation of quantum mechanics. *European Journal for the Philosophy of Science* 3: 19–32.

Esfeld, M., and V. Lam. 2008. Moderate structural realism about space-time. *Synthese* 160: 27–46.

Euler, L. 1748. Réflexions sur l'espace et le temps. *Histoire de l'Academie Royale des sciences et belles lettres* 4: 324–333.

Euler, L. 1761. *Letters to a German Princess*. Trans. H. Hunter. London: Murray and Highley.

Falkenstein, L. 2004. *Kant's Intuitionism*. Toronto: University of Toronto Press.

Ferguson, J. 1974. *The Philosophy of Dr. Samuel Clarke and Its Critics*. New York: Vantage Press.

Field, H. 1980. *Science without numbers*. Princeton: Princeton University Press.

Field, H. 1985. Can we dispense with space-time?. In *PSA 1984*, 2, ed. P. Asquith and P. Kitcher, 30–90. East Lansing: Philosophy of Science Association.

Field, H. 1989. Can we dispense with space-time? In *Realism, mathematics, and modality*, 171–226. London: Blackwell.

Foster, J. 1982. *The case for idealism*. London: Routledge & Kegan Paul.

Frasca-Spada, M. 2002. *Space and self in Hume's* Treatise. Cambridge: Cambridge University Press.

French, S. 1989. Identity and individuality in classical and quantum physics. *Australasian Journal of Philosophy* 67: 432–446.

French, S. 2006. Structure as a weapon of the realist. *Proceedings of the Aristotelian Society* 106: 167–185.

French, S. 2011. Metaphysical underdetermination: Why worry? *Synthese* 180: 205–221.

French, S. 2012. The presentation of objects and the representation of structure. In *Structural realism: Structure, object, and causality*, ed. M. Landry and D. Rickles, 3–28. Dordrecht: Springer.

French, S. 2014. *The structure of the world*. Oxford: Oxford University Press.

French, S., and J. Ladyman. 2003. Remodelling structural realism: Quantum physics and the metaphysics of structure. *Synthese* 136: 31–56.

French, S., and J. Ladyman. 2012. In defense of ontic structural realism. In *Scientific structuralism*, ed. P. Bokulich and A. Bokulich, 25–42. Dordrecht: Springer.

French, S., and J. Saatsi. 2006. Realism about structure: The semantic view and non-linguistic representations. *Philosophy of Science* 73: 548–559.

Friedman, M. 1981. Theoretical explanation. In *Reduction, time and reality*, ed. R. Healey. Cambridge: University of Cambridge Press.

Friedman, M. 1983. *Foundations of space-time theories*. Princeton: Princeton University Press.

Friedman, M. 1992. *Kant and the exact sciences*. Cambridge: Harvard University Press.

Friedman, M. 2000. *A Parting of the ways: Carnap, Cassirer, and Heidegger*. Chicago: Open Court.

Friedman, M. 2001. *Dynamics of reason*. Stanford: CSLI Publications.

Friedman, M. 2004. Introduction. In *Metaphysical foundations of natural science*, ed. I. Kant. Cambridge: Cambridge University Press.

Friedman, M. 2005. Ernst Cassirer and the philosophy of science. In *Continental philosophy of science*, ed. G. Gutting, 71–84. London: Blackwell.

Friedman, M. 2013a. Neo-Kantianism, scientific realism, and modern physics. In *Scientific metaphysics*, ed. D. Ross, J. Ladyman, and H. Kincaid, 182–197. Oxford: Oxford University Press.

Friedman, M. 2013b. *Kant's construction of nature*. Cambridge: Cambridge University Press.

Funkenstein, A. 1986. *Theology and the scientific imagination from the middle ages to the seventeenth century*. Princeton: Princeton University Press.

Furth, M. 1967. Monadology. *Philosophical Review* 76: 169–200.

Futch, M.J. 2008. *Leibniz's metaphysics of time and space*. Berlin: Springer-Verlag.

Garber, D. 1992. *Descartes' metaphysical physics*. Chicago: University of Chicago Press.

Garber, D. 2009. *Leibniz: Body, substance, monad*. Oxford: Oxford University Press.

Garber, D., J. Henry, L. Joy, and A. Gabbey. 1998. New doctrines of body and its powers, place, and space. In *The Cambridge history of seventeenth century philosophy*, vol. 1, ed. D. Garber and M. Ayers, 553–623. Cambridge: Cambridge University Press.

Gassendi, P. 1658. *Petri Gassendi Opera Omnia in sex tomos divisa*, 6 volumes. Lyon: Laurent Anisson and Jean-Baptiste Devenet.

Gassendi, P. 1972. *The Selected Works of Pierre Gassendi*, Ed. and Trans. C.G. Brush. New York: Johnson.

Gassendi, P. 1976. *The reality of infinite void*. In *The Concepts of Space and Time*. Trans. M. Capek and W. Emge, 91–96. Dordrecht: D. Reidel.

Glymour, C. 1977. Indistinguishable space-times and the fundamental group. In *Foundations of space-time theories, Minnesota studies in the philosophy of science*, vol. 8., ed. J. Earman, C. Glymour, and J. Stachel, 50–60. Minneapolis: University of Minnesota Press.

Gorham, G. 2003. Descartes's Dilemma of Eminent Containment. *Dialogue* 42: 3–25.

Gorham, G. 2011. Newton on God's relation to space and time: The Cartesian framework. *Archiv für Geschichte der Philosophie* 93: 281–320.

Grant, E. 1981. *Much Ado about nothing: Theories of space and vacuum from the Middle Ages to the acientific revolution*. Cambridge: Cambridge University Press.

Grant, E. 2010. *The nature of natural philosophy in the late Middle Ages*. Washington, DC: Catholic University of America Press.

Graves, J.C. 1971. *The conceptual foundations of contemporary relativity theory*. Cambridge: MIT Press.

Greene, B. 2011. *The hidden reality*. New York: Knopf.

Hall, A.R. 1990. *Henry more and the scientific revolution*. Cambridge: Cambridge University Press.

Hall, A.R. 2002. *Henry more and the scientific revolution*, 2nd ed. Cambridge: Cambridge University Press.

Harré, R. 1986. *Varieties of realism*. Oxford: Blackwell Publishers.

Hartz, G.A. 2007. *Leibniz's final system*. London: Routledge.

Healey, R. 1995. Substance, modality and spacetime. *Erkenntnis* 42: 287–316.

Hedrich, R. 2009. Quantum gravity: Has spacetime quantum properties?, philsci-archive.pitt.edu/archive/00004445. (arXiv:0902.0190).

Hellman, G. 1989. *Mathematics without numbers*. Oxford: Oxford University Press.

Henry, J. 1979. Francesco Patrizi da Cherso's Concept of Space and its Later Influence. *Annals of Science* 36: 549–575.

Hinckfuss, I. 1975. *The existence of space*. Oxford: Oxford University Press.

Hiscock, W.G. 1937. In *David Gregory, Isaac Newton, and their circle*, ed. W.G. Hiscock. Oxford: Oxford University Press.

Hobbes, T. 1839. *The English works of Thomas Hobbes of Malmesbury*, vol. 1, ed. W. Molesworth. London: John Bohn.

Hoefer, C. 1996. *The metaphysics of space-time substantivalism*, 5–27. XCIII: Journal of Philosophy.

Hoefer, C. 1998. Absolute versus Relational Spacetime: For Better or Worse, the Debate Goes On. *British Journal for the Philosophy of Science* 49: 451–467.

Hoefer, C. 2000. Energy conservation in GTR. *Studies in History and Philosophy of Modern Physics* 31: 187–199.

Holden, T. 2004. *The architecture of matter*. Oxford: Oxford University Press.

Hooker, C.A. 1971. The relational doctrines of space and time. *British Journal for the Philosophy of Science* 22: 97–130.

Horgan, T., and M. Potrc. 2008. *Austere realism*. Cambridge, MA: MIT Press.

Horwich, P. 1978. On the existence of time, space, and space-time. *Noûs* 12: 397–419.

Hübner, K. 1983. *Critique of Scientific Reason*. Trans. P.R. Dixon and H.M. Dixon. Chicago: University of Chicago Press.

Huggett, N. 1999. Why manifold substantivalism is probably Not a consequence of classical mechanics. *International Studies in the Philosophy of Science* 13: 17–34.

Huggett, N. 2006a. Can Spacetime Help Settle any Issues in Modern Philosophy?", philsci-archive. pitt.edu/id/eprint/3085.

Huggett, N. 2006b. The regularity account of relational spacetime, *Mind* 114: 41–73.

Huggett, N. 2008. Why the parts of absolute space are immobile. *British Journal for the Philosophy of Science* 59: 391–407.

Huggett, N., and C. Hoefer. 2015. Absolute and relational theories of space and motion. In *The Stanford encyclopedia of philosophy*, ed. E. Zalta. http://plato.stanford.edu/archives/spr2015/entries/spacetime-theories.

Huggett, N., and C. Wüthrich. 2013. Emergent spacetime and empirical (in)coherence. *Studies in History and Philosophy of Modern Physics* 44: 276–285.

Hume, D. 2000 [1739]. *A treatise of human nature,* ed. D.F. Norton and M.J. Norton. Oxford: Oxford University Press.

Husserl, E. 1997 [1907]. *Thing and Space*. Trans. and Ed. R. Rojcewicz. Dordrecht: Kluwer.

Huygens, C. 1993. *Codex Huygens7a in Penetralia Motus: la fondazione relativistica della meccanica nell'opera di Chr. Huygens*, ed. G. Mormino. Trans. M. Stan. Firenze: Nova Italia.

Jammer, M. 1993 [1954]. *Concepts of space*, 3rd ed. New York: Dover Publications.

Janiak, A. 2000. Space, atoms, and mathematical divisibility in Newton. *Studies in History and Philosophy of Science* 31: 203–230.

Janiak, A. 2008. *Newton as philosopher*. Cambridge: Cambridge University Press.

Jauernig, A. 2008. Leibniz on motion and the equivalence of hypotheses. *The Leibniz Review* 18: 1–40.

Jauernig, A. 2009. Leibniz on motion—Reply to Edward Slowik. *The Leibniz Review* 19: 139–147.

Jesseph, D. 1993. *Berkeley's philosophy of mathematics*. Chicago: University of Chicago Press.

Jesseph, D. 2016. Ratios, quotients, and the language of nature. In *The language of nature: Reassessing the mathematization of natural philosophy*, 160–177. Minnesota studies in the philosophy of science. Minneapolis: University of Minnesota Press.

Johnson, W.E. 1964 [1922]. *Logic*, vol. II. New York: Dover Publications.

Kant, I. 1992. *Theoretical Philosophy: 1755–1770*. Trans. and Ed. D. Walford and M. Meerbote. Cambridge: Cambridge University Press.

Kant, I. 1997. *Prolegomena to Any Future Metaphysics*. Trans. and Ed. G. Hatfield. Cambridge: Cambridge University Press.

Kant, I. 1998. *Critique of Pure Reason*. Trans and Ed. P. Guyer and A. Wood. Cambridge: Cambridge University Press.

Kant, I. 2004. *Metaphysical Foundations of Natural Science*. Trans. and Ed. M. Friedman. Cambridge: Cambridge University Press.

Kant, I. 2012. *Natural Science*, ed. E. Watkins. Trans. L.W. Beck, J.B. Edwards, O. Reinhardt, M. Schönfeld, and E. Watkins. Cambridge: Cambridge University Press.

Kaplunovsky, V., and M. Weinstein. 1985. Space-time: Arena or illusion? *Physical Review* D31: 1879–1898.

Khamara, E. 1993. Leibniz's theory of space: A reconstruction. *Philosophical Quarterly* 43: 472–488.

Khamara, E. 2006. *Space, time, and theology in the Leibniz-Newton controversy*. Dordrecht: Ontos Verlag.

Kitcher, P. 1981. Explanatory unification. *Philosophy of Science* 48: 507–531.

Korté, H. 2006. Einstein's hole argument and Weyl's field-body relationism. In *Physical theory and its interpretation*, ed. W. Demopoulus and I. Pitowsky, 183–212. Leiden: Springer.

Koslow, A. 1976. Ontological and ideological issues of the classical theory of space and time. In *Motion and time, space and matter*, ed. P.K. Machamer and R.G. Turnbull, 224–263. Columbus: Ohio State University Press.

Kosso, P. 1997. *Appearance and reality*. Oxford: Oxford University Press.

Koyré, A. 1957. *From the closed world to the infinite universe*. Baltimore: Johns Hopkins University Press.

Koyré, A. 1965. *Newtonian studies*. Cambridge, MA: Harvard University Press.

Koyré, A., and I.B. Cohen. 1962. Newton and the Leibniz-Clarke correspondence. *Archives Internationales d'Histoire des Sciences* 15: 63–126.

Kragh, H. 2011. *Higher speculations: Grand theories and failed revolutions in physics and cosmology*. Oxford: Oxford University Press.

Ladyman, J. 1998. What is structural realism? *Studies in the History and Philosophy of Science* 29: 409–424.

Ladyman, J. 2009. Structural realism. In *Stanford encyclopedia of philosophy*, ed. E. Zalta. http://plato.stanford.edu/entries/structural-realism/.

Ladyman, J., and D. Ross. 2007. *Every thing must go: Metaphysics naturalized*. Oxford: Oxford University Press.

Lam, V. 2011. Gravitational and nongravitational energy: The need for background structures. *Philosophy of Science* 78: 1012–1023.

Lam, V., and M. Esfeld. 2013. A dilemma for the emergence of spacetime in canonical quantum gravity. *Studies in History and Philosophy of Modern Physics* 44: 286–293.

Lang, H. 1998. *The order of nature in Aristotle's physics: Place and the elements*. Cambridge: Cambridge University Press.

Laudan, L. 1996. *Beyond positivism and relativism*. Boulder: Westview Press.

Leibniz, G.W. 1923. In *Sämtliche Schriften und Briefe*, ed. Akademie der Wissenschaften der DDR. Darmstadt/Berlin: Akademie-Verlag.

Leibniz, G.W. 1962. In *Leibnizens Mathematische Schriften*, ed. C.I. Gerhardt. Hildesheim: Olms.

Leibniz, G.W. 1965. In *Die Philosophischen Schriften von Leibniz*, ed. C.I. Gerhardt. Hildesheim: Olms.

Leibniz, G.W. 1969. *Leibniz: Philosophical Letters and Papers*, 2nd ed., ed. and trans. L.E. Loemker. Dordrecht: Kluwer.

Leibniz, G.W. 1989. *Leibniz: Philosophical Essays*. Ed. and Trans. R. Ariew and D. Garber. Indianapolis: Hackett.

Leibniz, G.W. 1995. *Leibniz: Philosophical Writings*. Ed. and Trans. M. Morris and G.H.R. Parkinson. Rutland, Vermont: C. Tuttle.

Leibniz, G.W. 1996. *New Essay on Human Understanding*, ed. and trans. by P. Remnant and J. Bennett. Cambridge: Cambridge University Press.

Leibniz, G.W. 2001. *The Labyrinth of the Continuum: Writings of 1672 to 1686*. Ed. and Trans. R. Arthur. New Haven: Yale University Press.

Leibniz, G.W. 2007. *The Leibniz-Des Bosses Ccorrespondence*. Ed. and Trans. B. Look and D. Rutherford. New Haven: Yale University Press.

Leibniz, G.W. and S. Clarke. 2000. *Leibniz and Clarke Correspondence*. Ed. and Trans. R. Ariew. Indianapolis: Hackett.

Leijenhorst, C. 2002. *The Mechanisation of Aristotelianism*. Leiden: Brill.

Leijenhorst, C., and C. Lüthy. 2002. The erosion of Aristotelianism: Confessional physics in early modern Germany and the Dutch Republic. In *The dynamics of Aristotelian natural philosophy from antiquity to the seventeenth century*, ed. C. Leijenhorst, C. Lüthy, and J. Thijssen, 349–374. Leiden: Brill.

Lennon, T. 1993. *The Battle of the Gods and Giants: The Legacies of Descartes and Gassendi, 1655–1715*. Princeton: Princeton University Press.

Lewis, D. 1993. Mathematics is megethology. *Philosophia Mathematica* 1: 3–23.

Liston, M. 2000. Review of M. Steiner. *The Applicability of Mathematics as a Philosophical Problem Philosophia Mathematica* 8: 190–207.

Locke, J. 1975 [1690]. *An essay concerning human understanding*, ed. P. Nidditch. Oxford: Oxford University Press.

Lodge, P. 2003. Leibniz on relativity and the motion of bodies. *Philosophical Topics* 31: 277–308.

LoLordo, A. 2007. *Pierre Gassendi and the birth of early modern philosophy*. Cambridge: Cambridge University Press.

Loux, M. 2006. *Metaphysics*, 3rd ed. Routledge: New York.

Lowe, E.J. 2002. *A survey of metaphysics*. Oxford: Oxford University Press.

Mach, E. 1906. *Space and geometry in the light of physiological, psychological and physical inquiry*. Chicago: Open Court.

Mach, E. 1960 [1893]. *The Science of Mechanics*, 6th ed. Tans. T. McCormack. London: Open Court.

Magalhães, E. 2006. Armstrong on the spatio-temporality of universals. *Australasian Journal of Philosophy* 84: 301–308.

Magirus, J. 1642. *Physiologiae peipateticae contractio*, Lib.1, Cap. VIII. Cambridge.

Maier, A. 1982. *On the Threshold of Exact Science*. Trans. S.D. Sargent. Philadelphia: University of Pennsylvania Press.

Malament, D. 1977. Observationally indistinguishable space-times. In *Foundations of space-time theories*, Minnesota Studies in the Philosophy of Science, vol. 8., ed. J. Earman, C. Glymour, and J. Stachel. Minneapolis: University of Minnesota Press.

Manders, K. 1982. On the spacetime ontology of physical theories. *Philosophy of Science* 49: 575–590.

Mates, B. 1986. *The philosophy of Leibniz: Metaphysics and language*. New York: Oxford University Press.

Maudlin, T. 1988. *The essence of space-time*, 82–91. II: Philosophy of Science Association.

Maudlin, T. 1993. Buckets of water and waves of space: Why spacetime is probably a substance. *Philosophy of Science* 60: 183–203.

Maudlin, T. 2007a. *The metaphysics within physics*. Oxford: Oxford University Press.

Maudlin, T. 2007b. Completeness, supervenience, and ontology. *Journal of Physics A: Mathematical and Theoretical* 40: 3151–3171.

Maudlin, T. 2012. *Philosophy of physics: Space and time*. Princeton: Princeton University Press.

McGuire, J.E. 1978a. Existence, actuality and necessity: Newton on space and time, *Annals of Science*, 35: 463–508. Reprinted in, *Tradition and innovation: Newton's metaphysics of nature*. Dordrecht: Kluwer, 1995, 1–51.

McGuire, J.E. 1978b. Tempus et Locus. In Newton on Place, Time, and God: An Unpublished Source. Trans. and Ed. J.E. McGuire, *British Journal for the History of Science*, 11: 114–129.

McGuire, J.E. 1982. Space, infinity, and indivisibility: Newton on the creation of matter. In *Contemporary Newtonian research*. ed. Z. Bechler, 145–190. Dordrecht: Reidel.

McGuire, J.E. 1983. Space, geometrical objects and infinity: Newton and Descartes on extension. In *Nature mathematized*, ed. W.R. Shea. Dordrecht: D. Reidel. Reprinted in *Tradition and Innovation: Newton's Metaphysics of Nature*. Dordrecht: Kluwer, 1995, 151–189.

McGuire, J.E. 1990. Predicates of pure existence: Newton on God's space and time. In *Philosophical perspectives on Newtonian science*, ed. P. Bricker and R.I.G. Hughes, 91–108. Cambridge, MA: MIT Press.

McGuire, J.E. 2000. The fate of the date: The theology of Newton's *Principia* Revisited. In *Rethinking the scientific revolution*, ed. M. Osler, 271–296. Cambridge: Cambridge University Press.

McGuire, J.E. 2007. A dialogue with Descartes: Newton's ontology of true and immutable natures. *Journal of the History of Philosophy* 45: 103–125.

McGuire, J.E., and E. Slowik. 2012. Newton's ontology of omnipresence and infinite space. *Oxford Studies in Early Modern Philosophy* 6: 279–308.

McGuire, J.E., and M. Tamny. 1983. *Certain Philosophical Questions: Newton's Trinity Notebook*. Ed. and Trans. J.E. McGuire and M. Tamny. Cambridge: Cambridge University Press.

McLaughlin, B. 2008. Emergence and supervenience. In *Emergence*, ed. A. Bedau and P. Humphreys, 81–98. Cambridge: MIT Press.

More, H. 1662. Epistolae Quatour ad Renatum Des-Cartes. In *A collection of several philosophical writings*, 2nd ed, 54–133. London: James Flesher.

More, H. 1925. In *Philosophical writings of Henry More*, ed. F.I. MacKinnon. New York: AMS Press.

More, H. 1995 [1679]. *Henry More's Manual of Metaphysics: A Translation of the* Enchiridium Metaphysicum, Parts I and II, Trans. A. Jacob. Hildesheim: Olms.

More, H. 1997a [1655]. *An antidote against atheism*, 2nd ed. Bristol: Thommes Press.

More, H. 1997b [1659]. *The immortality of the soul*. Bristol: Thommes Press.

Morganti, M. 2004. On the preferability of epistemic structural realism. *Synthese* 142: 81–107.

Morison, B. 2002. *On location: Aristotle's concept of place*. Oxford: Oxford University Press.

Mugnai, M. 1992. *Leibniz' theory of relations*. Stuttgart: Franz Steiner Verlag.

Mundy, B. 1983. Relational theories of Euclidean space and Minkowski spacetime. *Philosophy of Science* 50: 205–226.

Mundy, B. 1992. Space-time and isomorphism. In *PSA 1992*, vol. 1, ed. D. Hull, M. Forbes, and K. Okruhlik, 515–527. East Lansing: Philosophy of Science Association.

Murdoch, J.E., and E.D. Sylla. 1978. The science of motion. In *Science in the middle ages*, ed. D.C. Lindberg. Chicago: University of Chicago Press.

Nerlich, G. 2005. Can the parts of space move?: On paragraph six of Newton's Scholium. *Erkenntnis* 62: 119–135.

Neumann, C. 1976 [1870]. On the necessity of the absolute Frame of reference. In *The Concepts of Space and Time*. Trans. and Ed. M. Capek. Dordrecht: Reidel.

Newstead, A. and J. Franklin. 2012. Indispensability without Platonism. In *Properties, powers and structures*, ed. A. Bird, B. Ellis, and H. Sankey, 81–97. New York: Routledge.

Newton, I. 1962a. De gravitatione et aequipondio fluidorum. In *Unpublished Scientific Papers of Isaac Newton*. Trans. and Ed. A.R. Hall and M.B. Hall, 89–164. Cambridge: Cambridge University Press.

Newton, I. 1962b. *Mathematical principals of natural philosophy*. Trans. and Ed. A. Motte and F. Cajori. Berkeley: University of California Press.

Newton, I. 1983. *Certain Philosophical Questions: Newton's Trinity Notebook*. Ed. and Trans. J.E. McGuire and M. Tamny. Cambridge: Cambridge University Press.

Newton, I. 2004. *Philosophical Writings*. Trans. and Ed. A. Janiak and C. Johnson. Cambridge: Cambridge University Press.

Ney, A. 2010. The status of our ordinary three dimensions in a quantum universe. *Noûs* 46: 525–560.

Ney, A., and D. Albert. 2013. *The wave function: Essays on the metaphysics of quantum mechanics*. Oxford: Oxford University Press.

Norton, J. 2008. Why constructive relativity fails. *British Journal for the Philosophy of Science* 59: 821–834.

Nozick, R. 2001. *Invariances: The structure of the objective world*. New York: Belknap.

Ockham, W. 1967. *Philosophical Writings*. Trans and Ed. P. Boehner. London: Thomas Nelson.

Oresme, N. 1968. *Le Livre Du Ciel et Du Monde*. Ed. and Trans. A.D. Menut and A.J. Denomy. Madison: University of Wisconsin Press, 1968.

Pardies, I. 1670. *A Discourse of Local Motion (Discours du mouvement local)*. Trans. A. M. London: Moses Pitt.

Pasnau, R. 2011. *Metaphysical themes: 1274–1671*. Oxford: Oxford University Press.

Philoponus, J. 1992. *Commentaries on Aristotle's Physics*. Trans. J.O. Urmson. In *Corollaries on place and void*. Ithaca: Cornell University Press.

Pincock, C. 2012. *Mathematics and scientific representation*. Oxford: Oxford University Press.

Poincaré, H. 1905 [1952]. *Science and Hypothesis*. Trans. W.J. Greenstreet. New York: Dover.

Pooley, O. 2006. Points, particles, and structural realism. In *The structural foundations of quantum gravity,* ed. D. Rickles, S. French, and J. Saatsi, 83–120. Oxford: Oxford University Press.

Pooley, O. 2013. Substantivalist and Relationalist approaches to spacetime. In *Oxford handbook of philosophy of physics*, ed. B. Batterman, 522–586. Oxford: University of Oxford Press.

Popper, K. 1953. A note on Berkeley as precursor of Mach. *British Journal for the Philosophy of Science* 4: 26–36.

Priest, G. 2005. *Towards non-being*. Oxford: Oxford University Press.

Psillos, S. 2006. *The* structure, the *whole* structure and nothing *but* the structure? *Philosophy of Science* 73: 560–570.

Psillos, S. 2010. Scientific realism: Between platonism and nominalism. *Philosophy of Science* 77: 947–958.

Psillos, S. 2011. Living with the abstract: Realism and models. *Synthese* 180: 3–17.

Psillos, S. 2012. Anti-nominalistic scientific realism: A defense. In *Properties, powers and structures*, ed. A. Bird, B. Ellis, and H. Sankey, 63–80. New York: Routledge.

Quine, W.V.O. 1966. Carnap and logical truth. In *The ways of paradox*. New York: Random House.

Reid, J. 2007. The evolution of henry More's theory of divine absolute space. *Journal of the History of Philosophy* 45: 79–102.

Reid, J. 2012. *The metaphysics of Henry more*. Dordrecht: Springer.

Resnik, M. 1985. How nominalist is Hartry Field's nominalism. *Philosophical Studies* 47: 163–181.

Resnik, M. 1997. *Mathematics as a science of patterns*. Oxford: Oxford University Press.

Rickles, D. 2005. A new spin on the hole argument. *Studies in History and Philosophy of Modern Physics* 36: 415–434.

Rickles, D. 2008a. Who's afraid of background independence?. In *The ontology of spacetime II,* ed. D. Dieks, 133–152. Amsterdam: Elsevier.

Rickles, D. 2008b. Quantum gravity: A primer for philosophers. In *The Ashgate companion to contemporary philosophy of physics*, ed. D. Rickles, 262–365. Aldershot: Ashgate Publishing Limited.

Rickles, D., and S. French. 2006. Quantum gravity meets structuralism: Interweaving relations in the foundations of physics. In *The structural foundations of quantum gravity*, ed. D. Rickles, S. French, and J. Saatsi, 1–39. Oxford: Oxford University Press.

Roberts, J. 2003. Leibniz on force and absolute motion. *Philosophy of Science* 70: 553–573.

Rohault, J. 1969 [1723]. *A System of Natural Philosophy*. Trans. J. Clarke and S. Clarke. New York: Johnson Reprint Corp.

Rosen, G. 2010. Metaphysical dependence: Grounding and reduction. In *Modality: metaphysics, logic, and epistemology*, ed. B. Hale and A. Hoffman, 109–136. Oxford: Oxford University Press.

Rovelli, C. 1997. Halfway through the Woods: Contemporary research on space and time. In *The cosmos of science*, ed. J. Earman and J. Norton, 180–224. Pittsburgh: University of Pittsburgh Press.

Rovelli, C. 2001. Quantum spacetime: What do we know? In *Physics meets philosophy at the Planck Scale*, ed. C. Callender and N. Huggett, 101–122. Cambridge: Cambridge University Press.

Rovelli, C. 2004. *Quantum gravity*. Cambridge: Cambridge University Press.

Russell, B. 1938. *The principles of mathematics*. New York: W. W. Norton.

Russell, B. 1992 [1937]. *A critical exposition of the philosophy of Leibniz*, 2nd ed. London: Routledge.

Rutherford, D. 1990. Phenomenalism and the reality of body in Leibniz's later philosophy. *Studia Leibnitiana* 22: 11–28.

Rutherford, D. 1995. *Leibniz and the rational order of nature*. Cambridge: Cambridge University Press.

Rutherford, D. 2008. Leibniz as idealist. *Oxford Studies in Early Modern Philosophy* 4: 141–190.

Ryckman, T. 2005. *The reign of relativity*. Oxford: Oxford University Press.

Rynasiewicz, R. 1995. By their properties, causes, and effects: Newton's Scholium on time, space, place, and motion. *Studies in History and Philosophy of Science* 25(Part I): 133–153, 26(Part II): 295–321.

Rynasiewicz, R. 1996. Absolute versus relational space-time: An outmoded debate? *Journal of Philosophy*, XCIII: 279–306.

Rynasiewicz, R. 2000. On the distinction between absolute and relative motion. *Philosophy of Science* 67: 70–93.

Saatsi, J. 2010. Whence ontological structural realism? In *EPSA: Epistemology and methodology of science*, ed. M. Rédei, M. Dorato, and M. Suàrez, 255–265. Dordrecht: Springer.

Salmon, N. 1998. Nonexistence. *Noûs* 32: 277–319.

Sambursky, S. 1962. *The physical world of late antiquity*. Princeton: Princeton University Press.

Saunders, S. 1993. To what physics corresponds. In *Correspondence, invariance and heuristics: Essays in honour of Heinz post*, ed. S. French and H. Kamminga, 295–325. Dordrecht: Kluwer.

Schmaltz, T. 2008. *Descartes on causation*. Oxford: Oxford University Press.

Schroer, B., and H.-W. Wiesbrock. 2000. Modular theory and geometry. *Review of Mathematical Physics* 12: 139–158.

Sedley, D. 1987. Philoponus' conception of space. In *Philoponus and the rejection of Aristotelian science*, ed. R. Sorabji,140–154. London: Duckworth.

Sepkoski, D. 2007. *Nominalism and constructivism in seventeenth-century mathematical philosophy*. New York: Routledge.

Shapiro, S. 1997. *Philosophy of mathematics: Structure and ontology*. Oxford: Oxford University Press.

Shapiro, S. 2000. *Thinking about mathematics*. Oxford: Oxford University Press.

Shea, W. 1991. *The magic of numbers and motion: The scientific career of René Descartes*. Canton: Science History Publications.

Sklar, L. 1974. *Space, time, and spacetime*. Berkeley: University of California Press.

Sklar, L. 1985. *Philosophy and spacetime physics*. Berkeley: University of California Press.

Sklar, L. 1990. Real Quantities and Their Sensible Measures. In *Philosophical perspectives on Newtonian science*, ed. P. Bricker and R.I.G. Hughes, 57–76. Cambridge, MA: MIT Press.

Skow, B. 2007. Sklar's Maneuver. *British Journal for the Philosophy of Science* 58: 777–786.

Slowik, E. 2002. *Cartesian spacetime*. Dordrecht: Kluwer.

Slowik, E. 2005a. Spacetime, ontology, and structural realism. *International Studies in the Philosophy of Science* 19: 147–166.

Slowik, E. 2005b. On the cartesian ontology of general relativity: Or, conventionalism in the history of the substantival/relational debate. *Philosophy of Science* 72: 1312–1323.

Slowik, E. 2006. The 'dynamics' of Leibnizian relationism: Reference frames and force in Leibniz's plenum. *Studies in History and Philosophy of Modern Physics* 37: 617–634.

Slowik, E. 2009. Another go-around on Leibniz on rotation. *The Leibniz Review* 19: 131–138.

Slowik, E. 2010. The fate of mathematical place: Objectivity and the theory of lived-space from Husserl to Casey. In *Space, time, and spacetime*, ed. V. Petkov, 291–312. Dordrecht: Springer.

Slowik, E. 2014. Hobbes and the phantasm of space. *Hobbes Studies* 27(2014): 61–79.

Smart, J.J.C. 1964. *Problems of space and time*. New York: Macmillan.

Smolin, L. 2000. *Three roads to quantum gravity*. New York: Basic Books.

Smolin, L. 2006. The case for background independence. In *The structural foundations of quantum gravity*, ed. D. Rickles, S. French, and J. Saatsi, 196–239. Oxford: Oxford University Press.

Snelders, H. 1980. Christiaan Huygens and the concepts of matter. In *Studies on Christiaan Huygens*, ed. H. Bos et al., 104–125. Swetz & Zeitlinger: Linse.

Snobelen, S. 2001. 'God of gods, and Lord of lords': The theology of Isaac Newton's general Scholium to the *Principia*. *Osiris* 16: 169–208.

Sorabji, R. 1988. *Matter, space, and motion: Theories in antiquity and their sequel*. Ithaca, New York: Cornell University Press.

Stachel, J. 2002. 'The relations between things' versus 'The things between relations': The deeper meaning of the hole argument. In *Reading natural philosophy*, ed. D. Malament, 231–266. Chicago: Open Court.

Stachel, J. 2006. Structure, individuality, and quantum gravity. In *The structural foundations of quantum gravity*, ed. D. Rickles, S. French, and J. Saatsi, 53–82. Oxford: Oxford University Press.

Stan, M. 2012. Newton and Wolff: The Leibnizian Reaction to the *Principia*, 1716–1763. *The Southern Journal of Philosophy* 50: 459–481.

Stan, M. 2013. Kant's third law of mechanics: The long shadow of Leibniz. *Studies in History and Philosophy of Science* 44: 493–504.

Stan, M. 2014a. Once more onto the Breach: Kant and Newton. *Metascience*, 23: 233–242.

Stan, M. 2014b. Absolute space and the riddle of rotation: Kant's response to Newton. *Oxford Studies in Early Modern Philosophy* 7: 257–308.

Stein, H. 1967. Newtonian space-time. *Texas Quarterly* 10: 174–200.

Stein, H. 1977a. Some philosophical prehistory of general relativity. In *Foundations of space-time theories*, Minnesota studies in the philosophy of science, vol. 8, ed. J. Earman, C. Glymour, and J. Stachel, 3–49. Minneapolis: University of Minnesota Press.

Stein, H. 1977b. On space-time and ontology: Extract from a letter to Adolf Grünbaum. In *Foundations of space-time theories*, Minnesota studies in the philosophy of science, vol. 8, ed. J. Earman, C. Glymour, and J. Stachel, 374–403. Minneapolis: University of Minnesota Press.

Stein, H. 1989. Yes, but...some skeptical remarks on realism and anti-realism. *Dialectica* 43: 47–65.

Stein, H. 2002. Newton's metaphysics. In *The Cambridge companion to Newton*, ed. I.B. Cohen and G.E. Smith, 256–307. Cambridge: Cambridge University Press.

Suárez, F. 1965 [1597]. *Disputationes Metaphysicae*, vol. 2. Hildesheim: Verlag.

Susskind, L. 1995. The world as a Hologram. *Journal of Mathematical Physics* 36: 6377–6396.

Sylla, E. 2002. Space and spirit in the transition from Aristotelian to Newtonian space. In *The dynamics of Aristotelian natural philosophy from antiquity to the seventeenth century*, ed. C. Leijenhorst, C. Lüthy, and J. Thijssen, 249–287. Leiden: Brill.

Tahko, T.E. and E.J. Lowe. 2015. Ontological dependence. In *The Stanford encyclopedia of philosophy*, ed. E. Zalta. http://plato.stanford.edu/archives/spr2015/entries/dependence-ontological/.

Teller, P. 1987. Space-time as physical quantity. In *Kelvin's Baltimore lectures and modern theoretical physics*, ed. P. Achinstein and R. Kargon, 425–448. Cambridge, MA: MIT Press.

Teller, P. 1991. Substance, relations, and arguments about the nature of space-time. *Philosophical Review* 100: 362–397.

Toletus, F. 1589. *Commentaria una cum quaestionibus in Octo Libros Aristotelis de Physica Auscultatione*, 4th ed. Venice: Iuntas.

Torretti, R. 1999. *The philosophy of physics*. Cambridge: Cambridge University Press.

Toulmin, S. 1959. Criticism in the history of science: Newton on absolute space, time, and motion, I & II. *Philosophical Review* 68(1–29): 203–227.

van Fraassen, B. 1991. *Quantum mechanics: An empiricist view.* Oxford: Oxford University Press.

van Fraassen, B. 2006. Representation: The problem for structuralism. *Philosophy of Science* 73: 536–547.

van Fraassen, B. 2008. *Scientific representation: Paradoxes of perspective.* Oxford: Oxford University Press.

Wallace, D., and C. Timpson. 2010. Quantum mechanics on spacetime I: Spacetime state realism. *British Journal for the Philosophy of Science* 61: 697–727.

Watkins, E. 1997. The laws of motion from Newton to Kant. *Perspectives on Science* 5: 311–348.

Watkins, E. 2003. Forces and causes in Kant's early pre-critical writings. *Studies in History and Philosophy of Science* 34: 5–27.

Watkins, E. 2005. *Kant and the metaphysics of causality.* Cambridge: University of Cambridge Press.

Westfall, R. 1962. The foundation of Newton's philosophy of nature. *British Journal of the History of Science* 1: 171–182.

Westfall, R. 1971. *Force in Newton's physics.* New York: Elsevier.

Westfall, R. 1983. *Never at rest: A biography of Isaac Newton.* Cambridge: Cambridge University Press.

Westman, R.S. 1980. Huygens and the problem of Cartesianism. In *Studies of Christiaan Huygens*, ed. H. Bos et al. Linse: Swetz & Zeitlinger.

Weyl, H. 1949 [1927]. *Philosophy of mathematical and natural science.* Princeton: Princeton University Press.

Weyl, H. 1982 [1952]. *Symmetry.* Princeton: Princeton University Press.

Witten, E. 1989. The search for higher symmetry in string theory. *Transactions of the Royal Society of London* A329: 349–357.

Wolff, C. 2009 [1720]. *Rational thoughts on God, the World and the Soul of Human Beings, Also All Things in General.* Ed. and Trans. E. Watkins. In *Kant's* Critique of pure reason: *Background source materials*, 7–53. Cambridge: Cambridge University Press, 2009.

Worrall, J. 1989. Structural realism: The best of both worlds? *Dialectica* 43: 99–124.

Wüthrich, C. 2009. Challenging the spacetime structuralist. *Philosophy of Science* 76: 1039–1051.

Wüthrich, C. 2012. The structure of causal sets. *Journal for General Philosophy of Science* 43: 223–241.

Zepeda, J. 2014. The concept of space and the metaphysics of extended substance in Descartes. *History of Philosophy Quarterly* 31: 21–40.

Index

© Springer International Publishing Switzerland 2016   339
E. Slowik, *The Deep Metaphysics of Space*, European Studies in Philosophy of Science, DOI 10.1007/978-3-319-44868-8

Printed in the United States
By Bookmasters